D1075228
RDUE FINES:
er day per item
MATERIALS:

DEVELOPMENTS IN SEDIMENTOLOGY 28

HYPERSALINE BRINES AND EVAPORITIC ENVIRONMENTS

FURTHER TITLES IN THIS SERIES
VOLUMES 1, 2, 3, 5, 8 and 9 are out of print

4 *F.G. TICKELL*
THE TECHNIQUES OF SEDIMENTARY MINERALOGY
6 *L. VAN DER PLAS*
THE IDENTIFICATION OF DETRITAL FELDSPARS
7 *S. DZULYNSKI* and *E.K. WALTON*
SEDIMENTARY FEATURES OF FLYSCH AND GREYWACKES
10 *P.McL.D. DUFF, A. HALLAM* and *E.K. WALTON*
CYCLIC SEDIMENTATION
11 *C.C. REEVES Jr.*
INTRODUCTION TO PALEOLIMNOLOGY
12 *R.G.C. BATHURST*
CARBONATE SEDIMENTS AND THEIR DIAGENESIS
13 *A.A. MANTEN*
SILURIAN REEFS OF GOTLAND
14 *K.W. GLENNIE*
DESERT SEDIMENTARY ENVIRONMENTS
15 *C.E. WEAVER* and *L.D. POLLARD*
THE CHEMISTRY OF CLAY MINERALS
16 *H.H. RIEKE III* and *G.V. CHILINGARIAN*
COMPACTION OF ARGILLACEOUS SEDIMENTS
17 *M.D. PICARD* and *L.R. HIGH Jr.*
SEDIMENTARY STRUCTURES OF EPHEMERAL STREAMS
18 *G.V. CHILINGARIAN* and *K.H. WOLF*
COMPACTION OF COARSE-GRAINED SEDIMENTS
19 *W. SCHWARZACHER*
SEDIMENTATION MODELS AND QUANTITATIVE STRATIGRAPHY
20 *M.R. WALTER, Editor*
STROMATOLITES
21 *B. VELDE*
CLAYS AND CLAY MINERALS IN NATURAL AND SYNTHETIC SYSTEMS
22 *C.E. WEAVER* and *K.C. BECK*
MIOCENE OF THE SOUTHEASTERN UNITED STATES
23 *B.C. HEEZEN, Editor*
INFLUENCE OF ABYSSAL CIRCULATION ON SEDIMENTARY
ACCUMULATIONS IN SPACE AND TIME
24 *R.E. GRIM* and *N. GÜVEN*
BENTONITES
25A *G. LARSEN* and *G.V. CHILINGARIAN, Editors*
DIAGENESIS IN SEDIMENTS AND SEDIMENTARY ROCKS
26 *T. SUDO* and *S. SHIMODA, Editors*
CLAYS AND CLAY MINERALS OF JAPAN
27 *M.M. MORTLAND* and *V.C. FARMER*
INTERNATIONAL CLAY CONFERENCE 1978

DEVELOPMENTS IN SEDIMENTOLOGY 28

HYPERSALINE BRINES AND EVAPORITIC ENVIRONMENTS

Proceedings of the Bat Sheva Seminar on Saline Lakes and Natural Brines

Edited by

A. NISSENBAUM

The Weizmann Institute of Science, Rehovot, Israel

ELSEVIER SCIENTIFIC PUBLISHING COMPANY

Amsterdam — Oxford — New York 1980

ELSEVIER SCIENTIFIC PUBLISHING COMPANY
335 Jan van Galenstraat
P.O. Box 211, 1000 AE Amsterdam, The Netherlands

Distributors for the United States and Canada:

ELSEVIER/NORTH-HOLLAND INC.
52, Vanderbilt Avenue
New York, N.Y. 10017

Typesetting by Electrolyne Ltd., Jerusalem

Library of Congress Cataloging in Publication Data

Bat-Sheva Seminar on Saline Lakes and Natural Brines, Weizmann Institute of Science, 1977.
Hypersaline brines and evaporitiç environments.

(Developments in sedimentology ; v. 28)
"Held under the auspices of the Bat-Sheva de Rotschild Foundation for the Advancement of Science in Israel."
Bibliography: p.
Includes index.
1. Saline waters--Congresses. 2. Lakes--Congresses. 3. Evaporites--Congresses. 4. Sedimentation and deposition--Congresses. I. Nissenbaum, A. II. Batsheva de Rothschild Foundation for the Advancement of Science in Israel. III. Title. IV. Series.
GB1601.2.B37 1977 551.48'2 80-247

Library of Congress Cataloging in Publication Data

ISBN 0-444-41852-0

ISBN 0-444-41852-0 (Vol. 28)
ISBN 0-444-41238-7 (Series)

Printed in The Netherlands

FOREWORD

During the period from October 16–October 26, 1977, over fifty scientists from many countries gathered at the Weizmann Institute of Science, at Rehovot, Israel, for an international seminar on "Saline Lakes and Natural Brines." The seminar was held under the auspices of the Bat-Sheva de Rotschild Foundation for the Advancement of Science in Israel.* During the seminar about 30 papers were presented covering various aspects of evaporitic environments and hypersaline brines. The organizing committee for the seminar included: J. R. Gat (Chairman), J. Kushnir (Secretary), Y. Berman (Organizer), G. Assaf and A. Nissenbaum from the Weizmann Institute, D. Neev from the Geological Survey of Israel, M. Shilo from the Hadassa Medical School, I. Zak from the Hebrew University and A. Lerman from Northwestern University.

This particular subject was chosen because of the increasing interest in the hypersaline environment. Brine bodies and evaporitic environments represent one of the least understood environments in our world. For many years, such environments were considered a very marginal resource for human needs. With the advent of modern science and technology, the evaporitic environments have become of great geological, geochemical and practical significance. The economic importance of natural brines as a source for minerals, and as a possible collectors of solar energy, is matched by its inherent scientific interest. It poses a variety of fascinating problems in ecology (concerning adaptation of microorganisms to hostile environments), in hydrodynamics (stratification and mixing regimes), in physical chemistry (activity coefficients, ionic speciation and phase equilibria in brines), and geology (the formation of evaporite deposits).

The present volume includes some of the presentations given at the Bat-Sheva seminar. Although very far from a complete treatise on the subject we hope that it will present, to the initiate and unitiate alike, a taste of this fascinating environment.

J. R. Gat
J. Kushnir
A. Nissenbaum

Isotope Department, Weizmann Institute of Science, Rehovot, Israel

* The Bat-Sheva Foundation is a privat non-profit foundation incorporated in New York City in 1957. The foundation was established through the personal endowment of Bat-Sheva de Rotschild. The primary goal of the foundation is to support scientific research in Israel. Since 1963 the Foundation has sponsored 33 international School Seminars devoted to various fields of Science.

CONTENTS

Chapter 1

ISOTOPE HYDROLOGY OF VERY SALINE LAKES

J. R. GAT

Isotope Department, Weizmann Institute of Science, Rehovot, Israel

THE SALT EFFECT ON THE ISOTOPIC ENRICHMENT BY EVAPORATION

The heavier isotopic species of water (both H^2HO and $H_2{}^{18}O$) are enriched in a water body exposed to evaporation, such as a lake, as a result of the preferred evaporation of the lighter water molecule ($^1H_2{}^{16}O$).

In order to estimate the degree of enrichment of the heavy isotopic species during the course of the evaporation process, the latter was modelled by Craig and Gordon (1965) as a consecutive series of steps: water evaporates from a virtually saturated sublayer at the liquid surface, through a region where transport occurs predominantly by molecular diffusion, into a fully turbulent atmosphere where there is no further isotope separation. This model is based on the electrical resistance analogue model of evaporation discussed by Sverdrup (1951). The mathematical formulation for the case of a water body without inflow leads to the following equation* (upon neglect of possible resistance to transport in the liquid phase):

$$\frac{d\lambda}{d\,\ell n\,f} = \frac{h(\delta_L - \delta_V)/(1+\delta_L) - (1-\alpha*) - \Delta\epsilon}{(1-h)+\Delta\epsilon} = \frac{h(\delta_L - \Delta_V)/(1+\delta_L) - \epsilon}{(1-h)+\Delta E} \quad [1]$$

Similarly, in a well mixed lake open to inflow (I^+) and outflow (I^-), the change of isotope composition is given by Gat and Levy (1978):

$$\frac{d\lambda}{d\,\ell nN} = \frac{h(\delta_L - \delta_V)/(1+\delta_L) - \epsilon + \frac{I^+}{E}[(1-h)+\Delta\epsilon](\delta_L - \delta_I)/(1+\delta_L)}{[(1-h)+\Delta\epsilon]\,[1 - \frac{(I^+ - I^-)}{E}]} \quad [2]$$

* $R \equiv Ni/N$, where Ni and N are the number of moles of the isotopic (heavy) and dominant water species, respectively, $\lambda \equiv \ell n\, R/R_{std}$; $\delta = (R_{sample} - R_{std})/R_{std}$; $f \equiv N/N_o$, with subscript o referring to the initial state of the system; $\alpha* \equiv (R_V/R_L)$ at equilibrium, subscripts V,L denote the vapour and liquid phase, respectively; $\epsilon \equiv 1 - \alpha* + \Delta\epsilon$, where $\Delta\epsilon$ is the fractionation factor resulting from the non-equilibrium processes, $[\Delta\epsilon = (1-h)(\rho_i/\rho - 1)]$. $\rho_i, /\rho$ are the resistance to transport of the isotopic water species, as given by Craig and Gordon's model. h is a relative humidity, expressed relative to water vapour saturated at the temperature of the liquid.

In both these cases the change in isotope composition depends on similar ambient parameters, in particular on the humidity and the difference between the isotopic composition of the lake and its surrounding. The conditions (mechanism) of evaporation expresses itself through the value of the fractionation factor ϵ, whose value is also a function of temperature (α*) and of the degree of under saturation ($\Delta\epsilon$). The combination of all these factors fixes the "slope" of the evaporation line in the δ_D vs δ^{18}_O field of coordinates (Gat, 1971).

The isotopic composition of an evaporating water body approaches a steady state value in all cases. For the case of a closed water body (no inflow or outflow) this steady state value is (Craig and Gordon, 1965):

$$\delta_{L,S} = \frac{\epsilon}{h} + \delta_a \qquad [3]$$

or $\delta_{L,S} = \frac{\epsilon}{h'} + \delta_a$ for the case of a brine (Gat and Levy, 1978) where h' = h/a (see below). In the system open to inflow, the steady state value depends *inter alia* on the throughflow rate I^+/E and can hence be used as a measure of this property.

The enrichment in heavy isotopes during the evaporation process marks the surface waters in an unique way relative to the meteoric waters. These evaporated waters are characterised by a relative deuterium deficiency, resulting from the lower oxygen isotope fractionation during evaporation as compared to fractionation during equilibrium phase transitions, such as are involved in the rain processes. This is just another way of saying the slope of an "evaporation line" on a δ_D vs δ_{18} diagram to be less than that of the "meteoric water line" (which is the locus of the isotopic composition of rain waters (Craig,1961) and characterised to a good approximation by a slope of $\Delta\delta_D/\Delta\delta_{^{18}O} = 8$). This characteristic range of isotope composition is widely used to identify waters originating from surface water bodies.

The presence of high salt concentrations affects the evaporation and the accompanying isotope fractionation process in a number of ways:

a) the lower saturation-vapour pressure over the brine relative to a fresh water surface (the reduced water activity) reduces the saturation deficit over the solution, thereby reducing evaporation rates, the kinetic fractionation factor and the surface heat balance (temperature).

b) the isotopic thermodynamic fractionation factor α* becomes a function of both temperature and salinity (Sofer and Gat, 1972 and 1975).

c) changes in surface tension and in density, viscosity, etc. affect the wave structure, the formation of spray and the air-liquid interface conditions in general. All of these may affect the kinetic aspects of the water transport and fractionation.

In addition, high salinity results in changes in the vertical stability and in the stratification, which affects the mixing within the water column and thus the overall water balance, temperature, etc.; these in turn may influence the isotopic composition of the system under consideration.

The effects on the water activity and on the isotope separation factor between the brine and its equilibrium vapour can be expressed quantitatively, both "a" and "Γ" being known and *measurable* functions of the salinity (Gonfiantini, 1965; Sofer and Gat, 1972, 1975). Taking these factors into account, formulae 1 and 2 are slightly modified with h being replaced by h' = h/a and $\alpha^*_b = \Gamma\alpha*$ taking the place of $\alpha*$ of the pure water system (which was a function solely of temperature); α^*_b is now a function of both the concentration and nature (composition) of the solution.

To a good first approximation the salt effect manifests itself through its effect on the humidity. As the solutions evaporate and become more saline, the apparent humidity h' increases. As a result, δ_L does not increase monotonously towards a steady state value, but we observe a reversal in its value while the isotopic composition pursues an ever changing apparent steady state corresponding to its appropriate humidity. Except in extremely dry climates, evaporation will finally cease at the point where the ambient humidity approaches the saturation vapour pressure of the brine; isotopic exchange will however continue and will tend to bring the vapour and brine into isotopic equilibrium at a level of $\delta_{L,f} = \delta_a + \epsilon^*_b$; which is a fixed value for any given salinity and ambient humidity.

A typical example of the evolution of the isotopic composition in an evaporation brine body is shown in Fig. 1-1.

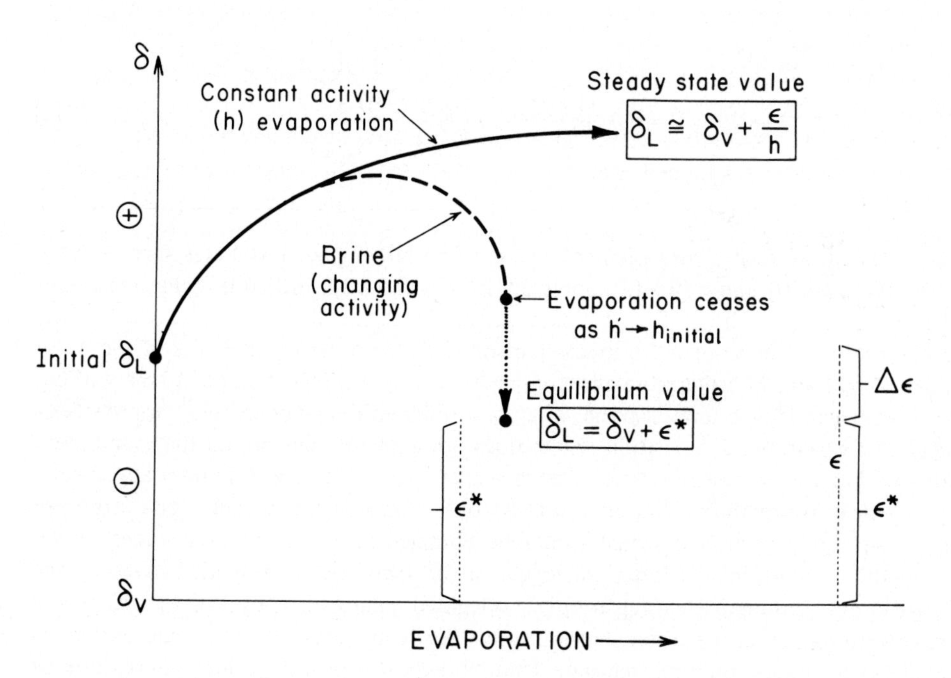

Fig. 1-1. Schematic of the evolution of the isotope composition of an evaporating water body at constant activity and apparent humidity (upper curve), and in a brine with changing activity (dashed line).

The magnitude of equilibrium and non-equilibrium fractionation terms are drawn on the lower bracket, which represents the δ value of the vapour. (See also Sofer and Gat, 1975).

STABLE ISOTOPES AS A TOOL IN EVAPORATION ESTIMATES AND WATER BALANCE STUDIES

Since the major changes in the isotopic composition of a water body occur as a result of the evaporation process, it is natural that isotope data are used as a measure of the evaporation term in the water balance of a lake. For this purpose one solves a material balance equation of the type

$$\frac{d(V\,\delta_L)}{dt} = I^+\,\delta_I - E\,\delta_E - I^-\,\delta_L \qquad [4]$$

in the transient state or under steady state conditions. This equation is equivalent to Equation 2 which is obtained on the introduction of the value for δ_e given by the Craig-Gordon theory.

Under steady state conditions one can get an expression relating δ_L to I/E, by equating the numerator of Equation 2 to zero:

$$\delta_{L,S} = \left((\frac{\epsilon}{h} + \delta_a) + \frac{I}{E}\frac{(1-h)}{h}\,\delta_I \right) / [1 + \frac{(1-h)}{h} \cdot \frac{I}{E}] \qquad [5]$$

or

$$E = I \cdot \frac{(1-h)}{h} \cdot \frac{(\delta_{L,S} - \delta_I)}{(\frac{\epsilon}{h} + \delta_a - \delta_{L,S})} \qquad [6]$$

The accuracy of such a measurement seems to be rather poor (estimates vary between ±10% (Gat, 1970) and ±50% (Zimmerman, 1970)), the main difficulty being the assessment of the value of h and δ_a. It should be noted that one obtains just the relative magnitude of the evaporation term in comparison with the other components of the water balance equation. Furthermore the composition of a terminal lake, i.e. a lake without outflow, where $I^+ = E$ (and also that of a pond which dries up completely), approaches a steady state isotopic composition which does not explicitly depend on the evaporation rate and the isotope composition of such a lake by itself does not contain any further hydrological information. *Strictum sensum* one compares the isotopic composition of any water body with that which would be obtained in a terminal lake under similar evaporation condition; the latter's δ value can be theoretically evaluated based on the Craig-Gordon formulation, using properly wieghted values of h, δ_a and δ_I and of the enrichment factor ϵ, the latter depending on the temperature and to some extent on details of the evaporation mechanism. Undoubtedly this procedure introduces quite an error into such a calculation. Most of these uncertainties can be circumvented whenever one can make measurements on a truly terminal lake in the area (a so-called "index lake," (Dincer, 1968)) or where isotopic measurements can be performed on a water body whose water balance is exactly known.

We have used such an approach to study the hydrology of playa lakes (sabkhas) in the Northern Sinai (Gat and Levy, 1978). These lakes are situated amidst high sand dunes close to the Mediterranean seashore, but are at present disconnected from the sea and its beach lagoons. They have the appearance of dry salt flats during the summer, with a hard cover of halite and gypsum; brines appear, however, just a few centimeters beneath the salt crust. In winter the sabkhas are flooded to a depth of about half a meter by meteoric water input from the surrounding dune areas. However, due to the occurence of evaporite deposits in the sabkhas the water remains saturated with respect to halite and calcium sulfate. As a consequence, it is impossible to establish the hydrological balance of the brine bodies through their salinity.

The evolution of the isotopic composition in four of these sabkhas, as shown in Fig. 1-2, indicates that there are considerable differences between them, in spite of their having similar salinities and being situated in close proximity one to another and hence exposed to similar evaporation rates. These differences must be interpreted as due to differences in the hydrological patterns operating in the system. The isotopic composition of sabkha "51" shoots up to highly enriched δ values within a few days, in a very similar manner to that of an exposed evaporation pan in the area. We then assume this sabkha to be a closed pond which dries up without further inflow, so that it can serve as an "index lake" for our further analyses. The steady state value of this sabkha, namely δ_L = +9.5‰ is taken to represent the steady state isotopic composition of a closed pond under the climate conditions which prevail in the area, as given by Equation [3]. Using Equation [6] one can then evaluate the value of I/E, which corresponds to the isotopic composition of the other sabkhas.

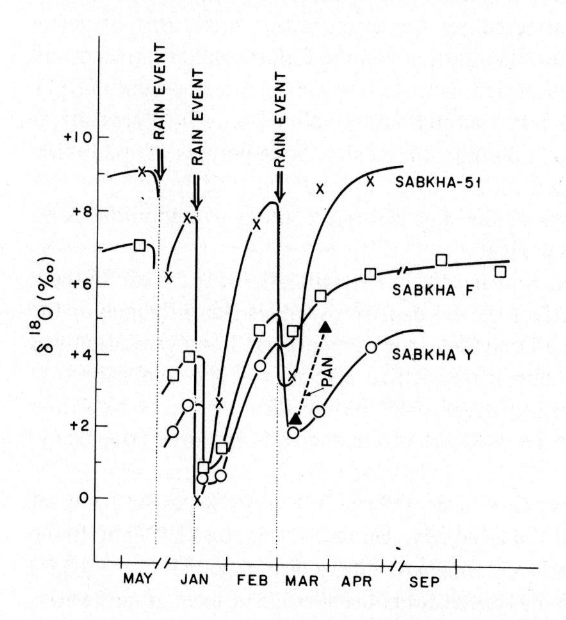

Fig. 1-2. The march of δ values in three inland sabkhas of the Bardawil area, following local rain events. Note the data from an evaporation pan exposed in the area.

The result obtained shows I/E to be 1 in the case of sabkha F, i.e. $I \sim E$. This sabkha appears to act as a terminal lake without outflow over quite some period; this finding is confirmed both by the almost constant water level in the lake for a few months and, more conclusively, by the identity of the value of $\delta_\epsilon = -0.07‰$ as calculated by solving Equation [2] = 0, with the composition of the meteoric waters in the area, δ_I which was independently estimated to be between −0.5‰ and 0‰. In the other sabkhas I/E>1; however water levels recede during the period of measurements indicating a subsurface outflow of brine into the groundwater systems.

The case of these sabkhas illustrates one situation where isotopic measurements yield hydrologic insight not readily obtainable through standard geochemical or hydrological measurements. In simpler systems, in which salinity (especially chloride content) is a conservative property of the aqueous phase, one may get a more accurate estimate of evaporation through salinity balance. However, in many situations, as for example in Lake Tiberias where there are unaccounted for salt seepages into the lake, or in other cases where industrial effluents upset the natural salt balance, stable isotopes remain the most reliable tracers in water balance studies. This is especially true in the case of the very saline water bodies such as the sabkhas described before, where the salt balance is upset by precipitation or dissolution of evaporites.

STRATIFICATION IN LAKES – THE ROLE OF ISOTOPIC STUDIES

A typical freshwater lake, such as Lake Kinnereth, is stratified during summer by virtue of the heating up of the top layer (typically the upper 15–25 m). Stratification then restricts the volume of water affected by the evaporative enrichment of stable isotopes (Gat, 1970), or, for that matter, the dilution volume for atmospheric tracers and gases such as tritium and oxygen. The transient isotope balance equations (equations [1] or [2]) must be formulated so as to take the epilimnion volume only into account; it should also include a term for changes in thermocline depth during periods of epilimnion growth. A steady state isotope concentration is not usually achieved because of this continuing change in depth of the thermocline, but isotopic content is a good check on homogeneity and mixing pattern in such a lake.

In a brine lake with freshwater inflow a permanent stratification can be established, based on a salinity gradients whose effect on the density overrides the influence of the seasonal heating and cooling cycle. The Dead Sea prior to its present homogenisation was an example for such a lake. In a permanently stratified lake the isotope composition of the deep water layers are a conservative property and the constancy of its composition with time can be taken as a very sensitive measure of the degree of isolation (stagnancy) of these layers.

Craig (unpublished) in 1960 gave a value of $\delta^{18}O = +4.39 \pm 0.02‰$ as the composition of deep waters (below 100 m) of the Dead Sea. On recent samples (1977) he found a uniform profile throughout the water column with the composition $\delta^{18}O = +4.38 \pm 0.02‰$. On the basis of these results one can exclude the seepage of local groundwaters into the abyssal brines in excess of a yearly contribution of 0.2% (v/v); this estimate is based on the assumption that local groundwaters are characterized by an isotope composition of $\delta^{18}O = -5.5‰$ (Gat and Dansgaard, 1972). We must be less dogmatic concerning contribution of surface water; although values measured over the years have ranged from +1.6‰ to +4.9‰ in an irregular seasonal manner, the average surface water composition has been close to that of the deep waters. Moreover the present depth profile was found to be quite uniform. As a result, the stable isotope content is not a sensitive measure of mixing with surface waters. In this respect the tritium content seems a more useful tool (Steinhorn, this volume); tritium being an attribute of atmospheric waters.

Prior to 1975, when the Dead Sea levels were higher, so that it consisted of two intercommunicating basins, one found horizontal differences in surface water composition. Southern basin waters were typically enriched by 0.4‰ relative to surface waters of the northern basin (Nissenbaum, 1969), thus providing a distinct marker for these water masses. The depth profiles of $\delta^{18}O$ in 1963–65, as reported by Nissenbaum (1969), peak at the pycnocline depth of 40–50 m; waters at this depth show the δ value characteristic for the southern basin. This lent support to the interflow model of the stratified Dead Sea which is still the accepted hydrodynamic model (Assaf and Nissenbaum, 1977). Unfortunately such an analysis is no longer possible since the Dead Sea, which is presently confined to the northern basin only, is much more homogenous in isotopic composition.

REFERENCES

Assaf, G. and Nissenbaum, A., 1977. "The evolution of the upper water mass of the Dead Sea 1805–1976", In: Desertic Terminal Lakes. (D. C. Greer, ed.), Univ. of Utah Water Research Lab, Utah, 61–72.

Craig, H, 1961. "Isotopic Variations in Meteoric Waters." Science 133, 1702–1703.

Craig, H. and L. I. Gordon, 1965. "Deuterium and Oxygen-18 Variations in the Ocean and Marine Atmosphere", in: Stable Isotopes in Oceanographic Studies and Paleotemperatures, Pisa, pp. 9–130.

Dincer, T., 1968. "The use of oxygen-18 and deuterium concentrations in the water balance of lakes". Water Resources Res., 4, 1289–1305.

Gat, J. R., 1970. "Environmental isotopes balance of Lake Tiberias", in: "Isotope in Hydrology 1970", IAEA, Vienna, 109–127.

Gat, J. R., 1970. Comments on the stable isotope method in regional groundwater investigations". Water Resources Res., 7, 980–993.

Gat, J. R. and W. Dansgaard, 1972. "Stable isotope survey of the fresh water occurences in Israel and the Jordan Rift Valley". J. Hydrology, 16, 177–211.

Gat, J. R. and Y. Levy, 1978. "Isotope hydrology of inland sabkhas in the Bardawil area, Sinai", Limnol. Oceanogr., 23, 841–850.

Gonfiantini, R., 1965. "effetti isotopici nell' evaporazione di acqua salate". Att. Soc. Tosc. Sc. Nat., Serie A72, 550.

Nissenbaum, A., 1969. "Studies in the geochemistry of the Jordan River–Dead Sea system". unpublished Ph.D. Thesis, UCLA, 288 pp.

Sofer, Z. and J. R. Gat, 1972. "Activities and concentrations of O-18 in concentrated aqueous solutions. Analytical and geophysical implications". Earth Planet. Sci. Lett., 15, 232–238.

Sofer, Z. and J. R. Gat, 1975. "The isotopic composition of evaporating brines: effect of the isotopic activity ratio in saline solutions". Earth Planet. Sci. Lett. 26, 179–186.

Sverdrup, H. V., 1951. "Evaporation from the oceans". Compendium of Meteorology. Am. Met. Soc., 1077.

Zimmermann, U. and D. Ehalt, 1970. "Stable isotopes in the study of the water balance of Lake Neusiedl, Austria". in: Isotope Hydrology 1970, Vienna, IAEA, 129–138.

Chapter 2

MIXING IN STRATIFIED FLUIDS

MELVIN E. STERN

Graduate School of Oceanography, University of Rhode Island, Kingston, R.I. 02881, USA

INTRODUCTION

This article is an exposition of certain aspects of fluid dynamics which may be useful in understanding and modifying the density structure in saline lakes. If these are small and strongly stratified, then they will have small rates of vertical mixing, and it seems feasible to alter and control the vertical density structure. A case in point is the Great Salt Lake. This has been divided by a causeway which has greatly increased the stratification in one half of the lake because of the excess evaporation in the other half (Assaf 1977), the two parts being connected only by culverts. A gravitationally driven circulation between the basins and through the culverts is thereby established, with heavy saline water flowing to the bottom of the fresher basin and thereby stratifying it.

At this conference we have also heard (Weinberger 1978) of the work being done to utilize salt stratified ponds and lakes as sources of energy. The principle here involves the (dynamical) trapping of the incoming solar energy, by the stable halocline, thereby realizing high equilibrium temperatures at the bottom of a "solar pond".

As a point of departure for this exposition, the aforementioned control action of the culvert will be illustrated with the aid of Fig. 2-1, which shows the free discharge between

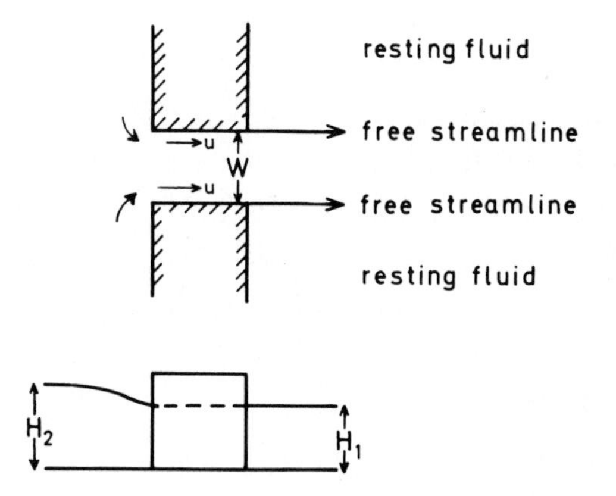

Fig. 2-1. The free discharge through a weir connecting two large bodies of water whose mean elevations are H_1, H_2 respectively. The top diagram is a plan view, and the bottom is a vertical section. u is the uniform horizontal velocity in the channel (weir), and W is the width measured transverse to the flow. The "free streamlines" denote the separation of the flow from the sides of the weir.

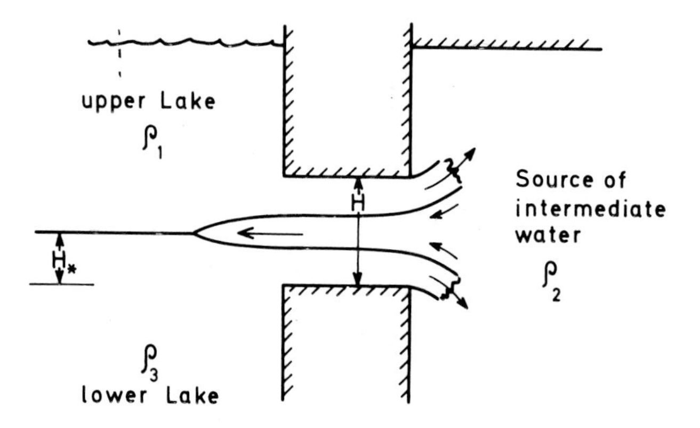

Fig. 2-2. An internal weir which is used to control the flow between a stratified lake (left) and a reservoir of intermediate water (right).

two *homogeneous* bodies of water connected by a weir. After considering this simplest case, we will indicate the kind of generalization which can be made for fluids having different densities, these fluids being connected by an internal weir (Fig. 2-2), such as may be used to establish and control the stratification in a solar pond.

We will then return to Fig. 2-1 in order to consider the effects which occur *after* the jet of water emerges from the weir and enters the right hand basin. The emerging jet (Fig. 2-3a) becomes unstable and turbulent, and it is necessary to understand these effects in the case of a homogeneous fluid, before we can proceed to the subject of mixing in a stratified fluid. In the latter case we shall see that a similar kind of instability occurs when a horizontal current has a vertical shear (Fig. 2-4). But here the entire process is

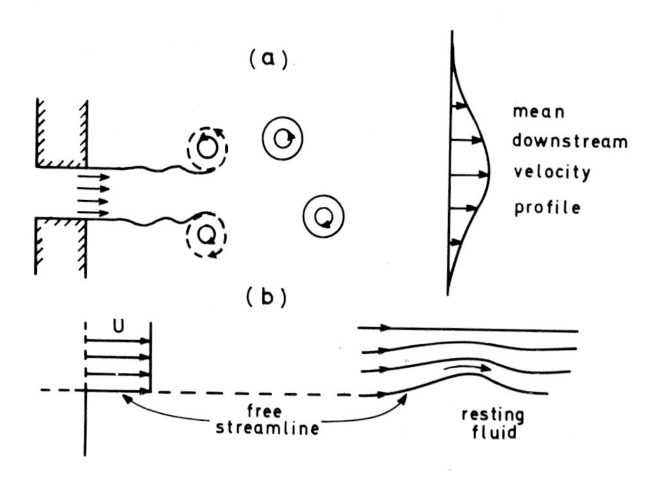

Fig. 2-3a. The instability of a jet with lateral shear. Top view of flow through a weir showing the amplification of waves, the formation of closed vortices and the modification of the mean jet flow.

2-3b. The Helmholtz instability problem showing the undistrubed flow and its perturbation.

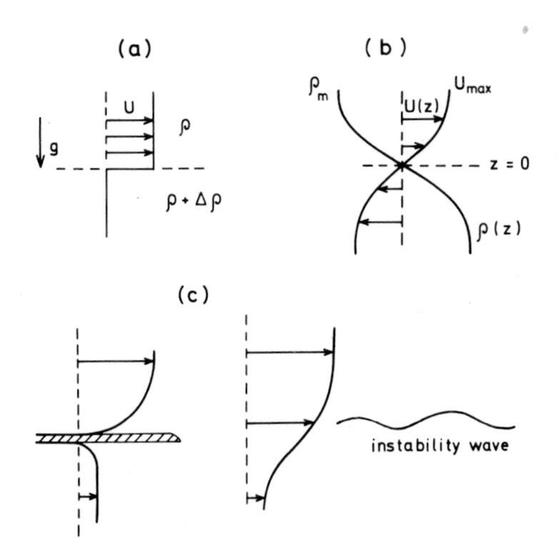

Fig. 2-4a. Basic state for instability problem in a two layer model.
2-4b. Basic state for a continuously stratified model.
2-4c. Splitter plate experiment for realizing the continuously stratified model.

substantially modified by the gravity (buoyancy) force, whereas this force is dynamically irrelevant in the flow with lateral shear (Fig. 2-3). The non-dimensional ratio of the gravity forces to the forces arising from the shear (Fig. 2-4) is called the Richardson number, and it is this which determines the degree of shear turbulence.

The course of the discussion then leads to the subject of entrainment, as illustrated by the deepening of the surface mixed layer under the action of a wind stress (Fig. -5). We will show that the wind generated turbulence is eventually confined to a surface mixed layer, beneath which there is essentially no mixing. This prodigious effect of the pycnocline raises the rather general oceanographic question as to how bottom water is "formed." This question will bring us to a consideration of convective effects, especially those due to evaporation. This produces a gravitationally unstable surface "microlayer", and the subsequent sinking of cold saline parcels produces large vertical velocities in the mixed layer which underlies the surface microlayer and overlies the stratified pycnocline. These large eddies entrain the fluid at the top of the pycnocline in a manner which is similar to the previously mentioned dynamical effect of the wind.

Since our fluid is stratified by temperature and salinity we note that the qualitative nature of the convection and mixing can be different from that which would occur in either pure thermal or pure haline convection. This subject is called "Double Diffusive" convection, in reference to the fact that the difference between the diffusivity of heat and salt (at the molecular level) is the underlying cause of a hierarchy of "large scale motions" which are produced in a fluid whose mean density field decreases upwards. Turner and Chen (1974) have demonstrated some sensational effects which occur in a fluid whose temperature, salinity, *and* density increase uniformly with depth. We shall see that these effects provide one mechanism by which bottom water can be formed without having a general overturn of the fluid.

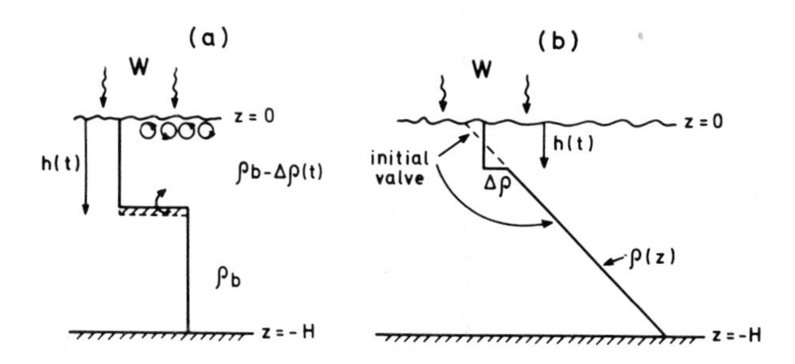

Fig. 2-5. The deepening of the mixed layer. W is proportional to the energy flux from the air to the water.
2-5a. Two layer model.
2-5b. Continuously stratified model.

THE CONTROL OF THE FLOW OF HOMOGENEOUS FLUIDS, THEIR INSTABILITY, AND THE TRANSITION TO TURBULENCE

a) The free discharge between two halves of a large lake connected by a Weir.

Suppose the left hand side (Fig. 2-1) of a lake has its free surface at the Height H_2, and the right hand side has a "slightly" lower head H_1. Upon opening the short channel which connects the two basins, the water will flow *smoothly* into the channel from the left hand basins, but the flow will *separate* (from the weir) as it emerges into the right hand basin. For our present purpose we may represent the separated flow by two free streamlines, on the exterior sides of which we have resting fluid whose height H_1 equals the original free surface height of the right hand basin. On the interior sides of the free streamline the flow has a uniform velocity u. Since the pressure must be continuous across the free streamline, it follows that the height of the free surface (inside) the emerging jet is also H_1. By proceeding upstream, in the region where the streamlines are straight and parallel, we deduce that the height of the free surface just inside the channel must also equal H_1, and the local velocity also equals u. Therefore the total kinetic plus potential energy (per unit volume) of a parcel on the free surface is $\rho u^2/2 + \rho g H_1$, where ρ is the uniform density and g is the acceleration of gravity. It is assumed here: i) that the downstream length of the weir is sufficiently large so that the flow is hydrostatic, ii) that the downstream length of the weir is sufficiently small so that wall friction can be neglected, iii) and that the two lakes are of infinite area, so that the flow in the weir is essentially steady. If we continue marching upstream on a given streamline, along which the total energy must be conserved (Bernoull's equation), we reach a region in the left hand lake where the horizontal velocity reduces to zero (compared to u), and the height of the free surface increases to H_2. Therefore, far upstream in the left hand lake the total energy of a parcel on the free surface is $\rho g H_2$. By equating upstream and downstream energies we have $\rho g H_2 = \rho u^2/2 + \rho g H_1$, or

$$u = [2g(H_2 - H_1)]^{1/2} \qquad [1]$$

This* solves the problem in terms of the given parameters, and the total volume transport uH_1W through the weir can then be computed from $(2g(H_2 - H_1))^{1/2} H_1 W$.

The foregoing is the simplest of a large class of interesting control problems related to flow through straits and estuaries (Anati, Assaf, and Thompson (1977). Fig. 2-2 shows a more complicated example, in which an internal weir is used to "build up" a pycnocline in the left hand lake. The latter originally has two homogeneous layers (ρ_1, ρ_3), with the interface location given by H_*. The lake is then connected to a source of intermediate water ρ_2 $(\rho_1 < \rho_2 < \rho_3)$ by means of an internal weir (channel) having vertical height H.. It is readily seen that intermediate water will flow between the upper and lower lake waters, and will spread laterally when it reaches the left hand basin. The volume compensating flows of the ρ_1, ρ_3 waters, on the other hand, will rise and descend (respectively) when they reach the right hand basin. The problem posed requires us to compute the rate at which intermediate water flows into the lake, and its thickness.

b) *On the instability of the jet in Fig. 2-1*. In accord with our program we return to the homogeneous fluid, and illustrate in Fig. 2-3a the observed sequence of events which occur at some distance downstream from Fig. 2-1. Waves of small amplitude appear on the free streamline, the amplitude increases downstream, and eventually a closed vortex "detaches" from the main flow. In this process some of the resting fluid which lies on the exterior side of the free streamline is incorporated or *entrained* into the jet, and thus the width of the mean velocity profile (shown on the right hand side of Fig. 2-3) increases downstream.

We will now give a highly simplified explanation of the instability process which eventually leads to the fully developed turbulence described above. The question is: why does a laminar equilibrium flow amplify small perturbations? The left hand side of Fig. 2-3b shows such an "equilibrium flow" (which has a certain resemblance to half of the laminar jet in Fig. 2-3a) in which the velocity U is uniform on one side of the free streamline and zero on the other side. Of course, this represents an equilibrium flow only to the extent that we neglect viscous forces, and such forces will also be neglected in the following instability considerations. We say that an equilibrium flow is unstable if any disturbance increases its energy at the expense of the energy associated with the mean flow. In nature, small amplitude random perturbations always exist as part of the noise background, and the question at hand is whether these will amplify with time. We do not propose to go into the large theoretical literature on this subject, and our point may be made with the following artifice. Consider the undisturbed flow on the left hand side of Fig. 2-3b, and then imagine that a small amplitude external force is steadily applied, but only in the vicinity of the free streamline. The latter is then displaced by a small amount from its equilibrium position, and the overlying streamlines are also deflected for reasons of mass continuity. At large perpendicular distances from the free streamline the fluid is undisturbed. Therefore the fluid moving over the "bump" in Fig. 2-3b must obviously accelerate, and the corresponding fluid pressure must drop, because of the "aspirator" (or Venturi) effect. It then follows that the pressure just above the free streamline in

* This is only valid if [1] is smaller than $(gH_1)^{1/2}$, at which value the flow becomes "critical" and we have a different problem.

Fig. 2-3b is lower than the pressure just below the free streamline (in the resting fluid). Were it not for the external force on the interface this differential pressure force would cause the free streamline to accelerate in the *same* direction in which it was displaced from its equilibrium level. If the external force is now removed we infer that this streamline displacement will amplify with time, and therefore the basic flow is unstable. Although the conclusion is correct, it must be emphasized that the foregoing argument is too artificial and too restricted, and the proper procedure requires the solution of an initial value problem. With a little stretch of the imagination we can apply the instability conclusion to Fig. 2-3a, thereby explaining the *downstream* amplification of the wave in terms of the temporal increase in the transverse velocity of each parcel in Fig. 2-3b.

It will be noted that the simple argument used above is independent of the horizontal dimension (e.g. wavelength) of the perturbation, and also independent of the magnitude U [cm/sec] of the basic current. These conclusions also follow on strictly dimensional grounds, since U is the only relevant parameter which describes the state of the basic current, and there is no external length or time scale in Fig. 2-3b. Therefore if the basic state is unstable for any U (or for any wavelength) then it must be unstable for all U (or for any wavelength).

SHEAR INSTABILITY IN STRATIFIED FLOW. THE RICHARDSON NUMBER

a) *Kelvin-Helmholtz instability in a two layered system.*

The equilibrium velocity profile in Fig. 2-4a is the same as previously considered except that the shear is now in the vertical direction, and the fluid may have a different density on either side of the free streamline. If $\Delta\rho = 0$, so that the density $\rho + \Delta\rho$ of the lower layer is identical with the density ρ of the upper layer, then the stability properties of our present model will be identical to those found previously (Fig. 2-3b), despite the presence of the gravity acceleration g in Fig. 2-4. The reason for this is that the potential energy is the same on both sides of the free streamline if $\Delta\rho = 0$, and we have here an illustration of the general rule, that the force of gravity is dynamically irrelevant (passive) unless there are density *variations* present. The gravity force enters the dynamical equations in the form of a buoyancy acceleration $g\Delta\rho/\rho$, and the density variations are otherwise unimportant ($\Delta\rho \ll \rho$). This is called the *Boussinesq approximation*.

From these remarks it is apparent that the stability of the equilibrium state in Fig. 2-4a can depend on only two parameters U [cm/sec] and $g\Delta\rho/\rho$ [cm/sec^2], with instability being assured as $g\Delta\rho/\rho \rightarrow 0$. Since no non-dimensional number can be constructed from U, $g\Delta\rho/\rho$ it follows that the present system must be unstable for *all* values of $(U, g\Delta\rho/\rho)$, if it is unstable for *any* particular value of these parameters. Since instability has been shown for $g\Delta\rho/\rho \rightarrow 0$, we conclude that Fig. 2-4a is also an unstable equilibrium. On the other hand, a length scale $U^2/(g\Delta\rho/\rho)$ can be constructed from the given parameters, and thus it does not follow that all wavelength perturbations will amplify. On intuitive grounds it is plausible that the importance of buoyancy forces, relative to shear forces, increases as the wavelength increases, and therefore wavelengths longer than some multiple of $U^2(g\Delta\rho/\rho)$ will not amplify.

b) *Instability in a continuous velocity and density profile.*

The next step is to consider a continuous velocity profile U(z) and a continuous density profile $\rho(z)$ (Fig. 2-4b), having a characteristic vertical length scale h, as well as a velocity scale U_{max}, and a given $g\Delta\rho/\rho_m$. From these given parameters we can now form a non-dimensional *Richardson Number*

$$Ri = \frac{g(\Delta\rho/\rho_m)h}{U_{max}^2} \qquad [2]$$

the value of which determines whether the equilibrium laminar flow is stable. In order to demonstrate this fact we will restrict the following discussion to the case in which the velocity and density profiles are geometrically similar in the sense

$$U(z) = U_{max}\, f(z/h)$$
$$\rho(z) - \rho_m = -\Delta\rho\, f(z/h) \qquad [3]$$

where f is an arbitrary (but fixed) form function with $f(o) = 0$, $f'(o) = 1$, and $h = (U'(o)/U_{max})^{-1}$. In the limit $h \to o$ with U_{max}, $U'g\Delta\rho/\rho_m$ fixed, the velocity profiles in Fig. 2-4b degenerate into those in Fig. 2-4a, and thus we conclude that Fig. 2-4b is unstable for sufficiently small Ri. An increase in h implies an increase in [2], which may also be interpreted as an increase in g (the gravitational restoring force on the perturbation). From the latter interpretation it then becomes intuitively plausible that the equilibrium flow in Fig. 2-5 is stable for sufficiently large values of [2]. This inference is verified by the Miles-Howard Theorem (see Stern, 1975), which Theorem states that if

$$\frac{-(g/\rho_m)\, \partial\rho/\partial z}{(\partial U/\partial z)^2} \qquad [4]$$

is greater than ¼ at all values of z then the equilibrium is stable. The expression in [4] is called theGradient Richardson Number. Upon substituting [3] in [4] we find that the latter equals $Ri/f'(z/h) > Ri/f'(o) = Ri$, since f is a monotonic decreasing function of height. Thus we conclude that if [2] is greater than ¼ then [4] is everywhere greater than ¼ and the stratified shear flow (Fig. 2-4b) is stable.

This result holds only if the density profile is geometrically similar to the velocity profile. If the length scale h is not the same in each of the two equations in [3], then the stability criterion depends on the detailed vertical variation of [4], rather than on one single overall Richardson Number [2].

c) *Experiments*

These theoretical results are confirmed by a number of experiments (Thorpe, 1973). If the density and velocity profiles are similar, and if the Richardson Number is below the theoretical value, then spontaneously amplifying wave perturbations of the laminar flow are observed. These grow to a definite finite amplitude, as determined by the external parameters, and then "break," thereby partially mixing both the density and the momentum in a definite height interval h_1. The waves and turbulence then dissipate, producing a new laminar equilibrium state whose profiles are somewhat similar to the initial state (Fig. 2-4b), the main difference being that the vertical scale of the shear layer has been increased from h to h_1, and the Richardson Number [2] has increased from its initial subcritical value to a final value which is approximately ¼.

The physics of this process can be illustrated further by means of "splitter plate experiments" (Browand and Winant, 1973) for which Fig. 2-4c is a schematic. Here we have two controlled streams of different salinity (or temperature) and different velocities which are fed onto opposite sides of the thin splitter plate, and which merge downstream to produce a velocity profile like that shown in the figure. If the two fluids have the same density, instability waves and turbulence will form downstream (as in Fig. 2-3a) and the width of the mean shear flow will continually increase. There is a remarkable difference, however, if the lower fluid (Fig. 2-4b) has a higher density than the upper fluid. Although waves and closed vorticies still develop in the lee of the plate (if the Richardson Number is subcritical) these produce a vertical mixing of *density* as well as momentum. Furthermore, the buoyancy restoring force on the transversely displaced fluid will inhibit the vertical penetration of the eddy motions. The net result is that the eddies die out after breaking, and the downstream flow then returns to a new stable laminar state. This is in marked contrast to the *persistent* downstream turbulence which occurs in the previously mentioned homogeneous shear flow. We can appreciate the difference by considering the energetics. In the homogeneous shear flow, energy is transferred from the mean flow into the eddies, and is eventually dissipated by molecular viscosity in the turbulent "cascade" to the smallest scales of motion. In the stratified flow, on the other hand, part of the mean flow energy goes immediately into potential energy as the width of the shear layer increases. The latter process cannot continue indefinitely downstream, because the potential energy would have to increase indefinitely, and more energy would be required than is available in the kinetic energy of the shear flow.

When the Richardson Number [2] is measured on a scale of ten meters or larger in the ocean pycnocline, the result is almost always much larger than the critical number. But many measurements of microstructure indicate subcritical Richardson Numbers (and instability) on smaller scales (Woods and Wiley, 1972), these being intermittent in space and time. We have seen that Kelvin-Helmoltz instability rapidly tends to remove the source of instability – should one exist – and thus the fundamental question arises as to the mechanism which produces intermittent regions of low Richardson Number. There are no "splitter plates" in the pycnocline, and no obvious reason why low Richardson Numbers should be produced, but the answer to this question obviously determines the average rate of mixing in the entire water body. What then are the prime energy sources for the production of small scale turbulence in the pycnocline, and what is the relation of this turbulence to the mean state of the pycnocline?

THE WIND MIXED SURFACE LAYER AND ENTRAINMENT AT ITS BOTTOM

A potent prime energy source is available at the surface of the wind driven ocean, and turbulent eddies are invariably produced immediately below the surface as the surf breaks etc. But a large fraction of the energy input is immediately dissipated in the local cascade, and only a fraction is available to mix the underlying pycnocline (Fig. 2-5) and to increase the total potential energy. Let $\rho_b W$ be the average rate at which the wind supplies energy to a unit surface area of the ocean in a unit time, where the nominal water density ρ_b is introduced here merely for normalizing convenience at a later stage. Thus W has the kinematic dimensions cm^3/sec^2, and a certain fraction of this – say $\Gamma<1$ – is available to increase the potential energy.

Consider first the case of *two* miscible layers (Fig. 2-5a) separated by a relatively sharp interface at $z = h(t)$, the initial value of which is $h(o)$. $\Delta\rho(o)$ denotes the initial value of the density deficit $\Delta\rho(t)$ of the upper layer relative to the density $\rho + \Delta\rho$ of the lower layer (N.B. $\Delta\rho \ll \rho$). Suppose that a wind stress is applied at $t = 0$, so that a turbulent shear flow will begin to diffuse downwards. The order of magnitude of the diameter of the eddies at any time will be comparable with the vertical dimension of the shear flow, and will increase accordingly, but the velocity of the shear flow and the velocity of the eddies decreases with depth. After some time the diameter of the eddies is such that they extend down to $z = -h$, and are deflected by the interface. In this process the eddies will drag along (entrain) a portion of the continuously stratified interface, and eventually mix it into the upper layer, so that $\Delta\rho(t)$ increases with time. The volume of the upper layer is thereby increased at the expense of the lower layer, but the density increase of the upper layer is not accompanied by any change in the density (or velocity) of the lower layer. The center of gravity of the entire fluid is clearly being raised by this entrainment process, and the rate of increase of potential energy is $\Gamma\rho_b W$. Moreover, it is intuitively obvious that this fraction (Γ) must decrease with time, because h(t) increases and because the intensity of the wind driven turbulent velocities at $z = -h$ (which is responsible for the entrainment) decreases.

From our present point of view $\rho_b W$ may be taken as a "given" parameter, whose value may be increased by merely increasing the wind speed. But Γ is by no means given, and its determination is an essential part of the problem at hand. Here as in so many other interesting oceanographical problems we simply lack the fundamental fluid dynamical knowledge which will allow us to proceed very far, and therefore the foregoing ideas will only be partially quantified in that which follows.

If $\rho(z)$ is the density of an element of height dz and unit cross sectional area, then $\rho(z)gzdz$ is its potential energy relative to the $z = -h$ datum. The total potential energy of the whole water column is then given by

$$\int_{-H}^{-h} \rho_b gzdz + \int_{-h}^{o} (\rho_b - \Delta\rho)gzdz = \frac{-\rho_b g H^2}{2} + \frac{g\Delta\rho}{2}h^2$$

The rate of change of this is less than $\rho_b W$, according to our energy principle, and thus we have

$$\frac{d}{dt}\frac{g\Delta\rho}{\rho_b}\frac{h^2}{2} < W \qquad [5a]$$

Moreover the total mass of the water column is

$$\rho_b (H - h) + (\rho_b - \Delta\rho)h$$

and since the rate of change of this must obviously be zero, we have

$$\frac{d}{dt}(h\Delta\rho) = 0 \qquad [5b]$$

From [4] and [5] we get

$$\frac{g}{2\rho_b}(h\Delta\rho)dh/dt < W \qquad [6a]$$

$$h(t) - h(o) < \frac{2\int_o^t W dt}{\frac{g\,\Delta\rho(o)}{\rho_b}h(o)} \qquad [6b]$$

which gives an upper bound on h(t). This may be converted to an equality by multiplying the right hand side of [6a] by the (unknown) fraction $\Gamma(t)$. This fraction decreases towards zero with time as mentioned above, because the turbulent velocities decrease with depth in the mixed layer. Therefore the rate of increase of potential energy will tend towards zero as $t \rightarrow \infty$, and the mixed layer depth h(t) tends to a certain limit, even with the continued application of a given wind stress (i.e. W = constant). For the qualifications and basis of these remarks see Kato and Phillips (1969).

Now consider the case where the fluid has initially a uniform vertical density gradient $\partial\bar{\rho}/\partial z<0$ with $\rho(o) = \rho_o$ denoting the initial surface density. After a certain time (Fig. 2-4b) the turbulence mixes the water conpletely in some depth h(t), leaving the underlying fluid unaltered. Consequently there is a density jump

$$\Delta\rho = \frac{h(t)}{2}(-\partial\bar{\rho}/\partial z) \qquad [7]$$

as is required by the conservation of mass (or mean density). By integrating $gz\bar{\rho}(z)$ for the profile shown in Fig. 2-5b we readily find that the rate of increase of potential energy is

$$g\left(-\frac{\partial\bar{\rho}}{\partial z}\right)\frac{h^2}{4}\frac{dh}{dt} \qquad [8]$$

where use has been made of [7]. This must be less than W, according to our energy principle and thus we find

$$h(t) < \left[\frac{12}{\frac{g}{\rho_b}\left(-\frac{\partial\bar{\rho}}{\partial z}\right)}\int_0^t W dt\right]^{1/3} \qquad [9]$$

For constant W the right hand side of this increases as $t^{1/3}$, which is an even smaller growth rate than that (t^1) which corresponds to (6b). The reason is that the mixing in the upper layer of Fig. 2-5b leads to increasing values of $\Delta\rho(t)$, thereby implying increasing interfacial buoyancy forces which effectively eliminate further mixing after a certain time interval.

CONVECTION

Certain aspects of Fig. 2-5 are similar, and certain aspects are different, if the source of energy W for mixing is due to evaporative cooling rather than a wind stress. Suppose that the upper layer in Fig. 2-5a is initially warmer than the lower layer, and that evaporation commences at time t = 0. Then the water will start to cool in a thin evaporative boundary layer of vertical thickness h_e, across which the density variation is $\alpha\Delta T_e$, where α is the coefficient of thermal expansion. The equilibrium value of these parameters is determined by the critical Rayleight Number (Stern, 1975; Katsaros et al, 1977):

$$\frac{g\alpha\Delta T_e h_e^3}{K_T\nu} \sim 10^3 \qquad [10]$$

where ν is the kinematic molecular viscosity and K_T is the thermal diffusivity. When ΔT_e and h_e correspond to the critical value given by Eq. [10], the evaporative boundary layer ceases to cool further and ceases to grow downwards, because the layer develops gravitational instabilities which immediately transports the heat of evaporation upwards. Another equation for $(\Delta T_e, h_e)$ is obtained by setting the molecular heat flux ($\sim K_T\Delta T_e/h_e$) at the surface of the water equal to the product of the evaporation rate and the latent heat of evaporation. For typical evaporation rates ΔT_e is of the order of 1°C. These convectively produced temperature fluctuations decrease with depth, because the descending elements are mixed as they sink in the upper layer. The gravitationally driven velocities, on the other hand, *increase* with distance below $z = 0$, and these velocities produce an entrainment at the base of the mixed layer which is similar to that discussed above. But there is a qualitative difference in so far as the mechanically driven turbulence *decreases* with depth, whereas convective turbulence increases with depth and is therefore more effective in increasing the mixed layer depth. Nevertheless the interface, or the pycnocline (Fig. 2-5), will still be eroded at the top and will not overturn until $h(t) \rightarrow -H$. This means that no bottom water is formed until such time as the *entire* stratification in the whole water column is destroyed.

This important conclusion depends critically on horizontal homogeneity, and the mechanism may (perhaps) be applicable to extratropical fresh water lakes, which exhibit a vernal overturning. But a complete winter time overturn does not occur in (for example) the polar pycnocline because the surface water is relatively fresh, and it is difficult to find any sizeable portion of that ocean where the density is ever uniform with depth. The question then is how the cold bottom temperature is actually maintained, and I suppose that similar problems must arise with tropical lakes and with saline lakes which do not exhibit a vernal overturn.

DOUBLE DIFFUSIVE CONVECTION

The subject under discussion now represents one way in which (winter time) surface cooling can result in a significant cooling of the bottom water without a general destruction of the pycnocline. Suppose the initial density stratification in Fig. 5b is due entirely to salt, and that the surface is then cooled at a constant rate (Turner, 1973). As previously mentioned, an upper homogeneous layer (beneath the evaporative microlayer) is once again formed, but now this layer ($0 > z > -h$) will be *cooler* than the (salt stratified) fluid below. Once again turbulent entrainment causes h to increase with time, and dh/dt decreases as h increases. At this stage, however, molecular diffusion of heat across the discontinuity surface at $z = -h$ becomes important, and results in a cooling of a thin layer on the *underside* of the discontinuity. The unstable temperature gradient here exceeds the stable salinity gradient, and convection and mixing is thereby initiated on the underside of $z = -h$. Thus a second homogeneous layer starts to grow downwards, with the overlying layer ($o > z > -h$) remaining in place and merely transporting heat upwards. The essential point here is that the molecular diffusivity of salt K_S is two orders of magnitude smaller than K_T, and therefore the conductive salt flux across $z = -h$ is small in comparison with the conductive heat flux. As the second mixed layer grows by eroding the underlying halocline, a stage is eventually reached where molecular heat diffusion across its lower boundary again becomes important and forms a third layer in the same way that the second layer was formed. By this mechanism it is possible to realize a series of convecting layers, separated by much thinner interfaces, such that the temperature, salinity, and density at the center of each layer is greater than that at the center of the overlying layer. The double diffusive mechanics allows the potential energy associated with the temperature field to be released without a general destruction of the pycnocline, and the upwards flux of heat is much larger than that which would be computed by the product of K_T and the *mean* vertical temperature gradient.

Such layers have been convincingly documented in the Arctic Ocean (Neshyba, Neal, and Denner, 1971), and in Lake Powell (Osborne, 1973), among other places. The mechanics at the interface between convecting layers has recently been discussed by Linden and Shirtclife (1978). For the inverse case in which a layer of hot, salty and light water overlies a cold, fresh layer (Stern, 1975) we find "salt fingers" in the interface, and these release the potential energy in the salinity stratification. Williams (1974) has obtained shadowgraph photos of arrays of such fingers in the ocean.

A number of exotic possibilities for water mass formation and transformation by these mechanisms have been explored by Turner and Chen (1974), and these authors kindly made available a film of their experiments, which was shown at this symposium. Since it is very hard to use heat as the working substance Turner and Chen used sugar and salt solutions instead (the diffusivity of sugar is one third that of salt). They produced an initial stratification in which sugar concentration decreased linearly upwards, the salt concentration increased linearly upwards, but the resultant density field decreased upwards. The magnitudes were chosen such that the system was stable to all *infinitesimal* perturbations, due account being taken of the diffusivities. But a small *finite* perturbation was then introduced in the form of a sloping sidewall boundary (Phillips, 1970), which causes the isopycnal surfaces to depart from the horizontal in a very thin boundary layer near the sloping wall. The ensuing motions are, however, completely different from what would occur if only one substance was used in producing the stratification. The experiment shows that the original double diffusive equilibrium is metastable, and that such a finite perturbation can release the potential energy in one of the two fields which contribute to the density stratification. The new stationary state, which evolves after a long time, consists of convecting layers separated by thin interfaces. But the manner in which this new state arises involves a complex system of "large scale" lateral eddies, or intrusive wedges, which originate at the inclined boundary and then spread accross the tank. The double diffusive effects occur at the interfaces of these wedges and provide the prime energy source for the motion. The experiment shows how significant heat exchange between the bottom and top of a salt stratified water body can occur without a complete overturn anywhere.

This lecture was prepared at the Max-Planck Institut für Meteorologie, and I would like to thank Prof. Klaus Hasselmann for inviting me to spend my sabbatical year there.

BIBLIOGRAPHY

Anati, A., Assaf, G., and Thompson, R., 1977. Laboratory Models of Sea Straits. J. Fluid Mech. 81, 341–351.

Assaf, G., 1977. The Great Salt Lake: An application of physical studies. Unpublished manuscript.

Browand, F. K. and Winant, C. D., Laboratory observations of shear layer instability in a stratified fluid. Boundary Layer Meteorology, 5, 67–77.

Kato, H. and Phillips, O. M., 1969. On the penetration of a turbulent layer into a stratified fluid. J. Fluid Mech, 37, 643–55.

Katsaros, K. B., Liu, W. T., Businger, J. A. and Tillman, J. E., 1977. Heat Transport and Thermal Structure in the interfacial boundary layer measured in an open tank of water in turbulent free convection. J. Fluid Mech, 83, 2, 311–355.

Linden, P. E. and Shirtcliffe, T. G., 1978. The diffusive interface in double-diffusive convection. J. Fluid Mech. (To be published).

Neshyba, S., Neal, V. T. and Denner, W., 1971. Temperature and Conductivity measurements under Ice Island. T–3, J. Geophys. Res., 76, 8107–8120.

Osborne, T. R., 1973. Temperature Microstructure in Powell Lake. J. Phys. Ocean., 3, 302–307.

Phillips, O. M., 1970. On flows induced by diffusion in a stably stratified fluid. Deep-Sea Res., 17, 435–443.

Stern, M. E., 1975. Ocean Circulation Physics. Academic Press, New York.

Thorpe, S. A., 1973. Turbulence in stably stratified fluids: A review of laboratory experiments. Boundary Layer Meteorology, 5, 95–119.

Turner, J. S. and Chen, C. F., 1974. Two-dimensional effects in double diffusive convection. J. Fluid Mech., 63, 577–592.

Weinberger, Z., 1978. Temperature distribution in a solar lake and its hydrodynamic stability (unpublished report).

Woods, J. D. and Wiley, R. L., 1972. Billow Turbulence and Ocean Microstructure. Deep Sea Res., 19, 87–121.

Williams, A. J., 1974. Salt Fingers Observed in the Mediterranean Outflow. Science 185, 941–943.

Chapter 3

ECOLOGY OF HYPERSALINE ENVIRONMENTS

HELGE LARSEN

Department of Biochemistry, Norwegian Institute of Technology, University of Trondheim, Trondheim, Norway

INTRODUCTION

About 30 years ago ZoBell (1946) reported some observations, parts of which are quoted in Table 3-1 and which might aptly serve as a starting point for the present discussion. ZoBell observed the ability of bacteria from habitats of different salinity to grow on nutrient media of different salinity. By salinity of the nutrient medium was meant the content of NaCl. Using sewage (low or no salt) as inoculum into the nutrient media, he found the highest number of bacteria developing in the media with the lowest salt concentrations. Increasing salt concentrations prevented the growth of an increasing proportion of the bacteria. These findings were not surprising: NaCl is well-known as an agent against bacterial growth; this is the basis for its widespread use as a preservative against bacterial spoilage. When using sea water as the source of bacteria, the highest number developed in the nutrient media at sea water salinity. Much lower numbers developed both at higher and lower salinities. At saturating salt concentration none of the bacteria from sea water developed. Samples from Great Salt Lake gave the highest viable count of bacteria on media saturated with NaCl. Lowering the NaCl concentration gave decreasing viable counts. At the lowest NaCl concentration tested (0.5%) was found 1/50 of the count at saturation. The findings when using marine saltern as the source of bacteria, were strikingly similar to those from Great Salt Lake.

TABLE 3-1.

Comparative bacterial counts obtained by ZoBell (1946) by plating from habitats of different salinity on nutrient media of different salinity.

Source of bacteria	Salinity of nutrient media (%)					
	30	15	7	3.5	2	0.5
	Average growth index					
Sewage	0	3	6	14	45	100
Sea water	0	7	38	100	62	19
Great Salt Lake	100	72	26	14	8	2
Marine Saltern	100	65	30	19	13	4

The figures of Table 3-1 deal with bacteria only, and even limited to those developing on a specified nutrient medium. The findings lead, nevertheless, to a postulate for which a good deal of additional evidence has accumulated since ZoBell made his observations more than 30 years ago: The indigenous population of hypersaline environments are, as a rule, rather special organisms adapted to live in the strong brine, and they even prefer, or require, the high salinity in their environment for growth and reproduction.

NaCl is the dominating salt component of the hypersaline environments, ponds and lakes, throughout the world. There are, however, some exceptions to this rule. The Dead Sea contains, in addition to Na^+, a high concentration of Mg^{++}. In some cases Na_2CO_3 may be a dominating component; so has also been found for $CaCl_2$. The hypersaline environments are mostly aerobic, but also anoxic situations are encountered. The acidity may differ considerably from one environment to another.

Through the years many observations have been reported on life in hypersaline environments. We are, however, still far from a thorough understanding of the ecology of such environments. The observations are sparse and scattered and often irrelative so that they are difficult to compare. Still a rough picture is emerging from somewhat more detailed studies of a few geographically well-known hypersaline environments, namely Great Salt Lake (Utah), the Dead Sea (Israel), and the alkaline lakes of Wadi Natrun (Egypt). A survey of the known ecological relations of these environments is given in the following, together with those of the marine salterns which are situated in coastal areas various places around the world. These environments have some major ecological properties in common, but there are also characteristic differences.

GREAT SALT LAKE

Great Salt Lake is predominantly a NaCl lake. The saline water also contains a relatively high, but not dominating, amount of SO_4. Twenty years ago Great Salt Lake contained about 20% solids throughout the lake, the only dominating chemical component being NaCl, as mentioned. In 1957 a rock-filled railroad causeway was completed across the lake, dividing it into a northern and a southern basin. The only connection between the two basins are two small culverts, each about 5 m wide and 3 m deep. About 95% of the water from watershed streams flow into the southern part of the lake. During the past 20 years the salt concentration has decreased to about 12–13% in the southern part, and is estimated to reach about sea water salinity in another 20 years. On the other hand, in the northern basin evaporation has caused the salt concentration to reach saturation (Table 3-2), and thus created an interesting ecological situation which has recently been studied in some detail by Post (1977).

In the southern part of the lake there is at present a fairly rich population of organisms comprising a wide variety of types, as compared to the northern part where only a few but characteristic types of organisms occur in conspicuous numbers.

In the northern basin the hypersaline water column is not more than 10 m deep. Most of the brine is aerobic although the concentration of O_2 in the brine is modest. At the bottom, however, and certainly in the sediment, the condition is for the most part anoxic

TABLE **3-2.**

Major ions of hypersaline lakes and brines, in g/l.

	Great Salt Lake, northern basin[1]	Dead Sea[2]	Wadi Natrun lakes[3]		Marine saltern Puerto Rico[4]
			Zugm	Gaar	
Na^+	105.4	39.2	142.0	137.0	65.4
K^+	6.7	7.3	2.3	1.4	5.2
Mg^{++}	11.1	40.7	0	0	20.1
Ca^{++}	0.3	16.9	0	0	0.2
Cl^-	181.0	212.4	154.6	173.7	144.0
Br^-	0.2	5.1	–	–	–
$SO_4^=$	27.0	0.5	22.6	48.0	19.0
$HCO_3^-/CO_3^=$	0.72	0.2	67.2	6.6	–
Total salinity	332.5	322.6	393.9	374.2	253.9
pH	7.7	5.9–6.3	11.0	10.9	–

1 Post (1977)
2 Nissenbaum (1975)
3 Imhoff *et al.* (1978)
4 Nixon (1970)

and H_2S can be shown to be present. The temperature varies from –5°C during the winter to +40°C during the summer. pH is about 7.7.

The dominating organisms of the northern basin, their numbers and biomass, are listed in Table 3-3; the data are taken from the interesting article of Post (1977).

The typical brine algae, *Dunaliella salina* and *D. viridis*, are the dominating primary producers. Although they seem to thrive better at lower salt concentrations (Brock, 1975) they apparently grow and reproduce quite well in the hypersaline brine of the northern basin. *D. salina* occurs as a planktonic form. 0.2 – 1 x 10^3/ml are common numbers observed, and peak blooms of 10^4/ml which corresponds to a maximum estimated biomass of 24 g/m^3. *D. virdis* is not common in the plankton but mostly occurs on the underside of wood and rocks where the cells are not exposed to direct sunlight, and altogether in relatively modest numbers. This is interesting in view of the fact that in the southern and less saline basin of Great Salt Lake *D. viridis* is found to be the principal planktonic alga, reaching levels as high as 2 x 10^5/ml, *i.e.* a biomass of about 150 g/m^3 (Stephens and Gillespie, 1976).

Post (1977) found that in the northern basin the chemoorganotrophic bacteria are by far the most conspicuous members of the biota. The dominating types are the extremely halophilic bacteria of the genera *Halobacterium* and *Halococcus*. Their numbers do not show much fluctuation through the year. An average of 7 x 10^7/ml was observed. This corresponds to a biomass of about 300 g/m^3, the figures being somewhat lower during the winter and somewhat higher during the summer. A biomass of 300 g/m^3 is indeed an impressive figure. The halobacteria and the halococci, being red in color due to carotenoids

TABLE 3-3.

Characteristic organisms of Great Salt Lake, northern basin (Post, 1977). Less conspicuous or less known types are shown in brackets.

	no./ml	g/m3
Algae		
Dunaliella salina (max.)	10^4	24
Dunaliella viridis (max.)	0.2×10^4	1.4
Chemotrophic bacteria		
Halobacterium *Halococcus* (av..)	7×10^7	300
(Sulfate reducing bacteria)		
(Others)		
(Cyanobacteria)		
(Protozoa)		
Brine shrimp	1	0.1
Brine fly		

in the cells, are present in the hypersaline brine in such quantities that they impress a red color upon the brine. Their biomass exceeds by at least a 10-fold that of the maximum observed algal bloom.

Brock (personal communication) has observed that also other chemotrophic bacteria may be present in high numbers, but these bacteria have not yet been closer identified. Sulfate reducing bacteria and methane producing bacteria are suspected to be present since hydrogen sulfide and methane production have been observed, but such bacteria have not been isolated. Bacteriophage specific for halobacteria have been shown to occur, and may play an ecological role by their ability to lyse the bacteria.

Cyanobacteria and protozoa have been encountered in the hypersaline water of the northern basin, but in modest numbers and may not be very active. Some of them are suspected to be brought in by the water flowing through the culverts.

The brine shrimp (*Artemia salina*) and the brine fly (*Ephydra* spp.) are characteristic organisms encountered during the summer. Both the brine shrimp and the brine fly larvae seem to do well in the extremely salty environment of the northern basin. The eggs of the brine fly are hatched in the strong brine, and the larvae apparently feed well on the microorganisms. The eggs of the brine shrimps have not been demonstrated to hatch in the strong brine of the northern basin. It is therefore possible that the larvae (nauplia) and the shrimp are carried into the northern basin with the water from the south (Post, 1977).

Turning now to the ecological aspects of the northern basin of Great Salt Lake it should first be pointed out that the temperature is quite low during the winter (<5°C in the period December–March), and not much metabolism seems to go on. The bacteria are probably resting, and the algae too. No brine shrimp or brine fly is there. Only when the temperature reaches above 10°C in April/May is there an increase in bacteria, but not much. The algae (*D. salina*) increase significantly in number, and eventually (June/July) the brine fly and the brine shrimp appear. The temperature keeps in the range 20–35°C most of the time from June through September.

The algae are, of course, the primary producers, and they provide food for the organotrophic community, both the brine shrimp, the brine fly larvae and the bacteria. The brine shrimp is known to feed directly on the algae, especially *D. viridis*, and this is also likely to be the case, but not proved, for the brine fly larvae. The bacteria are envisioned to feed on excretion products of the algae, possibly glycerol to a significant extent and autolysates of the algae and the small animals. That is the simple food chain of this almost closed ecosystem.

A point of interest is that the content of dead organic material is rather high in the strong brine of the northern basin. There is a good deal of insoluble organic material in the form of pupal cases, dead brine shrimp and egg masses. This insoluble material seems to be very slowly converted by the bacteria. The content of soluble organic material has not been determined directly, but has been estimated to be of the order 100 mg/l which is a high figure compared to other bodies of water. The chemical nature of the soluble organic material has not been determined (Post, 1977).

According to Post (1977) nitrogen seems to be a key element in the ecosystem of the northern basin. Nitrate and nitrite could not be detected in the brine. Ammonia seems to be the main source of inorganic nitrogen, being utilized as such by the algae, and may at times be a limiting factor since it is present in long periods in very small (undetectable) amounts. No nitrogen fixation, nitrification or denitrification have been detected. The N cycle may thus be very simple: NH_3 is converted by the algae to organic N which in turn is converted back to NH_3 by the metabolism of the organotrophs, the latter step possibly in significant part through uric acid, an excretion product of the brine fly larvae.

As mentioned above sulfate is an abundant, but not dominating, component in the brine. At the bottom, *i.e.* at the lower 1–2 m of the not more than 10 m deep water column, and in the sediment, anoxic conditions prevail, and H_2S can be shown to be present. This is a strong indication of the presence of halophilic or halotolerant sulfate reducing bacteria, but they have not been studied.

Post (1977) reported that when simply keeping lake water from the northern basin in the laboratory at 28°C for 6 months, the bacteria and algae counts rose by one or more orders of magnitude above the highest levels observed in the lake. Post suggested that this observation might give some clue to which might be limiting factors for the biota under natural conditions. One important factor may be the temperature; the summer period available for growth is short. In the laboratory experiment the grazing invertebrates were not present. It is just possible that under natural conditions these invertebrates put a constraint on the population size of the algae and thus limit their organic output which provides food for the bacteria. A third factor which at times probably limits the biota under natural conditions is ammonia.

THE DEAD SEA

The Dead Sea differs hydrologically from Great Salt Lake in a number of respects. The total salinity (*i.e.* the sum of salts in g/l) is about the same, but the salt composition differs (Table 3-2). Concerning the cations the major differences are a much lower content of Na^+ and much higher contents of Mg^{++} and Ca^{++} in the Dead Sea. The very high content of Mg^{++} is indeed a unique phenomenon.

The dominating anion in the Dead Sea is Cl^-, such as is generally encountered in hypersaline, natural brines. However, $SO_4^=$ is very low compared to other brines, as f.ex. Great Salt Lake.

Another characteristic of the Dead Sea is its great depth, with a maximum of 320 m. According to Nissenbaum (1975) one should from an ecological point of view distinguish between the upper water mass, from the surface to about 80 m, which is aerobic and has temperatures in the range 23–36°C, and the lower water mass below 80 m which is anoxic, contains H_2S (0.5–1 mg/l), and has a fairly constant temperature in the range 21–23°C. pH is in the range 5.8–6.4 throughout the water column; this is noticeably lower than in Great Salt Lake.

The Dead Sea was traditionally believed to be completely barren of indigenous life until Volcani (1940, 1944) published his observations almost 40 years ago. He described the occurrence of the salt alga *Dunaliella*, a variety of different bacteria among which the halobacteria seemed to be prominent, and he described cyanobacteria, a ciliate and an amoeba. Since that time little was done to gain further understanding of the life in the Dead Sea until the work of Kaplan and Friedman (1970) who were the first attempting to quantify the biota of the Dead Sea (Table 3-4). The microbiology and biogeochemistry of the Dead Sea was ably reviewed and discussed by Nissenbaum (1975).

In the upper, aerobic water mass (<80 m) *D. viridis* is the dominating alga and possibly the only alga indigenous to the sea. It occurs in surface water of the Dead Sea in numbers of about 4 x 10^4 cells/ml. This corresponds to a biomass of the same order as that observed by Post (1977) for *D. salina* in the northern basin of Great Salt Lake. The number of *D. viridis* decreases rapidly with depth in the Dead Sea. At 50 m are found only 1/100 of that at the surface; at 100 m no algal cell was detected.

The dominating bacteria of the upper water mass are, as in Great Salt Lake, members of the genera *Halobacterium* and *Halococcus*. The halobacteria are the most common but appears to occur in much smaller numbers than in Great Salt Lake. The highest counts were found at the surface, about 7 x 10^6 cells/ml, corresponding to a biomass of 30 g/m^3. This is only 1/10 of that observed for this bacterium in Great Salt Lake. At 100 m depth the bacterial count was decreased about tenfold; at 250 m the decrease was still another tenfold. The numbers did not seem to vary much through the year.

In the lower anoxic water mass of the Dead Sea (>80 m) the biological activity seems to be very modest. However, already Volcani (1944) at the early date demonstrated by the use of elective culture technique that in mud, sediments and deeper layers of water, there are the potential abilities to carry out glucose fermentation, peptone decomposition, denitrification, sulfur oxidation and cellulose decomposition. The responsible organisms, chemotrophic bacteria, were not isolated and characterized. In recent years sulfate reducing bacteria from deeper waters and sediments have been studied, and are believed to play a

TABLE 3-4.

Characteristic organisms of the Dead Sea (Kaplan and Friedmann, 1970). Less conspicuous or less known types are shown in brackets.

	no./ml	g/m^3
Surface water		
Algae		
Dunaliella viridis (max.)	4×10^4	ca. 50
Chemotrophic bacteria		
Halobacterium (av.)	7×10^6	ca. 30
(*Halococcus*)		
(Others)		
Deep water/sediments		
(Chemotrophic bacteria)		
(Sulfate reducing bacteria)		
(Others)		
(Cyanobacteria)		
(Amoebae)		
(Protozoa)		

significant part in the S-cycle. The indications are that the bacterial sulfate reduction is extremely slow in the water column, but considerably more intense in the sediment (Nissenbaum, 1975).

On the basis of the information available it seems that the biomass of the Dead Sea is completely dominated by two types of organisms only: *D. viridis* and halobacteria. *D. viridis* is the primary producer; the halobacteria are the dominating mineralizing agents. In interesting contrast to Great Salt Lake zooplankton seems to be absent from the Dead Sea. This means that there are no animals grazing on the alga and controlling the size of its population. The bacteria are envisioned to feed on excretion products and lysates of the alga plus organic matter brought into the lake with water flowing in. The major source of in-flowing water is the river Jordan.

Dissolved organic material in the Dead Sea is about 10 mg/l, which is a fairly high figure for water bodies, but only 1/10 of that of Great Salt Lake. In addition NH_3 (2–6 mg/l) and organic N (0.3–3 mg/l) are quite high, but phosphate is very low and may possibly be a limiting factor (Nissenbaum, 1975).

It has been suggested that light might be an important limiting factor for *Dunaliella*, and may possibly be the dominating one since a high content of suspended material prevents the light from penetrating deep into the lake. Measurements have shown that the light intensity is reduced to 1% at 30 m depth (Kaplan and Friedman, 1970).

No clear conclusions have yet been reached as to which are the factors limiting and controlling the biota of the Dead Sea, its size and turn-over rates. In addition to the factors mentioned above, O_2, temperature, the high concentration of Mg^{++} (and Ca^{++}), and also the high concentration of certain trace elements, have been suggested to play a limiting role.

ALKALINE HYPERSALINE LAKES

Hypersaline lakes of high alkalinity are found in many places. Baas Becking (1928) reported the occurrence of red-colored bacteria and cyanobacteria from such localities in western United States. The red-colored bacteria may be present in such numbers that they color the brine, and Jannasch (1957), studying "die bakterielle Rotfärbung" in the alkaline salines of Wadi Natrun in Egypt, found that the dominating red-colored bacteria there were phototrophic sulfur bacteria (*Thiorhodaceae*). Some of the organisms contained sulfur globules inside the cells and seemed to belong to the *Chromatium/Thiocystis* group. There were also other types: vibrios and spirilla. Jannasch reported that these bacteria seemed to be extremely halophilic (no development at salt concentration below 20%) and some even alkaliophilic (best development in the pH-range 8.5–10.5).

More recently pure cultures of phototrophic sulfide-oxidizing spirilla that store sulfur out-side the cells, have been isolated from alkaline, hypersaline lakes. These organisms belong to the genus *Ectothiorhodospira*. Raymond and Sistrom (1967, 1969) described an organism isolated from Summer Lake, Oregon, which they named *E. halophila*; Imhoff and Trüper (1977) described an organism isolated from Wadi Natrun, Egypt, which they named *E. halochloris*. Both organisms are extremely halophilic; best growth at around 20% NaCl and still good growth at 30% NaCl.

At present Trüper and his collaborators are engaged in an extensive study of the ecology of the alkaline hypersaline lakes of the Wadi Natrun area in Egypt (Imhoff *et al.*, 1978). The Wadi Natrun is located northwest of Cairo. The bottom of this area is 23 m below sea level and a chain of shallow lakes extends along the deepest part. Water is slowly supplied from the ground, originating from the Nile. An extensive evaporation takes place and the water becomes highly enriched with mineral components.

Imhoff et al. (1978) report on the chemical composition of six of the lakes of the Wadi Natrun. The data listed in Table 3-2 are from two of the lakes of extremely high salinity and show some characteristic patterns: The dominating cation is Na^+. The dominating anion is Cl^-, but also $SO_4^=$ is high and so may $CO_3^=$ be. All the six lakes investigated showed pH values of about 11. Some of the lakes contained sulfide in the water, others did not at the time of the sampling. It has, however, been shown that sulfate-reducing bacteria are quite active in the ground, supplying H_2S to the water and playing an important role in the development of alkalinity (Abd-el-Malek and Rizk, 1963a, b).

None of the six lakes investigated by Imhoff *et al.* (1978) contained animals ranking above unicellular protozoa. The biota of microbes differed somewhat from one lake to the next, but still there were characteristic patterns, and only few types seemed to be represented in significant numbers (Table 3-5).

Four of the six lakes had an extremely high salinity (>30% total salts) and these lakes were all red in color. The absorption spectra, supported by observations in the microscope, revealed that the color was mainly due to the presence in large numbers of halobacteria. This finding was in some contrast to that of Jannasch (1957) who reported that the red

TABLE 3-5.

Characteristic organisms of the hypersaline lakes of Wadi Natrun – Egypt (Imhoff et al., 1978). Less conspicuous or less known types are shown in brackets.

Algae
Dunaliella salina
Phototrophic sulfur bacteria
Chromatium
Ectothiorhodospira halophila
Ectohtiorhodospira halochloris
Cyanobacteria
Spirulina
Others
Chemotrophic bacteria
Halobacterium
Sulfate reducing bacteria
(Others)
(Protozoa)

color of the Wadi Natrun brines examined by him was due to phototrophic sulfur bacteria. However, phototrophic sulfur bacteria (*Chromatium, Ectothiorhodospira*) were found in large numbers in the extremely saline Wadi Natrun brines also by Imhoff *et al.*, but the latter investigators pointed out that in the cases examined by them masses of phototrophic sulfur bacteria, and also cyanobacteria, tended to grow attached to the sediment often forming dense mats, while the halobacteria more tended to occur evenly suspended in the brine. The phototrophic bacteria were, however, also found in large numbers suspended in the water of the lakes, preferably near the bottom. *Dunaliella salina* was reported to give mass development. Flagellated protozoa were observed feeding upon the bacteria, but seemed to be relatively few in numbers.

The findings of Imhoff *et al.* strongly supported Jannasch' contention of a simple food chain in the Wadi Natrun lakes, based to a large extent upon a cycling of sulfur. Decaying halotolerant grass, of which there is an ample supply around the lakes, provides suitable organic matter for sulfate reducing bacteria. These bacteria produce CO_2 and H_2S which, in turn, serve as substrates for the phototrophic sulfur bacteria producing new organic matter and oxidizing the sulfide via elemental sulfur to sulfate. In addition to the grass of the surroundings and the phototrophic sulfur bacteria, also cyanobacteria and *Dunaliella* serve as primary producers; the cyanobacteria may play a role in the oxidation of sulfide. In addition to the sulfate reducing bacteria, the halobacteria seem to play an important role in the mineralization of the organic matter.

Imhoff *et al.* (1978) reported biologically important compounds of carbon, sulfur, nitrogen and phosphate to be present in abundance in all the Wadi Natrun lakes studied, and these elements can therefore not be considered growth limiting, although certain compounds of certain of the elements, *e.g.* CO_2 and H_2S, may be limiting at certain times. The same holds true for O_2. Furthermore, the lake waters are deficient in metal ions other than sodium and potassium, and this may influence the growth of the biota.

MARINE SALTERNS

Marine hypersaline ecosystems are found in coastal areas many places in the world. Natural hypersaline lagoons may form from almost enclosed shallow coastal waters exposed to a bright sun so that an extensive evaporation takes place. More frequently such hypersaline lagoons or ponds are man-made for the purpose of commercial production of NaCl from sea water. The manufacture of salt by solar evaporation of sea water is an ancient art (Baas-Becking, 1931a), and the way of making it is in principle the same today as in the ancient times: Sea water is led into a system of shallow ponds arranged in series. The water is retained in each pond for a certain period while evaporation by the sun takes place. In this way a brine results with an increasing concentration of salts from one pond to the next, and a fractional precipitation of the salts of the sea water occurs in the various ponds. The least soluble salts, $CaCO_3$ and $CaSO_4$, will first precipitate. Then NaCl, being the dominating component of sea water, is crystallized out in the following ponds, and is harvested, often washed, and dried in the sun. The remaining brine, which is strongly enriched in $MgCl_2$ (bitter salt) and KCl, may go to other ponds for precipitation upon further evaporation.

Few thorough studies have been carried out on the ecology of marine salterns. It seems, however, from the observations reported that a certain pattern is common to many, if not most, of them and largely independent of geographical locations. The pattern is the following.

The biological productivity in the dilute brine of the first evaporation pond(s) is quite high (Carpelan, 1957). A number of different primary producers, notably cyanobacteria but also green algae, develop. The cyanobacteria (*Spirulina, Oscillatoria, Coccochloris, Lyngbya* and others) tend to form a mat at the bottom of the pond, while the green alga, *Dunaliella viridis*, grows planktonic. The brine shrimp, *Artemia salina*, and the brine ciliate, *Fabrea salina*, are characteristic organisms of the zooplankton.

As the water evaporates not only do the concentrations of total salts increase, but the relative proportions of the salt components in solution also change. The number of species able to develop decreases rapidly as the brine becomes more concentrated, and so does the biological productivity of the ecosystem (Copeland and Jones, 1965). The cyanobacteria eventually die at the higher salt concentrations. The brine algae, *Dunaliella viridis* and *D. salina* become the dominating primary producers (Gibor, 1956a, b; Nixon, 1970). At the higher salt concentrations also a considerable population of bacteria develop, notably members of the genera *Halobacterium* and *Halococcus* which are red in color.

In the ponds where NaCl crystallizes out the concentration of total salt is about 10 times that of the sea water. The brine in these ponds is frequently red in color, mainly due to the presence of a very large number (appr. 10^8/ml) of the above-mentioned red-colored, halophilic bacteria. Also *D. salina* may be present in the concentrated brines in such numbers that it contributes to the red color, but the alga appears to be considerably

TABLE 3-6.

Characteristic organisms of concentrated brines from marine salterns (Nixon, 1970). Less conspicuous or less known types are shown in brackets.

Algae
Dunaliella salina
Dunaliella viridis
Chemotrophic bacteria
Halobacterium
Halococcus
(Others)
(Brine shrimp)

hampered in its metabolism in concentrated salt solution (Gibor, 1956a). *D. viridis* is less dominating at the highest salinities (Table 3-6).

The red-colored bacteria of the genera *Halobacterium* and *Halococcus* have been considered the dominating types of concentrated brines in marine salterns (Larsen, 1962). When the salt precipitates myriads of the bacteria adhere to the crystals and may stay alive in the salt for a long time if no special precautions are taken to kill or remove them. Much of the salt is later used as a preservative against bacterial spoilage of proteinaceous material such as fish, hides, etc. This salt may contain as many as $10^5 - 10^6$ viable cells/g of the halophilic, red-colored bacteria which, when given the proper conditions (high humidity and temperatures) may develop in the proteinaceous materials and spoil them. This is a well-known phenomenon in many parts of the world, and a reason why these bacteria have been so extensively studied.

Dundas (1977) reported recently that he had found the majority of bacteria in red-colored brines from the marine salterns to be colorless. A description of these organisms is needed to complete our knowledge of the biota of marine salterns.

It seems as if aerobic conditions normally prevail in the brine of marine salterns, although dissolved oxygen may be low. The bottom floor of the shallow ponds, beneath the mat of cyanobacteria if present, often contains black mud regardless of the concentration of the brine in the pond. This indicates anaerobiosis in the mud beneath the brine and activity of sulfate reducing bacteria, but they have not been studied.

The brine shrimp, *Artemia salina*, is found also in the concentrated brines of marine salterns. Nixon (1970) reports 3500 individuals/m^2 from a marine saltern of 35% salinity in Puerto Rico. This is a modest number.

Dissolved organic matter in the concentrated brines of marine salterns is high. Nixon (1970) reported figures of about 100 mg/l of organic C, which should roughly correspond to 200 mg/l of organic matter. The high content of dissolved organic matter obviously stems from the primary producers abounding in the brines at the lower salt concentrations. As the salt concentration increases most of the algae die, they autolyze and leave in the brine organic material which serves as food for the population of chemoorganotrophic organisms developing, notably the red-colored, halophilic bacteria, but also other bacteria.

Nixon (1970) found a ratio of organic C to organic N of 14:1 in the concentrated brine of the Puerto Rico marine saltern investigated by him. This indicates that proteinaceous material, including amino acids, makes up a modest part of the soluble organic matter in the brine. The organic components of the brines of marine salterns have not been further identified, but Nixon (1970) remarks that excretion products of the algae may make up a substantial part.

The gradual increase in salt concentration from sea water level to at least 10 times sea water, makes the brine of the marine salterns a considerably more complex ecosystem than for example the brines of Great Salt Lake and the Dead Sea which are of relatively constant composition. The increase in the concentration of salt results in a strict selection of organisms, as outlined above. In addition, the removal of water by evaporation causes shifts in the relative proportions of the components of the brine, which may also affect the biota. Upon evaporation Ca^{++} of the sea water is precipitated as $CaCO_3$ and $CaSO_4$ so that the strong brines contain relatively little of this element. On the other hand Mg^{++}, K^+ and $SO_4^{=}$ become relatively enriched in the strong brines.

Nixon (1970) reported that the levels of inorganic phosphate and nitrate were very low in brines from marine salterns. Enrichment experiments indicated that nitrogen was a major limiting factor in the brines studied by him, meaning that lack of nitrogen in suitable form limited growth of the primary producers.

GENERAL CONSIDERATIONS

From the specific cases discussed in the foregoing, it appears that life may not be sparse in hypersaline brines and lakes. On the contrary, such brines, even at the highest salt concentrations, may be as densely populated and contain a biomass as high as only rarely encountered in sea water or fresh water bodies. However, in contrast to the dilute and fresh water environments, the hypersaline environments are characterized by a low species diversity. In the most concentrated brines the number of types of dominating organisms is, indeed, limited to very few. This is clearly seen from Tables 3-3–6 which list the characteristic organisms of four different, extremely saline environments, namely Great Salt Lake, the Dead Sea, Wadi Natrun and marine salterns. In this connection should be particularly emphasized that the quantitatively dominating organisms of the different brines are, quite regularly and consistently, the same, or closely related, organisms, notably the red-colored bacteria of the genera *Halobacterium* and *Halococcus* and the algae *Dunaliella salina* and *D. viridis*. Admittedly, our knowledge of the biota of the brines are still far from complete, and future work may well reveal additional organisms indigenous to such environments. Still, it seems justified to conclude from what we know at present that the ability to live in strong brines is limited to few organisms, and that these specialized organisms are ubiquitously distributed in such environments all over the world.

The red-colored bacteria of strong brines have attracted considerable attention, and some insight in their physiological and biochemical peculiarities has been gained (Larsen, 1973; Dundas, 1977). These organisms are truly halophilic in the sense that generally they require at least 10–15% NaCl in their environment for growth, and best growth is attained in almost saturated salt solutions, *i.e.* 20–30% NaCl. Their requirement for NaCl is specific. NaCl cannot be replaced by other salts. These organisms have overcome the osmotic problem created by the strong salt solution in the way that they take up salt in the cells to a concentration roughly the same as that in the environment. The internal salt is mainly KCl, but also NaCl. Surprisingly, the metabolic machinery is not adversely affected by the extremely high salt concentration inside the cells. On the contrary, the enzymes are stimulated by the salt; in its absence the enzymes become inactivated and irreversibly denatured. This exceptional behavior of the metabolic units appears to be due to some very special properties of the proteins of these organisms. Most, if not all, proteins are strongly acidic in nature, and the high concentration of salt seems to be required for neutralization of the proteins so that they are kept in a conformational state at which they are metabolically active. In the absence of salt a denaturation takes place; the proteins may even go into solution and the cells may lyse. In addition to ionic interactions hydrophobic phenomena of the proteins seem to play an important role in relation to the salt (Lanyi, 1974).

Most of the halobacteria and halococci seem to have the same extreme salt requirement, regardless of the extremely halophilic environment they live in. An exception – or possibly better, a variant – to this rule is described from the Dead Sea. *Halobacterium volcanii* (Mullakhanbhai and Larsen, 1975) isolated from Dead Sea mud has an optimum requirement for NaCl of only about 10%, and this is less than half of that required optimally by other halobacteria. As a matter of fact other halobacteria have a minimum NaCl requirement considerably higher than the optimum requirement for *H. volcanii*. On the other hand, *H. volcanii* is seriously hampered in its development at NaCl concentration being optimum for the other halobacteria. This behavior of *H. volcanii* is interesting in view of the fact that the NaCl content of the Dead Sea is relatively modest compared to other extremely saline water bodies (Table 3-2). The extreme salinity of the Dead Sea is to a large extent due to its content of $MgCl_2$, and it could be shown that *H. volcanii* is extremely tolerant to $MgCl_2$. In other words, *H. volcanii* is remarkably well adapted for life in the Dead Sea.

In all cases investigated the halobacteria and the halococci have been found to have a considerable requirement also for Mg^{++} (1–5% $MgCl_2$), and this has been reported as possibly a general property of these organisms (Larsen, 1967). Recently has been described mass occurrence of halobacteria in the lakes of Wadi Natrun which are extremely low in Mg^{++} and Ca^{++} (Table 3-2, Imhoff *et al.*, 1978). A further characterization of the salt requirements of the Wadi Natrun halobacteria might thus reveal variants, and may confirm the importance of the apt remark by Dundas (1977) that "it might be fruitful to pay more attention to the original habitat in future work on extracellular salt relationship of halobacteria".

The frequent red coloration of the strong brines due to the mass occurrence of bacteria is one of the most striking features of these ecosystems, and has led to speculations as to the possible function of the pigment in these organisms. Colorless mutants display the same behavior towards salt as the red-colored parent strains. There is thus no direct relation between coloration and halophilism. However, the colorless mutants are quite sensitive towards light. At the light intensity of a bright sun, the colorless mutants are strongly hampered in their development. The red-colored parent strains are not affected by the light (Dundas and Larsen, 1962). A physiological role of the red pigment is thus to protect the cells against the bright sun to which these organisms are so often exposed in their natural habitats. On the other hand, the absorption of light by the pigment may increase the temperature of the brine which, in turn, may stimulate the metabolism of these bacteria having an optimum temperature as high as 40–50°C.

It has been shown that halobacteria under oxygen limiting conditions form a protein-pigment with properties similar to that of rhodopsin, and that light energy absorbed by this chromoprotein, called bacteriorhodopsin, may be converted to chemical energy within the cells, and without the participation of chlorophyll of which these cells are devoid. It has been suggested that the halobacteria use this system as an auxiliary device for ATP production in the cell under oxygen limiting conditions where a limited amount of ATP is synthesized via the respiratory chain (Oesterhelt, 1976).

When freshly isolated from nature the halobacteria often produce gas-filled vacuoles which provide buoyancy to the cells (Walsby, 1975). Petter (1932) who first described gas vacuoles in the halobacteria, put forth the reasonable hypothesis that a function of these structures is to lift the obligate, aerobic bacteria, which live in an environment of low oxygen solubility, towards air.

It is as yet not understood how the red-colored, extremely halophilic bacteria have evolved in nature. The halobacteria and the halococci both have their non-halophilic counterparts to which they seem closely related except for the halophilic character. This character seems to be a stable one. The most convincing experiments to demonstrate an adaptation to higher or lower salt concentrations have failed. Biochemical considerations on the basis of our present knowledge imply an extremely large number of mutations in order to convert a non-halophile to an extreme halophile or *vice versa* (Larsen, 1962, 1967, 1973). Also the way these bacteria spread in nature is obscure. From numerous tests in the laboratory we know that, with few exceptions, they are killed when the salt concentration goes somewhat below 10%. Still they seem ubiquitous in nature where the salt concentration is high. These findings have nourished an idea that the extremely halophilic character might come about by some special gene-transfer mechanism (Dundas, 1977). This remains, however, to be shown.

Also the brine alga, *Dunaliella*, is regularly found in the strongest brines. Two species are referred to, *D. viridis* and *D. salina*. Authors are sometimes in doubt about species designation which indicates the need for a closer taxonomic study. Both species mentioned are strongly halophilic, but with a optimum salt requirement somewhat lower than the halobacteria and the halococci.

Dunaliella belong to the green algae and is thus an eukaryote. The halophilic representatives are equipped with a mechanism to overcome the osmotic strain of the hypersaline environment which is quite different from that of the extremely halophilic bacteria. *Dunaliella* quite effectively excludes the salt from the cells, and compensates for the osmotic strain by accumulating inside the cells a corresponding amount of glycerol. For example, *D. viridis* growing in 4.25 M NaCl was reported to accumulate glycerol to a concentration of 4.4 molal. At lower salt concentrations correspondingly lower contents of glycerol were found inside the algal cells (Ben-Amotz and Avron, 1973; Borowitzka and Brown, 1974). A similar mechanism is known to operate in other eukaryotes withstanding high salt concentrations, e.g. holotolerant yeasts accumulating glycerol when exposed to a hypersaline environment (Gustafsson and Norkrans, 1976).

Mass occurrence of phototrophic sulfur bacteria has this far only been reported from strong brines of high alkalinity, *e.g.* Wadi Natrun in Egypt. There is at present no particular reason to believe that a high pH is essential for the occurrence in strong brines of this type of organisms. The phototrophic sulfur bacteria require H_2S, CO_2 and the simultaneous presence of light; a requirement for a high pH is not a general characteristic of these organisms. As pointed out by Imhoff *et al.* (1978) a mass development of phototrophic sulfur bacteria cannot be expected in the Dead Sea because of the lack of H_2S in the photic zone. The same holds true for the shallow ponds of the marine salterns where aerobiosis prevails throughout the brine. It is more difficult to explain the absence of such bacteria in Great Salt Lake where H_2S is present near the bottom at less than 10 m depth. May be phototrophic sulfur bacteria will be found in Great Salt Lake when carefully looked for. Or could it be that the salt influences these particular organisms so that they come to prefer an alkaline environment? These problems and also the general problem of the biochemical basis for their salt requirement, remain to be elucidated.

A number of halophilic cyanobacteria are known, but their halophilism is not known to be so extreme as in the case of the halobacteria, the halococci, and the phototrophic sulfur bacteria. They are therefore not so frequently encountered, and appear unhealthy, in the strongest brines (Nixon, 1970). There is, however, one outstanding exception to this rule, namely the mass occurrence of apparently healthy cyanobacteria in the strongly saline lakes of Wadi Natrun (Imhoff *et al.*, 1978). A possible explanation may be sought in the old report by Baas Becking (1931b) that while cyanobacteria are tolerant of Ca^{++} and Mg^{++} at lower salinities, they become sensitive to these ions at high salt level. At 4 M NaCl the cyanobacteria tested would grow only when no Ca^{++} or Mg^{++} was present. The brines of Wadi Natrun are precisely characterized by their low contents of Ca^{++} and Mg^{++}, whereas other strong brines contain considerable amounts of one or both of these ions (Table 2).

The biochemical basis for the halophilic character of the cyanobacteria has only recently been looked into, and our knowledge is limited. When grown at 3 M NaCl *Aphanothece halophytica* accumulates K^+ to a concentration of somewhat less than 1 M. Little or no Na^+ is taken up (Miller *et al.*, 1976). It thus appears that K^+ plays the role of a osmoregulator as in the halobacteria and the halococci, but the intracellular concentration of K^+ is far from adequate to compensate for the external salt. An organic solute of a polyol nature, and possibly free amino acids in additon, seem to be the main osmoregulators, thus displaying a regulatory mechanism reminding of *Dunaliella* and other eukaryotes (Tindall *et al.*, 1977).

It appears from the foregoing that some insight has been gained in the ecology of hypersaline environments, but much remains to be elucidated. In some cases suggestions have been set forth as to which are the dominating or limiting ecological factors for the display of life in such environments. One feels the need for more data to support these suggestions. Possibly will more information about these ecosystems lead to different explanations.

A special point of interest is the fact that hypersaline environments often contain a relatively large amount of dissolved organic matter, and thus a potentiality to support a population of organotrophs far larger than actually found. It would be of considerable interest to have some information about the chemical composition of the dissolved organic matter.

Sulfur metabolism represents another area which may prove interesting to study. Many of the hypersaline brines contain a considerable amount of $SO_4^=$ (Table 3-2) and the sediment of the bottom floor of the hypersaline lakes and ponds is often characteristically black and contains H_2S, indicating activity of sulfate reducing bacteria. Very little is known about these bacteria, but in view of the findings that about half of the organic matter in marine sediments may be mineralized by anaerobic sulfate respiration (Fenchel and Jörgensen, 1977), such a process may possibly be of considerable importance also in the element cycling of the hypersaline ecosystems.

REFERENCES

Abd-el-Malek, Y. and Rizk, S. G., 1963a. Bacterial sulphate reduction and the development of alkalinity. II. Laboratory experiments with soils. J. appl. Bacteriol. 26, 14–19.

Abd-el-Malek, Y. and Rizk, S. G., 1963b. Bacterial sulphate reduction and the development of alkalinity. III. Experiments under natural conditions. J. appl. Bacteriol. 26, 20–26.

Ben-Amotz, A. and Avron, M., 1973. The role of glycerol in the osmotic regulation of the halophilic alga *Dunaliella parva*. Plant Physiol. 51, 875–878.

Borowitzka, L. J. and Brown, A. D., 1974. The salt relations of marine and halophilic species of the unicellular green alga, *Dunaliella*. Arch. Microbiol. 96, 37–52.

Brock, T. D., 1975. Salinity and the ecology of *Dunaliella* from Great Salt Lake. J. gen. Microbiol. 89, 285–292.

Baas-Becking, L. G. M., 1928. On organisms living in concentrated brine. Tijdschrift der Nederl. Dierkundige Vereenigung 3de Serie, Deel I, Aflevering 1, p. 6–9.

Baas-Becking, L. G. M., 1931a. Historical notes on salt and salt-manufacture. Scient. Monthly 32, 434–446.

Baas-Becking, L. G. M., 1931b. Salt effects on swarmers of *Dunaliella viridis* Teod. J. gen. Physiol. 14, 765–779 (cited from Nixon, 1970).

Carpelan, L. H., 1957. Hydrobiology of the Alviso salt ponds. Ecology 38, 375–390.

Copeland, B. J. and Jones, R. S., 1965. Community metabolism in some hypersaline waters. Texas J. Sci. 17, 188–205.

Dundas, I. E. D., 1977. Physiology of *Halobacteriaceae*. Adv. Microbial Physiol. 15, 85–120.

Dundas, I. D. and Larsen, H., 1962. The physiological role of the carotenoid pigments of *Halobacterium salinarium*. Arch. Microbiol. 44, 233–239.

Fenchel, T. M. and Barker Jörgensen, B., 1977. Detritus food chains of aquatic ecosystems: the role of bacteria. Adv. Microbial Ecol. 1, 1–58.

Gibor, A., 1956a. The culture of brine algae. Biol. Bull. 111, 223–229.

Gibor, A., 1956b. Some ecological relationships between phyto- and zooplankton. Biol. Bull. 111, 230–234.

Gustafsson, L. and Norkrans, B., 1976. On the mechanism of salt tolerance. Production of glycerol and heat during growth of *Debaromyces hansenii*. Arch. Microbiol. 110, 177–183.

Imhoff, J. F. and Trüper, H. G., 1977. *Ectothiorhodospira halochloris* sp. nov., a new extremely halophilic phototrophic bacterium containing bacteriochlorophyll b. Arch. Microbiol. 114, 115–121.

Imhoff, J. F., Sahl, H. G., Soliman, G. S. H., and Trüper, H. G., 1978. The Wadi Natrun: chemical composition and microbial mass developments in alkaline brines of eutrophic desert lakes. Geomicrobiol. J., in press.

Jannasch, H. W., 1957. Die bakterielle Rotfärbung der Salzseen des Wadi Natrun (Ägypten). Arch. Hydrobiol. 53, 425–433.
Kaplan, I. R. and Friedman, A., 1970. Biological productivity in the Dead Sea. Part 1: Microorganisms in the water column. Israel J. Chem. 8, 513–528.
Lanyi, J. K., 1974. Salt-dependent properties of proteins from extremely halophilic bacteria. Bacteriol. Rev. 38, 272–290.
Larsen, H., 1962. Halophilism. In: The Bacteria (Gunsalus, I. C. and Stanier, R. Y. (eds.) Vol. IV p. 297–342. Academic Press, New York.
Larsen, H., 1967. Biochemical aspects of extreme halophilism. Adv. Microbial Physiol. 1, 97–132.
Larsen, H., 1973. The halobacteria's confusion to biology. Ant. v. Leeuwenhoek J. Microbiol. Serol. 39, 383–396.
Miller, D. M., Jones, J. H., Yopp, J. H., Tindall, D. R. and Schmid, W. E., 1976. Ion metabolism in the halophilic blue-green alga, *Aphanothece halophytica*. Arch. Microbiol. 111, 145–149.
Mullakhanbhai, M. F. and Larsen, H., 1975. *Halobacterium volcanii* spec. nov., a Dead Sea halobacterium with a moderate salt requirement. Arch. Microbiol. 104, 207–214.
Nissenbaum, A., 1975. The microbiology and biogeochemistry of the Dead Sea. Microbial Ecol. 2, 139–161.
Nixon, S., 1970. Characteristics of some hypersaline ecosystems. Doctor's thesis, University of North Carolina, Chapel Hill, North Carolina.
Oesterhelt, D., 1976. Isoprenoids and bacteriorhodopsin in halobacteria. Progr. Molec. Subcell. Biol. 4, 134–166.
Petter, H. F. M., 1932. Over roode en andere bacteriën van gezouten visch. Doctor's thesis, University of Utrecht, Utrecht.
Post, F. J., 1977. The microbial ecology of the Great Salt Lake. Microbial Ecol. 3, 143–165.
Raymond, J. C. and Sistrom, W. R., 1967. The isolation and preliminary characterization of a halophilic photosynthetic bacterium. Arch. Mikrobiol. 59, 255–268.
Raymond, J. C. and Sistrom, W. R., 1969. *Ectothiorhodospira halophila:* a new species of the genus *Ectothiorhodospira*. Arch. Mikrobiol. 69, 121–126.
Stephens, D. W. and Gillespie, D. M., 1976. Phytoplankton production in the Great Salt Lake, Utah, and a laboratory study of algal response to enrichment. Limnol. Oceanog. 21, 74–87.
Tindall, D. R., Yopp, J. H., Schmid, W. E. and Miller, D. M., 1977. Protein and amino acid composition of the obligate halophile *Aphanothece halophyta* (Cyanophyta). J. Phycol. 13, 127–133.
Volcani, B. E., 1940. Studies on the microflora of the Dead Sea. Doctor's thesis, Hebrew University, Jerusalem.
Volcani, B. E., 1944. The microorganisms of the Dead Sea. In: Papers Collected to Commemorate the 70th Anniversary of Dr. Chaim Weizmann, pp. 71–85. Daniel Sieff Research Institute. Rehovoth, Israel.
Walsby, A. E., 1975. Gas vesicles. Ann. Rev. Plant Physiol. 26, 427–439.
ZoBell, C. E., 1946. Marine Microbiology. Chronica Botanica, Waltham, Massachusetts.

Chapter 4

DISTRIBUTION AND PURE CULTURE STUDIES OF MORPHOLOGICALLY DISTINCT SOLAR LAKE MICROORGANISMS

PETER HIRSCH

Institut für Allgemeine Mikrobiologie, Universität Kiel, Germany

The small hyperthermal, monomictic "Solar Lake" near Elat (Gulf of Aqaba, Sinai) is characterized by a rather homogeneous epilimnion of nearly ambient temperature and a hypolimnion with gradients of various environmental parameters (Cohen et. al., 1974; Cohen, 1975). The locations, size and form, physical, chemical and limnological conditions, as well as microbiological observations have been described recently by Cohen et. al. (1977a,b,c), Krumbein et. al. (1977) Jørgensen and Cohen (1977). While the epilimnion has a lower salinity (45–90‰), the hypolimnion has a gradually increasing salinity of up to 190‰, increasing anaerobiosis and increasing sulphide concentration. Due to heliothermal heating during stratification a temperature maximum is built up in the upper hypolimnion that may reach 60°C, depending on the annual weather conditions. Lower down the temperature drops again to approximately 40°C at the sediment surface. These conditions – fairly stable throughout winter, spring and early summer – produce a variety of ecologically well-definable locations which could be expected to be inhabited by highly-adapted, specific microorganisms.

Primary production data reported by Cohen (1975) and Cohen et. a. (1977b) indicate extremely high values for the early stratification period (4,960 mg C $m^{-3} d^{-1}$), measured at 4 m in the hypolimnion (November 20, 1970). This highest productivity ever recorded for non-polluted natural waters is in contrast to the low epilimnion production of 50 mg C $m^{-3} d^{-1}$. Cohen et. al. (1977c) discuss rates determined for dark CO_2 incorporations. The value found (1,014 mg C $m^{-3} d^{-1}$) could correspond to a chemoorganotrophic bacterial production of 16,900 mg C m^{-3} d^{-1}, assuming that these chemoorganotrophs assimilated approximately 6% CO_2 of their total C uptake. Thus, chemoorganotrophic production by far exceeded primary production, indicating vigorous heterotrophic bacterial activity in the Solar Lake.

The annual accretion rate for shallow water mats observed by Krumbein et. al. (1977) indicated with 5 cm 100 yr^{-1} an extremely high degradation (remineralization) rate of the organic product, probably due to a most active bacterial population. All of these observations point to a physiologically highly diverse microbial population in this unusual ecosystem. First data on the presence of distinct microorganisms were given by Krumbein and Cohen (1974) Cohen (1975), Hirsch (1977), Cohen et. al. (1977b,c), Krumbein et al. (1977), Krumbein and Cohen (1977), Hirsch et. al. (1977), and Hirsch (1978). Most of these observations, however, had focussed on photosynthetic bacteria and cyanobacteria.

The present investigation was undertaken as part of a program to study the potential use of morphologically recognizable microorganisms as bioindicators for environmental conditions. The Solar Lake with all its microhabitats was thought to be ideally suited for studying unusual bacteria that are highly adapted to a variety of extreme conditions.

A survey of all morphologically recognizable microorganisms was made, therefore, and first isolation attempts yielded pure cultures of new, highly unusual bacteria with great promise to be used as bioindicators.

MATERIALS AND METHODS

Temperature, salinity and sulphide determination. The methods employed have been described previously (Hirsch, 1978).

Sampling, direct microscopy and enrichments. Techniques for measuring temperature and pH, for sampling, direct microscopy, and for enrichments have also been described previously (Hirsch, 1978). In direct microscopy the sample (or slide, plastic foil, enrichment) was studied exhaustively until new morphotypes could no longer be found.

Identification and documentation. Samples, enrichments or pure cultures were observed light-microscopically and – when possible – by use of transmission electron microscopy. Morphologically distinct microorganisms as well as others were recorded as drawings, light- and/or electron micrographs. Size measurements were made, and morphological features such as form, pigmentation, storage granules, mode of division or budding, etc. were noted. Behavioral features such as motility, attachment, predation, aggregation etc. were also recorded. Every morphologically distinct type ("morphotype") of microorganism found was registered in a filing system with consecutive number and documented with pictures to facilitate later recognition. Thus, assuming that developing (differentiating) microorganism stages could be recognized as belonging together, the numbers of *different* microorganisms could only be greater, not smaller if a morphotype was overlooked. Identification of the morphotypes with known genera was often possible. Identification with known species was not attempted except for diatoms. Pure cultures isolated were studied for their range of morphological variability and attempts were made to identify these with previously observed morphotypes.

Distribution studies. Presence or absence of a morphotype in a network of sampling stations was recorded by combining data obtained from direct microscopy of water samples, from the study of exposed artificial or natural surfaces, sediments, enrichments, or pure cultures. Except for observations on behavior and pure-culture growth no data were collected so far on viability or activities of these morphotypes.

Pure culture studies. Medium PYGV was employed as described by Hirsch (1978). For experiments in Kiel the medium was prepared with 2 x concentrated artificial sea water (Lyman and Fleming, 1940).

Temperature gradient experiments. An aluminum block temperature gradient (constructed and kindly donated by Dr. P. A. Klein, Fort Collins, Colorado, USA) was used to study temperature-dependent growth response and morphological changes in selected pure culture isolates. At each temperature the block had 4 parallel borings for test tube cultures. Of these, 3 borings were used as triplicates and one for temperature measurements. One end of the block was heated to 64°C, the other end was cooled to 13°C.

RESULTS

Physical and chemical parameters. For an initial survey the Solar Lake was visited during the first two weeks of March 1976. A general profile and some parameters measured in the deepest part (center) during this time are shown in Fig. 4-1. While the epilimnion was homogeneous with respect to ambient temperature, low salinity and high aeration, there was a dramatic increase in both salinity and temperature between 0.5 m and 1 m depth. Between 1.5 and 2.5 m (the temperature maximum!) intensive rotten odors indicated actively decaying organic matter. These layers did not contain sulphide.

The condition were rather stable. The following distinct environments where microbial types and associations could be expected to be different could, therefore, be recognized:

1. An aerobic neuston layer and epilimnion of low salinity, ambient temperature, and very high light intensity.

2. An aerobic zone characterized by rapid increase with depth of temperature and salinity; light intensity still high.

3. A microaerophilic to anaerobic zone at the temperature maximum (52°C) with reduced light intensity and still increasing salinity

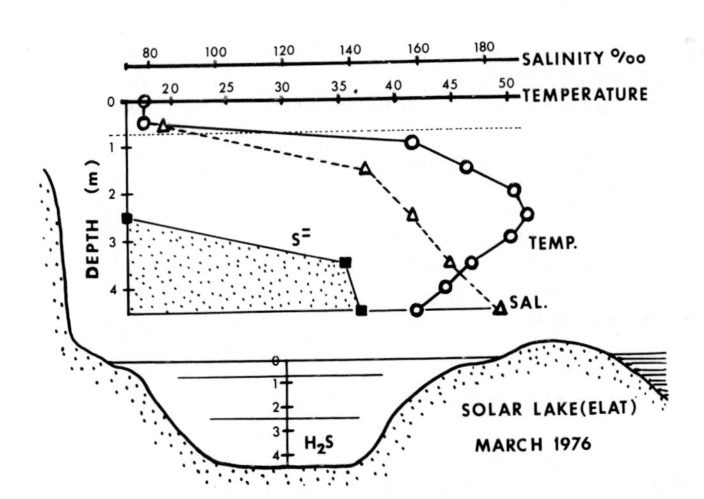

Fig. 4-1. Below: Solar Lake profile showing protection from wind (left) and barrier toward the Gulf of Aqaba (right). Epilimnion and hypolimnion indicated by horizontal separation lines.
Above: Salinity, temperature and sulphide profiles for the central sampling station on March 3, 1976. Sulphide concentrations ranged from zero (2.5 m) to 4.16 mg/l (4.5 m).

4. An anaerobic zone of low light intensity where sulphide and salinity increase further but with temperature now decreasing to 43°C
5. An anaerobic sediment surface of 40–43°C
6. Mat formations along the shore (Hirsch, 1978)

Diversity of microorganisms. Sampling procedures (Hirsch, 1978) and investigation of samples, enrichments and pure cultures resulted in the observation of 149 morphologically different, distinct types of microorganisms ("morphotypes"). These could be placed in 22 groups according to their main morphological or behavioral features such as shape, size, capsule formation, mode of reproduction, aggregation, spore formation, pigmentation, flexibility or gliding mobility (Table 4-1). The main microbial representatives in this hypersaline, heliothermal lake were the obviously most diverse bacteria (Fig. 4-2; 121 morphotypes, 81% of the total), while only a few eukaryotic microorganisms could be found. Fungi were totally missing. A complete list of morphotypes (working names) with data on their distribution is given in Table 4-2.

TABLE 4-1.

Morphotype grouping of 149 different, distinct microorganisms observed in the Solar Lake, March 1976.

Group	Characteristics	Number of morphotypes
1	*Cocci:* single or tetrads, ± polymer, s. t. irregularly-shaped	4
2	*Cocci:* budding	1
3	*Rods:* straight, length varying, single or in groups, ± sheath or polymer capsule	13
4	*Rods:* straight and in chains, ± sheath or polymer capsule	6
5	*Rods:* vibroid or resembling bananas	8
6	*Filaments:* long, flexible, s.t. tapering, like Mycoplasma, not gliding	6
7	*Filaments:* gliding, flexible, s. t. in chains	5
8	*Spindles* and pointed short filaments	3
9	*Sporulating* rods, filaments or spindles	4
10	*Budding* rods, or budding and tapering rods	6
11	*Spirilla,* screws, saprospiras	5
12	*Stalked, prosthecate bacteria*	7
13	*Hyphal and budding bacteria*	9
14	*Spirochaetes*	5
15	*Unusually-shaped* or irregular bacteria, s. t. branched	11
16	*Photosynthetic bacteria* w/o cyanobacteria	8
17	*Cyanobacteria*	20
18	*Diatoms*	11
19	*Colorless flagellates*	7
20	*Green or golden flagellates*	5
21	*Non-flagellated algae*	2
22	*Ciliates*	3
	Total	149

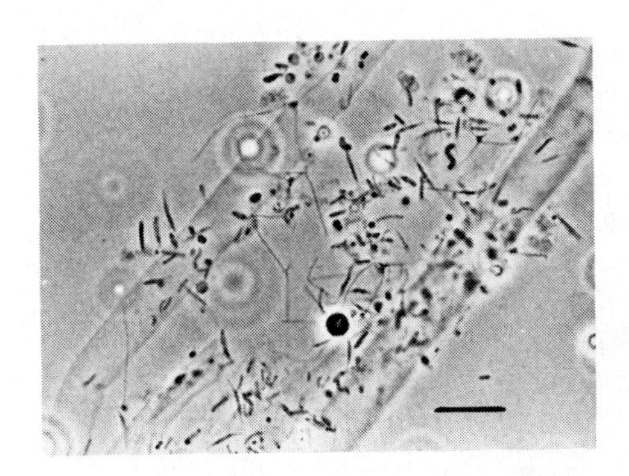

Fig. 4-2. Solar Lake bacteria from the sediment surface (4.5 m). Most forms are attached; Several types are prosthecate. Magnification bar 10 μm.

Among the microorganisms observed were some morphotypes of greater interest. Fig. 4-3a shows long, flexible bacterial filaments (no. 34) from 0.5 m which were present in large numbers and could be found throughout almost the whole water column. The *Spirulina* sp. 115 (Fig. 4-3b) came only from 0.5 m depth and was a rare organism. *Macromonas* sp. 50 (Fig. 4-3c) showed intensive motility in the 1.5 m sample as well as in enrichment cultures. This bacterium can be expected to oxidize sulphide as evident from the large sulphur globules present in most cells. *Thiodendron* sp. 64 (Fig. 4-3d) was found at 0.3 m in a discarded centrifuge glass containing cyanobacteria and purple sulphur bacteria. It is a rarely-seen prosthecate organism and does not yet exist in pure culture (Hirsch, 1974). The irregularly shaped coccus no. 29 usually showed aggregates reminescent of *Geodermatophilus* spp. (Fig. 4-3e). This encapsulated bacterium was present in most of the water column below the epilimnion. The micrograph was taken from a 4 m sample (44°C). *Chromatium* spp. were present in large numbers. Morphotype 40 (Fig. 4-3f) came from 3.5 m (49°C). Also seen on this micrograph is the filamentous cyanobacterium (*Oscillatoria* sp.) no. 43 which was likewise distributed between 3 and 4.5 m.

Additional unusual bacteria are shown in Fig. Spindel-shaped forms were most often found attached to diatom frustrules which appeared in various shapes of degradation (Fig. 4-4a). Spirochetes of various types were found in locations of vigorous breakdown of microorganisms. Fig. 4-4b shows morphotype no. 19 seen in a sediment surface sample together with cyanobacteria, *Prosthecobacter* 1 (arrow) and apple seed-shaped, budding bacteria no. 18.

A most interesting organism with branched hyphae, tetrahedral buds and tetrahedral or dichotomously-forked cells ("Dichotomicrobium 12") was observed throughout the whole Solar Lake profile. Fig. 4-4c is a micrograph of it from an anaerobic 2.5 m sample. Fig. 4-4d shows a pure culture (IFAM 953) obtained from 3.5 m (43°C) with hyphal network and tetrahedral shape. Hirsch et. al. (1977) have published electron micrographs of developmental stages of this organism. Figs. 4-5a and 5b show hyphal branches, terminal buds and short, stubby pili present on most (but not the younger) hyphae.

The apple seed-shaped bacterium no. 18 (Fig. 4-4b) was subsequently isolated in pure culture without loss of its distinctive morphological features (Fig. 4-6a). A screw-shaped

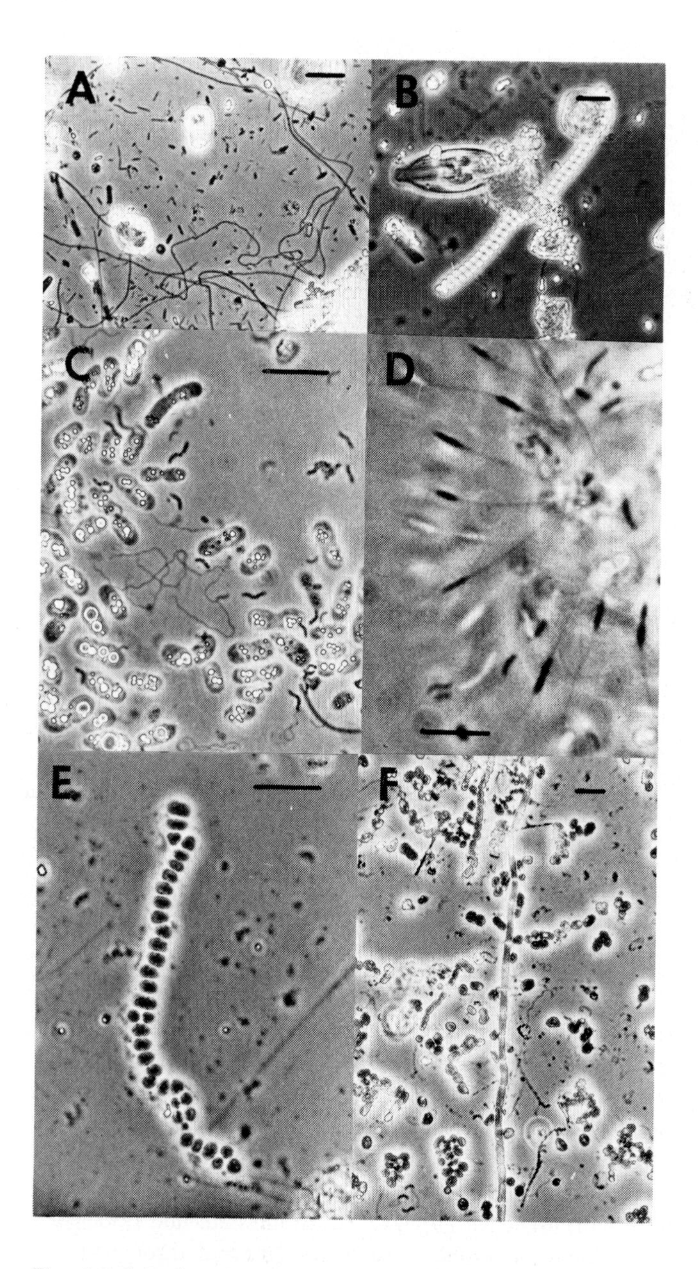

Fig. 4-3. Selection of Solar Lake bacteria observed in different depths. Magnification bars 10 μm.

4-3a. Flexible, filamentous bacterium 34 from 0.5 m.

4-3b. ***Spirulina*** sp. 115, a cyanobacterium from 0.5 m.

4-3c. ***Macromonas*** sp. 50 with large S^{o} inclusions. From 1.5 m.

4-3d. ***Thiodendron*** sp. 64, a prosthecate bacterium from 0.3 m.

4-3e. Irregularly-shaped coccus 29 from 4 m (44°C).

4-3f. ***Chromatium*** sp. 40, a purple sulfur bacterium from 3.5 m.

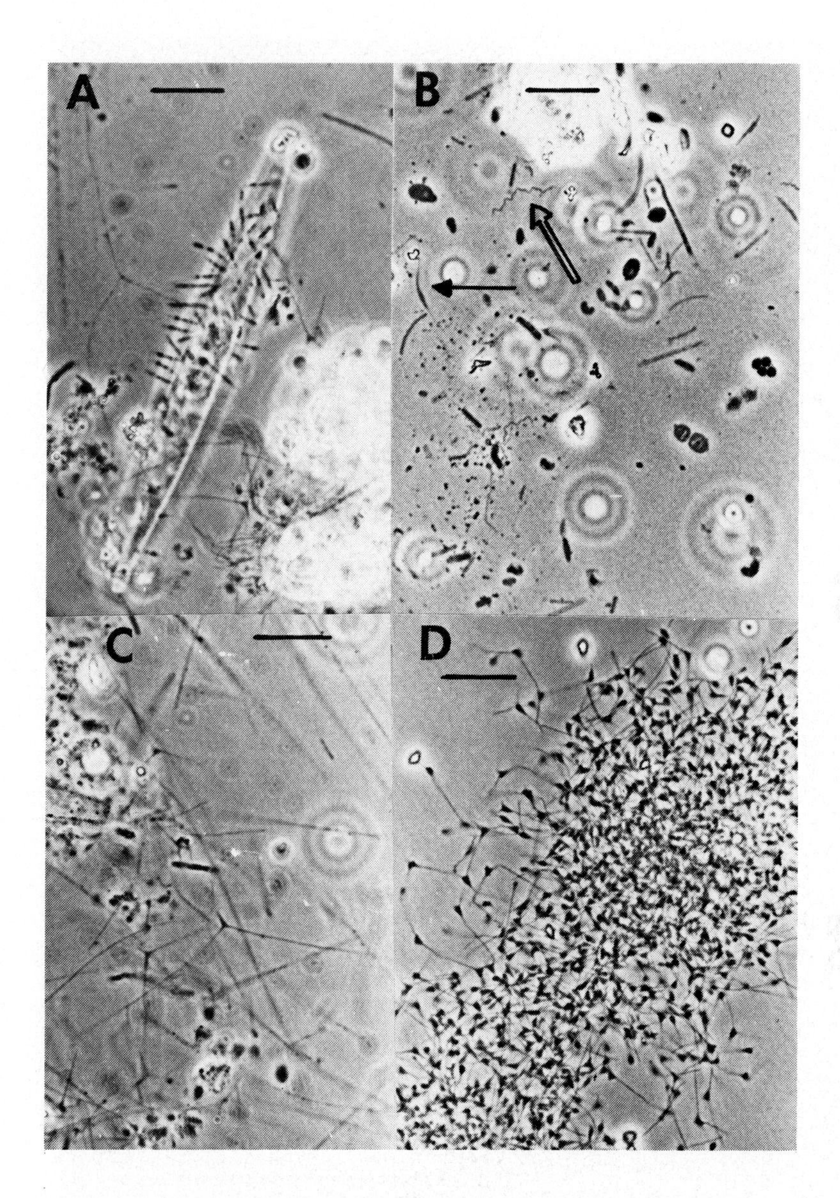

Fig. 4-4. Solar Lake bacteria: additional selection. Bar 10 μm.

4-4a. Spindle-shaped bacterium 13, attached in dense groups to diatom 67 and possibly involved in frustrule breakdown. Sample from 1 m (42.5°C).

4-4b. A sample from the 4.5 m sediment surface abounds in spirochetes 19 (open arrows) and also contains apple seed-shaped bacterium 18 (lower right; Fig. -6a) as well as ***Prosthecobacter*** 1 (arrow).

4-4c. An anaerobic sample from 2.5 m (52°C) with a new hyphal, budding tetrahedral bacterium ("Dichotomicrobium 12").

4-4d. Pure culture of "Dichotomicrobium 12" (strain IFAM 953) isolated from 3.5 m (43°C). This is the same morphotype as shown in Fig. -4c.

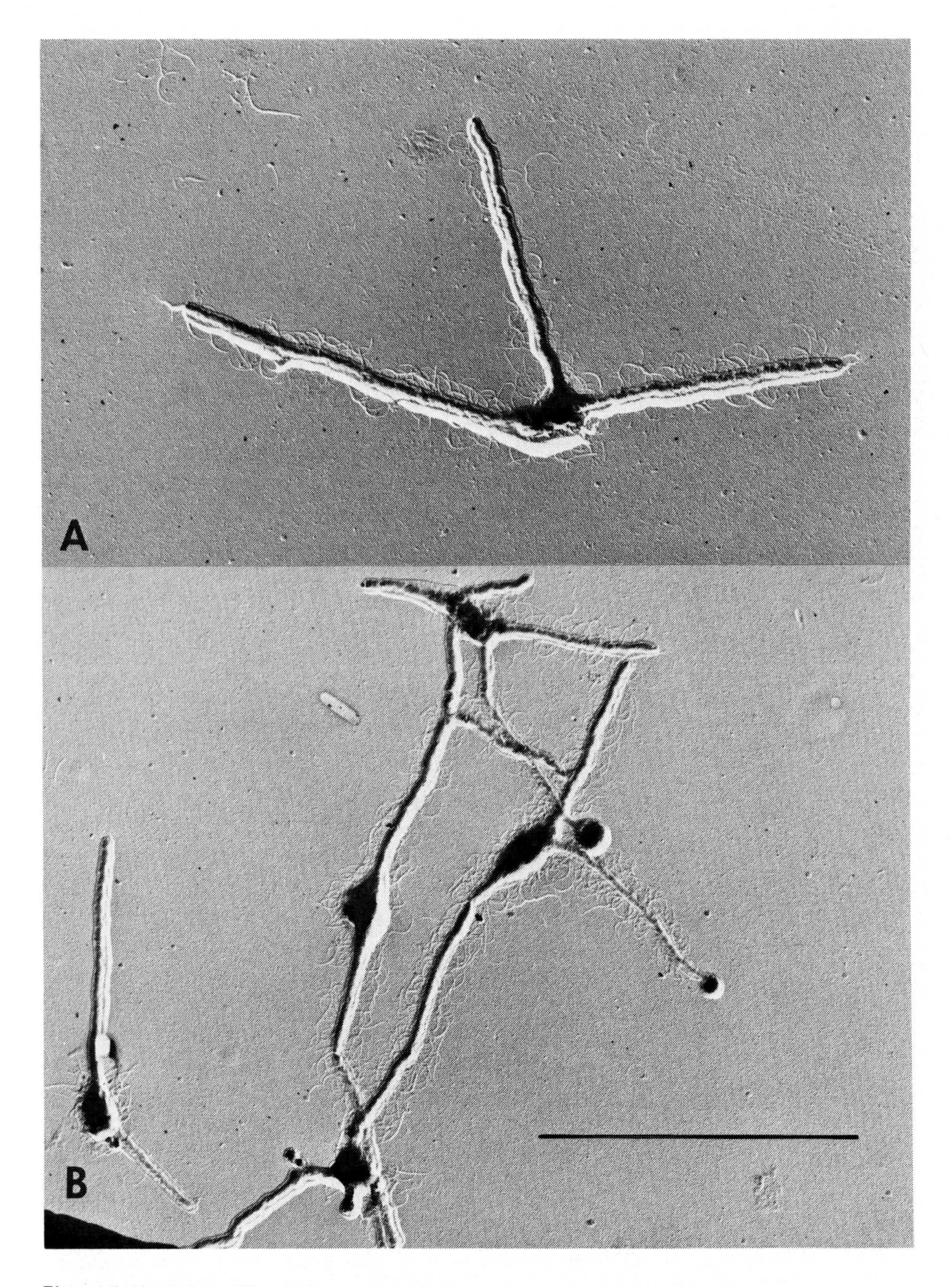

Fig. 4-5. Hyphal, budding "Dichotomicrobium 12" (strain IFAM 953) isolated from 3.5 m. Electron micrographs of shadow-cast preparations. Bar = 5 μm.

4-5a. Young cells with short, stubby pili on the hyphae.

4-5b. Cells with hyphal branches and terminal buds. Young hyphae do not show the short, stubby pili.

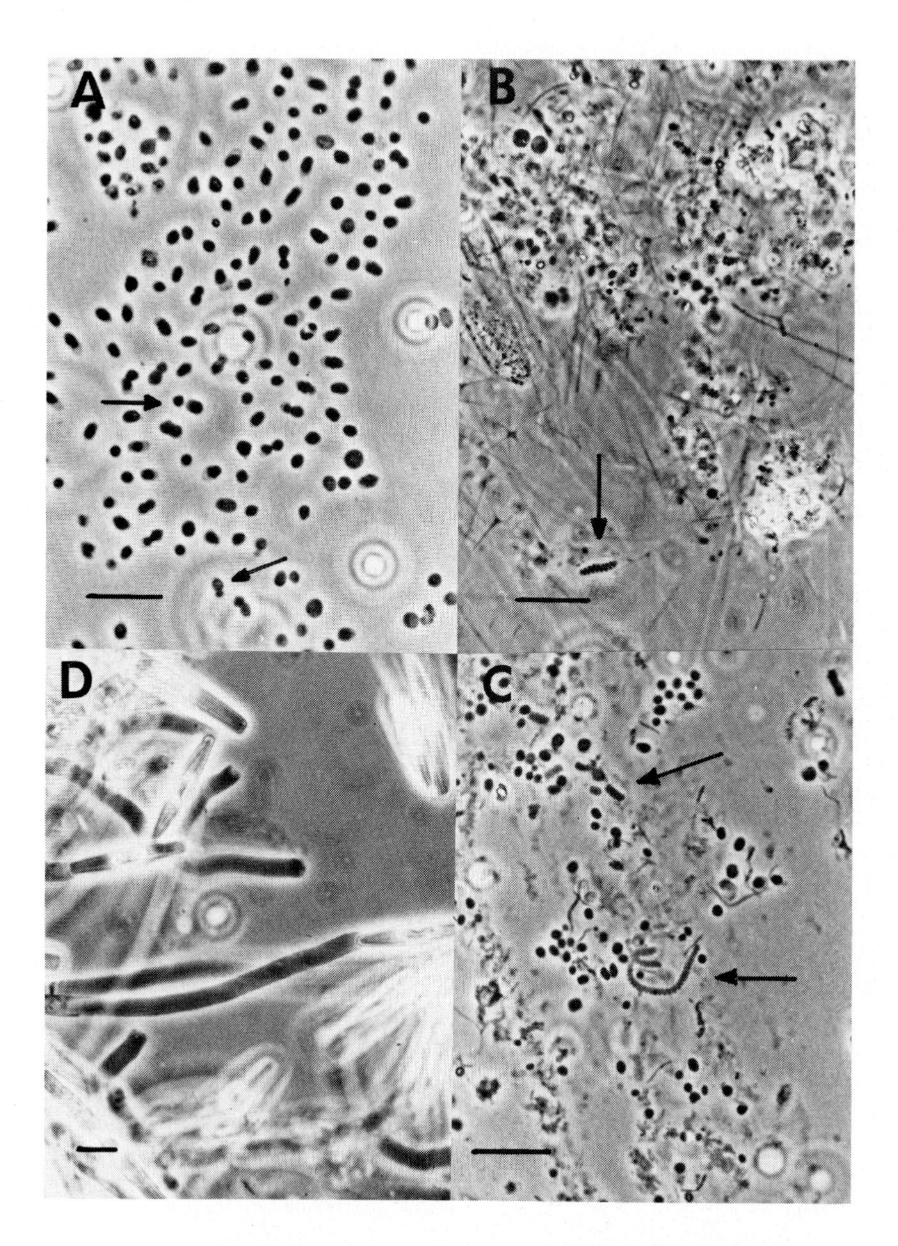

Fig. 4-6. Morphologically distinct Solar Lake microorganisms. Magnification bar 10 μm.

4-6a. Bacterium 18 (apple seed-shaped). Strain IFAM 1002 with buds (arrows).

4-6b. Screw-shaped bacterium 23 with often decreasing, polar turns was found between 0 and 2.5 m with greatest numbers below 0.5 m. Natural sample from 2.5 m, anaerobic.

4-6c. The screw-shaped bacterium (arrows) in two-membered culture with apple seed-shaped organism 18.

4-6d. Stalked diatom 67 from a submerged soda bottle (0.7 m). The organism (***Gomphonema*** sp. ?) occurs in the lake also without polymer stalk.

TABLE 4-2.

Complete list and depth distribution of all 149 Solar Lake morphotypes observed in march 1976. The names used are preliminary working names and subject to changes after obtaining pure culture data. Distribution types: see Fig. 4-7.

No.	Working name	distribution type	No.	Working name	distribution type
4	Diatom	1	136	Filaments (flexible)	2
5	Diatom	1	138	Rods (irregular, in chains)	2
6	Rod-shaped bacterium, bent	1	147	Tetrads	2
7	"Tetramicrobium"	1	1	***Prosthecabacter*** sp.	3
8	***Nevskia*** sp.	1	10	***Caulobacter*** sp.	3
9	***Hyphomicrobium*** sp.	1	18	Apple seed-shaped bacterium	3
14	Flagellate	1	47	Rod-shaped bacterium (straight)	3
17	Coccus	1	50	***Macromonas*** sp.	3
20	Rod (large, w. capsule)	1	55	***Oscillatoria*** sp.	3
24	***Caulobacter*** sp. (bacteroid)	1	58	Rod-shaped bacterium (small)	3
51	***Chromatium*** sp.	1	62	***Diplococcus*** sp. (small)	3
54	***Thiospirillum*** sp. (yellow)	1	64	***Thiodendron*** sp.	3
57	Diatom	1	73	Pointed rods	3
59	***Chroococcus*** sp.	1	81	***Caulobacter*** sp. (lemonoid)	3
60	Flagellate	1	86	***Bacillus*** sp. (filaments)	3
63	Ciliate	1	87	Spirochete (small)	3
65	Flagellate (cryptophyte)	1	90	Spirillum sp. (small, fast)	3
66	***Saprospira*** sp.	1	102	Diatom (asymmetrical)	3
68	Choanoflagellate	1	23	Screw-shaped bacterium	4
69	Rods (in chains, gliding)	1	25	***Aphanothece*** sp.	4
70	Filaments (gliding)	1	31	Flagellate (colorless)	4
75	***Beggiatoa*** sp. (thin)	1	93	***Gloeocapsa*** sp.	4
76	Spirochete (large, straight)	1	123	Budding rods	4
77	Flagellate (colorless)	1	146	Cyanobacteria (cocci, slime)	4
78	Flagellate (green)	1	12	"Dichotomicrobium" sp.	5
79	Rod-shaped bacterium (bent)	1	13	Spindle-shaped bacteria (attached)	6
80	Ciliate (large)	1	19	Spirochete	6
82	***Caulobacter*** sp. (round)	1	34	Flexible filaments (thick)	6
91	Diatom	1	74	***Oscillatoria*** sp.	6
92	Flexibacterium (large)	1	89	***Achromatium*** sp. (round)	6
95	Horse-shoe bacterium	1	105	***Bdellovibrio*** sp.	6
96	Ciliate	1	33	Spirillum	7
97	***Oscillatoria*** sp. (rostrum)	1	35	Filaments (long, thin)	7
98	Diatom	1	39	Bacterium (star-shaped)	7
99	Cyanobacterium (filaments)	1	120	Algal cyst (yellowish)	7
100	***Thiospirillum*** sp.	1	121	Coccus (w. gas vesicles)	7
103	Rods (in sheath, attached)	1	129	Rods (w. external sulphur)	7
104	Polyspheroid bacteria	1	141	***Blastobacter*** sp.	7
106	***Hyphomicrobium*** sp.	1	143	Rod-shaped bacterium (large)	7
107	Cyanobacterium sp. (large)	1	144	Rods (large, pointed)	7
108	Alga (Chrysophyte)	1	28	Cyanobacterium (filaments)	8
109	Flagellate	1	29	Coccus (irregular)	8
110	***Sphaerotilus*** sp.	1	32	Spirochete (long)	8
112	Diatom	1	37	Banana-shaped bacterium	8

TABLE 4-2. (Continued)

No.	Working name	distribution type	No.	Working name	distribution type
113	***Lineola*** sp.		40	***Chromatium*** sp.	8
115	Spirulina sp.	1	124	Rods (with polar gas vesicles)	8
116	***Mycobacterium*** sp.	1	126	Spirochetes (small, thin)	8
118	Flagellate	1	142	Filaments (long, straight)	8
119	Flagellate	1	36	Cyanobacterium (short, filam.)	9
131	Rod-shaped bacterium (rows)	1	38	***Prosthecochloris*** sp.	9
132	Flagellate (large, colorless)	1	42	***Desulfovibrio*** sp.	9
134	Diatom (large)	1	43	***Oscillatoria*** sp.	9
139	***Mycoplasma*** sp.	1	44	Rod-shaped bacterium (dark)	9
140	***Siderococcus*** sp.	1	46	Rods (straight and thin)	9
145	"Henricia" (stalked)	1	48	***Ancalomicrobium*** sp.	9
2	Diatom (Amphora)	2	49	Rod-shaped bacterium (red)	9
3	***Pasteuria*** sp.	2	53	Purple bacterium (yellow)	9
11	Bacterial filament (gliding)	2	83	Cyanobacterium (filaments, gas vesicles)	9
15	***Spirillum*** sp. (small)	2	84	***Thiospirillum*** sp. (red)	9
16	Rod-shaped bacterium (sheath)	2	85	***Ochrobium*** sp. (reddish)	9
21	Rod-shaped bacterium (thin)	2	127	Flagellate	9
22	***Vibrio*** sp. (small)	2	128	Rods (thick, w. sulfur)	9
26	***Hyphomicrobium*** sp.	2	148	Filaments (pointed, w. deposits)	9
27	***Pedomicrobium*** sp.	2	149	***Oscillatoria*** sp. (thin, dark)	9
30	Rod-shaped bacteria (groups)	2	41	***Caulobacter*** sp. (vibrioid)	10
56	***Pasteuria*** sp.	2	45	***Hyphomicrobium*** sp.	10
67	Diatom (stalked)	2	52	***Chromatium*** sp.	10
71	Cocci (budding)	2	61	***Thiotrix*** sp.	10
94	Rod-shaped bacteria (gas vesicles)	2	72	Tetrahedral bacteria (budding)	10
101	***Clostridium*** sp.	2	88	Rods (motile)	10
111	Rod-shaped bacterium (budding)	2	114	Cyanobacterium (round, capsule)	10
117	"blastomicrobium" sp.	2	125	Sporeformer (plectridium)	10
122	Rod-shaped bacteria (rosettes)	2	130	***Bacillus*** sp.	10
133	Diatom	2	137	Cyanobacterium (filaments, pointed)	10
135	Spindle-shaped bacteria	2			

bacterium, however, which in natural samples showed a polar decreasing amplitude of turn (Fig. 4-6b) lost this feature when grown in two-membered culture in the laboratory (Fig. 4-6c). So far all attempts to obtain growth in the absence of the contaminating bacterium (the apple seed-shaped organism 18) have failed.

Figure 4-6d shows a stalked diatom found to be quite common in the epilimnion. Confined to a submerged coke bottle or quiet water habitats, this diatom produced long stalks. But samples from the open water had only non-stalked cells.

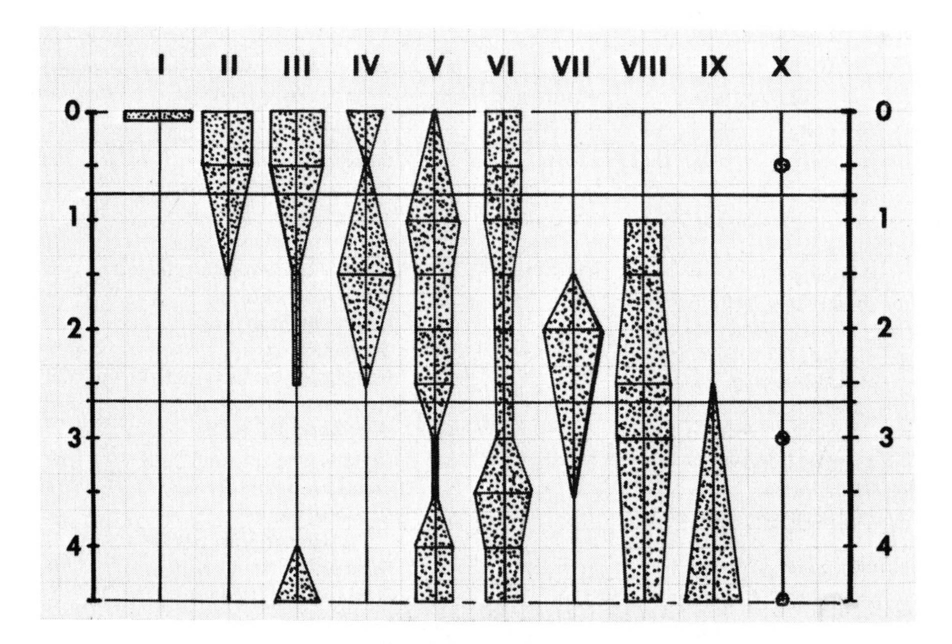

Fig. 4-7. All 149 Solar Lake microorganisms could be attributed to one distribution type (I–X). The drawing shows estimated frequencies with depths (m). Compare to Tables -2 and 3.

Distribution of morphotypes observed. The distribution of all 149 organisms (types) observed is given in Table 4-2. Generally, these microorganisms could be attributed to one of ten "distribution types" (Fig. 4-7, Table 4-2). While some of the patterns observed were quite distinct (types I, II, IV, VII and IX), others were less clear and may have included distribution of partially inactive or even dead cells (types III, V, etc.).

The depth distribution pattern organized according to different taxonomic groups (Table 4-3) shows high eubacterial diversity in the upper 2.5 m (types I–IV) with 64 morphotypes below 2.5 m. Eukaryotes were essentially absent from 2.5 m, with the possible exception of a flagellate. The most frequent distribution type observed was type I (water surface/neuston and mat surface along the shore) with 55 morphotypes (= 37% of total).

Pure culture isolation. Direct subculturing of samples or various enrichment techniques (Hirsch et. al., 1977) resulted in pure culture isolation of 28 strains (Table 4-4). This should not be understood, however, as the total yield of all efforts to isolate every morphologically distinct microorganism from the Solar Lake, but rather as an attempt to obtain *some interesting* forms. Hence, 8 strains of "Dichotomicrobium 12" were isolated from three different depths and by using three different techniques (Table 4-4). Additionally, other budding and some stalk-forming bacteria were obtained. A vitamin solution, Bacto-Yeast-Extract, or Bacto Peptone (0.025% w/v) was the main added compound in some enrichments. No addition to the samples resulted in 7 pure cultures.

TABLE 4-3.

Depth distribution pattern of microbiological groups attributed according to the distribution types shown in Fig. 4-7.

Distribution type	Predominant distribution	Number of different morphotypes					Total number of different microorganisms
		Bacteria	Cyanobacteria	Diatoms	Flagellates and other algae	Ciliates	
I	Water surface/neuston; shore mat surface	29	5	7	11	3	55
II	Epilimnion/thermocline	20	—	3	—	—	23
III	Epilimnion, down to the temperature maximum and again below 4 m	13	1	1	—	—	15
IV	Below or around the thermocline	2	3	—	1	—	6
V	Throughout most of the water column (mainly around 1 m)	1	—	—	—	—	1
VI	Hypolimnion and mainly epilimnion	5	1	—	—	—	6
VII	Thermophiles: around the temperature maximum	8	—	—	$(1)^{+}$	—	9
VIII	Hypolimnion throughout	7	1	—	—	—	8
IX	Lower hypolimnion, sulphide region	11	4	—	1	—	16
X	Sporadic distribution throughout column	8	2	—	—	—	10
	Total	104	17	11	14	3	149

TABLE 4-4.

Types and origin of 28 Solar Lake pure cultures and methods for their enrichment. BP – Bacto Peptone, YE – Bacto Yeast Extract. Enrichment techniques: A – sample directly incubated; B – sample incubated together with exposed and overgrown glass slide; C – sample incubated after addition of low amount of organic compounds.

Morphotype	Sample origin (m)	Enrichment Technique			Number of pure cultures
		Method	Additions (%)	Temperature °C	
"Dichotomicrobium"	1.5	B	–	43	2
"Dichotomicrobium"	1.5	C	BP – 0.025	43	2
"Dichotomicrobium"	2.5	C	YE – 0.025 + vitaminsol.	43	2
"Dichotomicrobium"	3.5	B	–	43	2
Budding bacteria (apple seed-shape)	0.5	C	Vitaminsol.	20	2
Budding bacteria (apple seed-shape)	1.5	C	YE – 0.025	43	3
Short, pelomorphic, budding rods	0.5	A	–	20	1
Short, pleomorphic, budding rods	0.5	C	Vitaminsol.	20	1
Short, pleomorphic, budding rods	1.5	C	YE – 0.025	43	5
Short, pleomorphic, budding rods	1.5	B	–	43	1
Bacillus (sporulating)	1.5	C	BP – 0.025	43	2
Irregular cocci, aggregated	2.5	C	YE – 0.025 + vitaminsol.	43	1
Budding cocci	1.5	C	YE – 0.025	43	1
Pasteuria (Slender, budding)	0.5	A	–	20	1
Caulobacter (slender, bacteroid)	0.5	C	Vitaminsol.	20	1
Caulobacter (drop-shaped)	0.5	C	Vitaminsol.	20	1

The 28 bacterial cultures obtained corresponded to 12 of the 149 microbial morphotypes seen (i.e. 8% of the total). Compared to the 104 *bacterial* morphotypes observed, the efficiency of isolation increases to 11.5%.

Morphological stability of morphotypes. It was already pointed out that the distinctive features of many morphotypes were qualitatively present in enrichments and pure cultures (compare apple seed-shaped bacteria in Figs. 4-2 and 6a). Other morphotypes change their features quantitatively when grown under laboratory conditions (compare screw-shaped bacteria Figs. 4-6b and 6c). The hyphal and budding bacterium no. 12 ("Dichotomicrobium") appeared morphologically different in samples taken from various depths. Differences existed in hyphal length, degree of branching, as well as in the mother-cell size and shape. Strain IFAM 954 was therefore grown in a temperature gradient created by an aluminum block, one end of which was cooled and the other end heated. The cells survived from between 13°C and 64°C, with shorter hyphae at 13°C, with smaller cells at 64°C, and with the most natural appearance between 29 and 48°C. Fig. 4-8 is a culture grown at 34°C.

Cell growth (increasing O.D.) occurred within 7 days between 38 and 64°C and with a maximum cell density at 47°C. Protein determinations revealed significant increases between 47 and 64°C. Further detailed experiments are in progress. Strain IFAM 954 was originally isolated from an anaerobic 3.5 m sample of 43°C and kept at 43°C during the isolation procedures.

DISCUSSION

The observation of a total of 149 different morphotypes in the Solar Lake makes one want to compare this number to those obtained from other aquatic habitats. Table 4-5 gives such data obtained with the same observation and isolation techniques as employed here. Thus, the Solar Lake had a great diversity of morphotypes, but it yielded the smallest number of morphotype pure cultures. In contrast, more than half of the 43 morphotypes of the Forest Pond could be isolated. This points to a high degree of physiological adaptation to distinct ecological niches in the Solar Lake which were difficult to reproduce in the laboratory.

The predominance of prokaryotes and low diversity of eukaryotes could be explained by greater prokaryotic resistance to environmental stress conditions such as high temperature and high salt concentrations. The complete absence of fungi could also be due to a lack of suitable substrates. Moderately halotolerant aquatic fungi known to live on pollen, submerged wood, macrophytes, algae or sediment organic matter would either not find their substrates in the Solar Lake or the conditions would be unsuitable (anaerobiosis, sulphide!). Likewise, an analysis of the Great Salt Lake biota (Post, 1977) did not reveal any fungi.

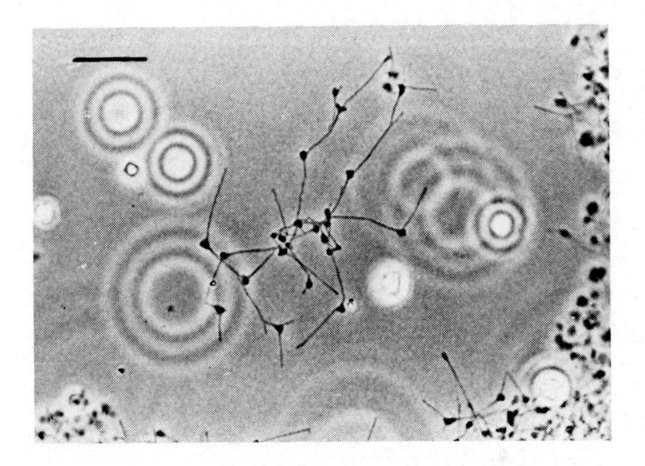

Fig. 4-8. "Dichotomicrobium" strain IFAM 954 isolated from a 3.5 m sample (43°C) grown at 34°C and showing "normal" morphology. Bar 10 μm.

TABLE 4-5.

A comparison of four aquatic habitats with respect to their numbers of bacterial morphotypes that could be observed during a short study period and possibly isolated in pure culture.

Site	No. of different bacterial morphotypes observed	Pure culture isolations		Morphotypes isolated in % of morphotypes observed
		No. of strains	No. of morphotypes	
Solar Lake (Sinai) (monomictic, heliothermal)	104	28	12	11.5
Forest Pond (Augusta, Mich.) (eutrophic, shallow)	43	72	24	55.8
Lake Plußsee (Holstein) (eutrophic, dimictic)	39	35	12	30.8
Lake Blunkersee (Holstein) (eutrophic, dimictic)	24	4	4	16.7

A central question of the present study was the certainty by which morphotypes could be unequivocally recognized. While some of the microorganisms were sufficiently distinct (*Macromonas, Caulobacter, Thiodendron,* "Dichotomicrobium", *Prosthecobacter, Spirulina* etc.) others were less well definable. This makes some morphotypes more valuable than others for interpretations in this study. The range of morphological variation observed in pure cultures often allowed better recognition of such morphotypes under natural conditions. A greater certainty can be otained from *in situ* growth and morphology studies with pure cultures to detect their range of morphological variability.

Another important, but still largely unsolved question is, if morphologically identical microorganisms (especially bacteria) would really be physiologically alike in nature. We already do know that rod-shaped bacteria of a certain size (diameter, length) and without other morphological features may be physiologically extremely different. Thus, "simple rods" may have little indicator value but comprise several different "physiotypes", and the numbers of different bacteria may in fact increase rather than decrease, if one could recognize the "physiotypes." More enrichment experiments may thus result in additional forms not seen with direct observation techniques. Working with morphotypes for a characterization of a habitat thus can only be the beginning. Eventually individual bacteria have to be investigated for their *in situ* activities.

Relatively few Solar Lake bacteria could be identified with cultivated, known genera. Bacteria such as *Macromonas* 50 or *Thiodendron* 64 have only been described on morphological grounds. A budding, drop-shaped *Pasteuria* (56) shows similarities to a thermophilic, euryhaline organism described by Kahan (1961) from the hot springs of Tiberias (Israel). But while his bacterium did not multiply below 30°C, *Pasteuria* 56 grew well at 20°C, and in the Solar Lake this morphotype occurred only between 0 and 0.5 m depth at ambient temperature.

It was thought, at first, that the occurrence of bacteria such as the screw- or apple seed-shaped organisms could be restricted to the Solar Lake. But this turned out to be wrong. Recently, both of these have been found in stratified beach sand interstitial water ("Farbstreifen-Sandwatt") near Surendorf (Baltic Sea, Holstein). Some odd forms such as the "Dichotomicrobium" seem to be unique to the Solar Lake. Their origin is uncertain, and it may well be possible that they have become highly adapted to the local specific conditions. Since this organism is actively involved in the formation and coherence of algal mats, its ecological role should be studied further. As concerns the diatoms, most forms found in the lake seem to arrive from outside with the inflowing water. The specificity of Solar Lake organisms will become clearer as soon as other, similar habitats can be studied.

The long, flexible filaments (34) resemble *Chloroflexus* spp. and could well represent new types of these photosynthetic bacteria, although pigmentation could not be observed. Their depth distribution pattern indicated the possibility of more than one population.

The presence of spindle-shaped bacteria (13) on degrading diatom shells poses the question if they may be the cause, or their presence the result of frustrule breakdown. Here experiments with pure cultures are urgently needed.

The number of photosynthetic primary producers (46 morphotypes = 31%) was quite high. Their distribution was generally found as expected (Cohen et. al., 1977b): diatoms and the few other algae lived at or near the surface water (neuston, epilimnion, shallow water mats). Photosynthetic bacteria occurred only within the hypolimnion or deeper mat layers (Hirsch, 1978). Occasional finds of purple bacteria on the water surface probably result from floating up of mat flocs. About half of the cyanobacteria (mostly unicellular) were surface organisms, while the other half (filamentous) was found either in the whole profile or only in the hypolimnion. This indicates that the potential for primary production is spread throughout the whole profile, as already found by Cohen et. al. (1977b). It does, of course, not provide any data on the real activity of these microorganisms.

About two-thirds of all morphotypes were non-photosynthetic. Some of these may have been chemolithotrophs but this was not studied. One could safely assume, however, that considerably more than just half of the morphotypes seen were participating in the vigorous bacterial heterotrophic activities postulated by others (Cohen et. al., 1977c). The predominance of bacteria in distribution zones I to III indicates that high heterotrophic activity should be found here. In March/April 1971, Cohen et. al. (1977c) found the highest CO_2 dark uptake to occur in about 1 m depth, while in March 1974 this maximum was found at 2.5 m. Annual, seasonal and diurnal variations of this pattern can be expected, and the present investigation of March 1976 only describes a momentary constellation.

The new bacterium 12 ("Dichotomicrobium") was found in all layers of the Solar Lake. Higher cell densities were observed between 1 and 2.5 m, and between 4 and 4.5 m. Pure cultures were obtained from 1.5, 2.5 and 3.5 m. This distribution pattern could be verified by laboratory growth studies and temperature gradient experiments with strain IFAM 954, a pure isolate. This strain remained viable over a temperature similar to that from which the organism had been isolated (43°C) and grown during the enrichment and isolation. Under conditions of lower and higher temperature a noticeable quantitative change of morphology was observed; hyphal length and branching varied just as did the cell size and shape. Further studies to quantitate these effects and to investigate the influence of high salt concentrations on growth and morphology are in progress. The properties observed so far indicate a potential use of this bacterium as bioindicator in future Solar Lake ecological studies.

SUMMARY

Morphologically distinct microorganisms of the monomictic, hyperthermal Solar Lake near Elat (Sinai) were investigated during early March 1976. Direct microscopic techniques yielded in samples, enrichments and pure cultures a total of 149 "morphotypes" which in some cases could be identified as known genera. Several bacteria found have only rarely been seen in other locations (*Macromonas, Thiodendron*). Some forms were completely new, and of these a dichotomously branching, hyphal and budding bacterium (no. 12) appears to be specific for this Solar Lake. The 149 different microorganisms could be attributed to one of 10 distribution patterns which in part corresponded to 6 environments (zones) characterized by differences in physical and chemical parameters.

28 Pure cultures were obtained; these corresponded to 12 of the 149 morphotypes observed. This low isolation efficiency is explained by a high degree of adaptation of Solar Lake microorganisms to extreme environments.

The dichotomously branched bacterium no. 12 (strain IFAM 954) was tested for growth and morphological changes in a laboratory temperature gradient. Optimal growth occurred at about the temperature from which the organism had been isolated. Temperatures below or above resulted in quantitative morphological changes (hyphal length, degree of branching, cell size, but the qualitative features needed for *in situ* recognition were still present. The validity of the morphotype concept for an initial description of ecosystem species diversity is discussed.

ACKNOWLEDGEMENTS

This work was made possible through the generous hospitality shown to me by Prof. M. Shilo and Drs. M. Varon and M. Kessel, Department of Microbiological Chemistry, Hebrew University, Jerusalem. The Steinitz Biological Laboratory, Elat has been helpful during the visits at the Solar Lake. Excellent and responsible technical assistance by B. Doose and E. Heiden (Kiel) as well as by I. Levanon (Elat) is gratefully acknowledged. I also thank Profs. S. Golubić (Boston) and T. Anagnostidis (Thesaloniki) for helpful discussions. The investigations were supported by the Deutsche Forschungsgemeinschaft.

REFERENCES

Cohen, Y., 1975. Dynamics of prokaryotic photosynthetic communities of the Solar Lake. Ph.D. Thesis, Hebrew University, Jerusalem; 149 pp.

Cohen, Y., W. E. Krumbein and M. Shilo, 1974. The Solar Lake: Limnology and microbiology of a hypersaline, monomictic heliothermal heated sea-marginal pond (Gulf of Aqaba, Sinai). Contrib. Sympos. XXIV. Congr. of C.I.E.S.M., Monaco.

Cohen, Y., W. E. Krumbein, M. Goldberg and M. Shilo, 1977a. Solar Lake (Sinai). I. Physical and chemical limnology. Limnol. Oceanogr. 22, 597–608.

Cohen, Y., W. E. Krumbein and M. Shilo, (1977b). Solar Lake (Sinai). II. Distribution of photosynthetic microorganisms and primary production. Limnol. Oceanogr. 22, 609–620.

Cohen, Y., W. E. Krumbein and M. Shilo, 1977c. Solar Lake (Sinai). III. Bacterial distribution and production. Limnol. Oceanogr. 22, 621–634.

Hirsch, P., 1974. Thiodendron. In: Bergey's Manual of Determinative Bacteriology, 8th Edition (R. E. Buchanan and N. E. Gibbons, eds.). The Williams and Wilkins Com., Baltimore, Md.

Hirsch, P., 1977. Unusual bacteria of a Solar Lake: distribution and pure culture studies. Amer. Soc. Microbiol., Abstr. N. 33.

Hirsch, P., 1978. Microbial mats in a hypersaline Solar Lake: types, composition and distribution. In: Environmental Biogeochemistry and Geomicrobiology 1, Chapt. 16: The aquatic environment (Proc. 3rd Int. Sympos. Wolffenbüttel, 1977; W. E. Krumbein (ed.) Ann Arbor Sci.

Hirsch, P., M. Müller and H. Schlesner, 1977. New aquatic budding and prosthecate bacteria and their taxonomic position. In: "Aquatic Microbiology" (F. A. Skinner and J. M. Shewan, Editors), Soc. Appl. Bacteriol. Sympos. Ser. 6, 107–133, Academic Press, London.

Jørgensen, B. B. and Y. Cohen, 1977. Solar Lake (Sinai). 5. The sulfur cycle of the benthic cyanobacterial mats. Limnol. Oceanogr. 22, 657–666.

Kahan, D., 1961. Thermophilic micro-organisms of uncertain taxonomic status from the hot springs of Tiberias (Israel). Nature 192, 1212–1213.

Krumbein, W. E., and Y. Cohen, 1974. Biogene, klastische und evaporitische Sedimentation in einem mesothermen, monomiktischen ufernahen See (Golf von Aqaba). Geolog. Rundschau 63, 1035–1065.

Krumbein, W. E. and Y. Cohen, 1977. Primary production, mat formation and lithification: contribution of oxygenic and facultative anoxygenic cyanobacteria. In: Fossil Algae (E. Flügel, Edit.) Chapt. 3, Springer-Verlag, Berlin.

Krumbein, W. E., Y. Cohen and M. Shilo, 1977. Solar Lake (Sinai). 4. Stromatolithic cyanobacterial mats. Limnol. Oceanogr. 22, 635–656.

Lyman, J. and R. H. Fleming, 1940. Composition of sea water. J. Marine Res. 3, 134–146.

Post, F. J., 1977. The microbial ecology of the Great Salt Lake. Microbial Ecology 3, 143–165.

Chapter 5

TERMINAL LAKE LEVEL VARIABILITY AND MAN'S ATTEMPTS TO COPE IT WITH THEM

D. C. GREER

Department of Geography-Geology, Weber State College, Ogden, Utah, USA

The average terminal lake is a somewhat exotic feature to most because it is usually located in the arid, lightly populated regions of our planet. Even to those who live near them terminal lakes are very often enigmatic natural features that are usually avoided, overlooked or simply ignored. Unfortunately this attitude often leads to a lack of understanding about their importance, reason for being and peculiarities. In terms of priorities the populations of areas near terminal lakes usually place their own waterbody considerably down the priority list of important or desirable natural resources unless they are directly affected in some way by the vagaries of such features. The lack of understanding about the nature of terminal lakes is not only unfortunate but potentially dangerous and costly to man's structures, agricultural undertakings and utilization of such lakes as chemical sources, transportation links or recreation areas. As with many of nature's perfidies the failure of man to understand as fully as possible the problem inherent in terminal lakes can lead to numerous and potentially dangerous consequences.

Because closed lakes have no outlet, their level, surface area and volume is dependent upon the drainage and evaporation rate of their particular closed environment. The size of terminal lakes is thus dependent upon the weather or more accurately its long term effect, climate. Since the essence of weather and climate is one of change it follows that the essence of terminal lakes is also change. Unfortunately the relationship between climate and lake level is indirect and often delayed for months or years thereby obscuring this fact from the uninformed observer. In addition man interferes with terminal lakes and their mechanisms, upsetting the balance regulated by nature often to his own detriment. In effect terminal lakes can and do change their level, size and volume constantly and man should be cognizant of this fact in all of his undertakings concerning terminal lakes. He should never be so naive as to believe that he understands completely the tempo of rising or falling lake level trends which he perceives to be taking place. Man observes nature in terms of microscopic history and makes his judgments on this basis, but a falling lake stage in one century can give way to a rising lake stage in another. Thus, decisions about the utilization of, construction of structures near, or the formulation of strategies concerning terminal lakes are usually made in the light of probabilities and costs viewed within a constricted time frame. A more lengthy and informed view of terminal lakes would save man considerable time, effort and expense.

Two of the best known examples of problem-plagued terminal lakes and man's efforts to utilize, control and regulate them are found today in the Soviet Union's Caspian Sea and the Great Salt Lake in the United States. Both of these lakes are currently being studied by the respective governments of the countries in which they are located in an effort to better understand their peculiarities, utilize their resources and maintain them as viable entities. In each case man has contributed to the lake's ecological problems and this, coupled with natural changes in the environments of both, has resulted in a critical situation for their biota, ecological viability and usefulness to man. Due to economic priorities the restoration of both lakes to "natural" conditions is neither feasible nor perhaps desirable, thus as a consequence various strategies have and are being devised to cope with the man and naturally created problems of both lakes. Ironically the problems confronting each of these two lakes is at present due to opposite circumstances. In the case of the Caspian it is plagued by a falling lake stage while the Great Salt Lake has been rising during the past decade. The condition of both lakes is not necessarily naturally irreversible but the strategies presently being developed by the respective governments relating to them must of necessity assume that their problems will not be quickly remedied by natural means.

THE CASPIAN SEA

The Caspian Sea basin contains about 40% of the population of the U.S.S.R. and an even greater amount of the industrial capacity of the country. This means that the water resources available to the population and industry of the basin are more heavily taxed than for any other region of the Soviet Union. Most of the water used in the Basin is supplied by the Volga and its system which is also the principal source of water for the Caspian Sea. Since World War II the Soviet Government has embarked on a program of dam building, diversion and irrigation which has reduced the flow into the Caspian Sea. Some of this diversion has been of a temporary nature (in order to fill reservoirs) but increased utilization of the water resources of the Basin's rivers has resulted in a marked reduction in the amount reaching the Caspian Sea. This man-made reduction coupled with a drier climatic regime in the catchment basin of the sea has resulted in a rapidly declining surface level as shown by the following table.

Year	Withdrawals in Cubic Kilometers	Surface Level in Meters
1930	nil	-26 m below sea level
1940	10 km^3	-28 m below sea level
1970	30 km^3	-29 m below sea level
2000	60 km^3	-30 m or more

The amount of water withdrawn during the past five year plan (1970–75) increased by 20% and long term planning for industrial and agricultural development in the U.S.S.R. point to the consumption of from two to three times the current rate by the end of this century. This means that the flow from the Volga and other rivers in the Caspian Basin will be more and more heavily utilized which will, in turn, reduce the inflow and level of the Caspian Sea.

The effect of these withdrawals has had a marked impact upon the level of the Caspian Sea because the basin of the lake like that of most terminal lakes is shallow, at least in its northern third. Depths in the southern part approach 1000 m but the extremely shallow northern third of the lake is the portion which has been most effected by the falling lake

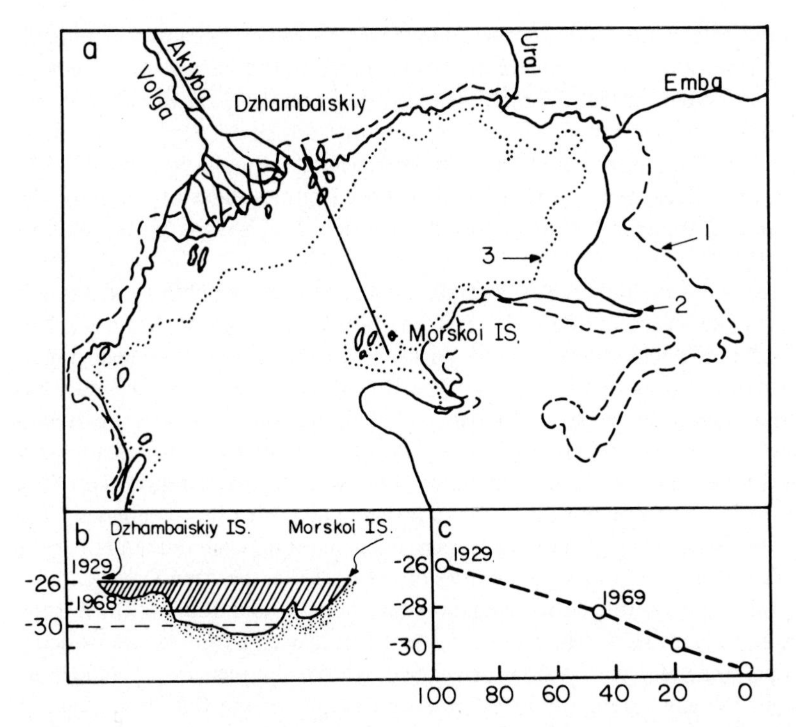

Fig. 5-1. The Caspian Sea Shorelines: 1 – 1929; 2 – 1968–69; 3 – minus 30 m isobath. b. Cross-section of sea from the Ozhambiaskiy to Morskoi Islands. c. Area of sea N.E. of cross-section comparing surface area in % with depth in m below sea level. (From: Marti, Biologicheskata Puoductivnost Kaspiskovo Morya, 1974.).

stage. The total surface area of that part of the Sea, north-east of a line extending from Dzhumbaiskiy Island, near the mouth of the Volga River to Morskoi Island to the southeast has been reduced more than one half during the period 1929 to 1969 (Fig. 5-1). A further decline in level of four meters would eliminate this portion of the sea completely. The southern part of the Caspian has not been effected as severely by the falling lake stage because it is much deeper and has a steeper shoreline. It is in the northern part of the lake where the greatest amount of sea life is produced and it is, therefore, important to the survival of the fishing industry and particularly the sturgeon and caviar industry of the Caspian that the northern arm of the lake be saved. The conditions in the northern part are best for the breeding and reproduction of sturgeon and other fish due to the input of nutrient rich waters from the Volga, Ural and Emba Rivers that flow into this part of the sea plus an enhanced photosynthesis in the shallow water. The prolific plant and animal life of the northern arm is thus more threatened by the falling lake stage than it is in any other part of the lake.

The threat to sea life in the Caspian is in and of itself serious enough, due to the amount and type of fish production which comes from this sea (the Russian population is supplied with about 80% of its consumption of caviar and sturgeon from this source) but this is not the only thing threatened by the falling level, however. The Caspian serves as an important transportation link with more inland water traffic carried on it and its tributaries than any system in the U.S.S.R. The decline in sea level on the Caspian has reduced the usefulness of or eliminated entirely some of the ports along its shoreline. The deepwater ports of Baku, Krasnovodsk and Makhachkala, though not immediately threatened, have been deepened and some of their port facilities moved because of the declining lake level. If the lake continues to fall some of the sea's many ports would be threatened with closure.

The Caspian Sea is not the only water body threatened by man's activities and nature's drier conditions in the southern U.S.S.R. in recent years. The Sea of Azov which is connected through the Kerch Straits with the Black Sea is also in danger of changing its regime due to a reduced flow of fresh water. The Don River, its principal tributary has been changed to a cascade of reservoirs by dams all along its course in recent years with much of its flow diverted for irrigation and industry. The sea of Azov, which formerly contained more fish per square mile than any other of the world's seas, is losing this prolific sea life because saltwater from the Black Sea is invading as fresh water decreases. The Aral Sea east of the Caspian is also threatened, perhaps even more severely than either the Caspian or the Sea of Azov. The two tributaries of the Aral Sea, the Amu Darya and Syr Darya are both very heavily utilized for irrigation and industry and could in the not too distant future be entirely utilized. The fate of the Aral Sea, in such a case, would be desiccation to a residual brine reservoir with unforseen ecological consequences. It is the hope of Soviet planners to ameliorate or reverse the desiccation of the Southern slope of the U.S.S.R. by maintaining the level of flow into these southern terminal lakes through water diversion, damming or a combination thereof.

In the course of many years a number of alternative solutions have been proposed which would divert streamflow from north to south and east to west. As early as 1932 the general meeting of the U.S.S.R. Academy of Sciences unanimously agreed on the urgent need to divert northern rivers into the Volga basin. The principal drawback then as now was economic resources but a number of other problems have been recognized in recent years complicating the situation even further. Almost all of these problems are ecological considerations which need to be thoroughly studied before any large scale water diversion projects can be contemplated. The XXV Party Congress of the U.S.S.R. has set a high priority for the diversion of water from north to south within the country and it is the charge of the Academy of Science to determine the feasibility, practicality and environmental impact of such an undertaking. The importance of this undertaking is underscored by the increasing threat to the viability and existence of the terminal lakes and seas on the southern margins of the U.S.S.R. caused by the growing water withdrawals from the streams which feed these water bodies. Targets need to be set and goals reached before conditions deteriorate to the point that the lakes' environments cannot be saved.

The proposals offered up to now have been numerous but it appears these can be grouped into four basic projects of which a variety of combinations is possible. The basic proposals are: 1) A canal across the North Caucasian Foreland bringing water from the Sea of Azov or Black Sea to the Caspian; 2) A dam across the Caspian Sea dividing the more shallow north from the deeper south; 3) Diversion of the Ob River into the Aral and Caspian Basins; and 4) Diversion of northern European rivers into the Volga and its tributaries. Each of these proposals has been considered in the past with arguments presented for and against each by their proponents and detractors. The idea of a canal connecting the Sea of Azov with the Caspian was attacked because the Sea of Azov is already becoming hyper-saline and such a canal would further increase its salinity. Such a canal could also introduce exotic biota into the Caspian that might damage existing species. A dam across the Caspian would help to maintain the level of the northern half of the lake but would cause an increase in salinity in the south and a freshening of the north which could threaten the species it is designed to save. Diversion of the Ob River into the Aral and Caspian Basins which was earlier proposed to go through the Northern Urals from a huge reservoir created by damming the lower Ob now appears to be impractical. The largest oilfield in the U.S.S.R. is now being developed in Western Siberia and such a lake would flood these important resources. A dam higher up the system at the confluence of the Ob and Irtysh could supply water through the Turgay Sink in sufficient quantities to meet the present and short-term future needs of the Aral Sea Basin, however. This project along with the diversion of northern European Rivers into the Volga-Kama System, supplying the needs of the Caspian and Azov Sea Basins, appears to be the most favored projects at the present time.

There are basically four strategies presently being considered by the Academy of Sciences' Water Problems Institute as solutions to the decline in surface level of the Caspian Sea as well as the increase in salinity of the Sea of Azov (Voropaev, 1976). The first alternative envisions a separate water supply for the European and Asian parts of the U.S.S.R. Water from the Northern Dvina and Pechora Rivers would be diverted into the Volga Basin through a series of canals some 1200 km in length. This project would require the raising of the diverted waters a maximum of 160 m which would then supply about 1000 m^3 of water per second to the Volga system. Part of this flow would be transferred across the Volga-Don Isthmus to the Don River increasing its flow into the Sea of Azov (The remainder would flow into the Caspian). The Asiatic side of this variant would shunt Ob-Irtysh waters into the Aral Sea Basin through a system of canals about 300 km in length requiring an increase in head of 150 m and would supply about 2400 m^3 per second of water to irrigate new lands.

A second variant would combine the water needs of both European and Asiatic U.S.S.R. into one massive project designed to transfer water from the Ob and Northern European Rivers through the Volga system into the basins of the Caspian, Aral and Azov Seas. Some of the flow would come from northern European rivers but the principal source of water would come from the lower Ob. Because the large estuary of the Ob River is fresh and lies below existing oilfields, water could be pumped through a tunnel under the Urals into the Pechora basin and from there into the Kama-Volga system. This transfer would require construction of 4,400 km of canals along which the water head would be raised 260 m supplying 2400 m^3 per second into the Volga system. The redistribution of this water in the south would require a second project on the lower Volga to divert flow into the Aral Sea Basin. For this an additional 2400 km of canals would have to be constructed to carry and raise the head 210 m so that 2000 m^3 per second of water could be diverted into the systems of the Syr and Amu Darya Rivers. Part of the flow would also be shunted through the Volga-Don Canal into the Don River basin thereby increasing the input of fresh water into the Sea of Azov.

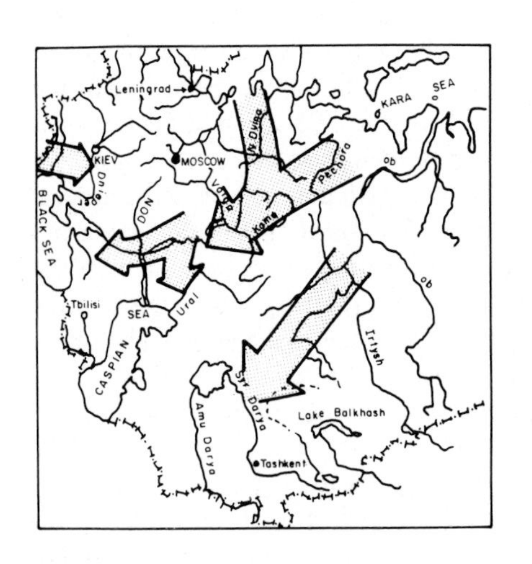

Variant I

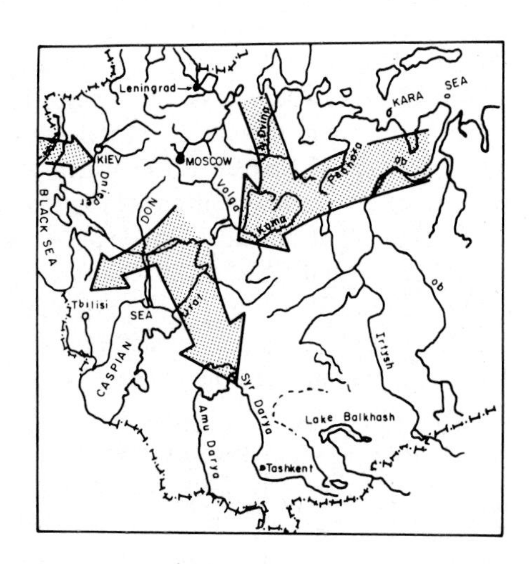

Variant II

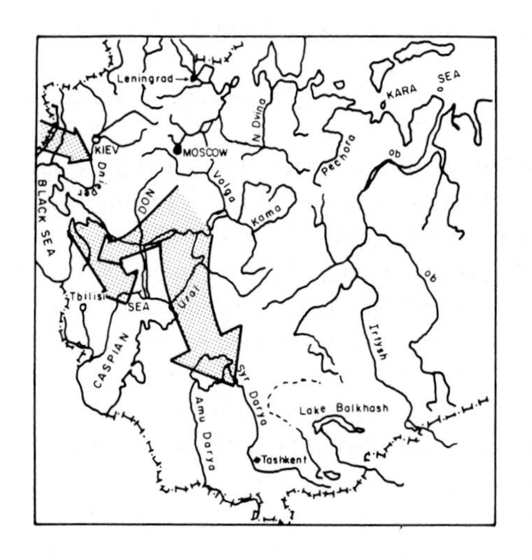

Variant III

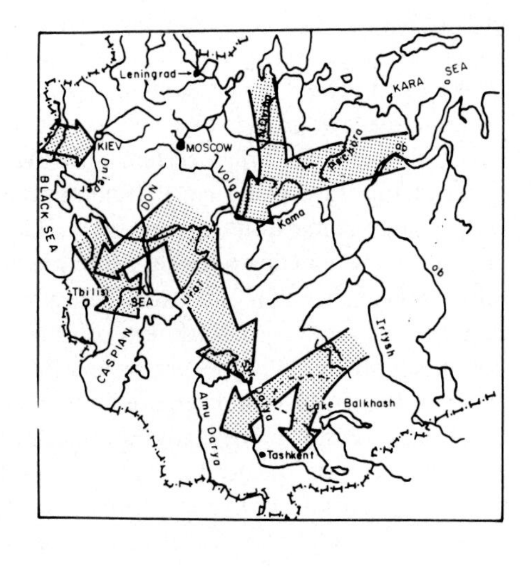

Variant IV

A third alternative assumes that the flow of the Volga River supplemented by water from the Black Sea to maintain the level of the Caspian could meet the needs of the southern slope of the U.S.S.R. This alternative would require the construction of a barrier dam across the Caspian Sea in order to control the water and salt balance in both halves of the lake. Fresh water from the Volga flowing into the northern part of the sea would be mixed in proper proportions with the sea water from the Black Sea to maintain the salinity level of the northern arm. The two parts of the Caspian would be connected by a sluce to allow the water in the northern basin to communicate with that in the southern. Water saved from the Volga by the diversion of Black Sea into the Caspian would be transferred into the Aral and Azov Seas to compensate for losses in those basins created by agricultural, hydroelectrical and industrial uses. This variant would require the construction of 2400 km. of canals and a rise in head of 210 m to get 2000 m^3 per second of Volga River water into the Aral Sea Basin. A second canal system some 1100 km in length would be required to deliver 3000 m^3 per second of Black Sea water over a rise of 28 m into the Caspian Sea.

A fourth alternative would combine aspects of all the previously mentioned variants into one comprehensive system known as the "Single Water Economy System" or SWES. This variant would be a country-wide comprehensive solution to the water needs of the U.S.S.R. The first three variants could be considered as comprising parts of the greater whole and could also be viewed as stages in the development process toward the completion of the SWES. This comprehensive system assumes that river runoff regulation should be accomplished not only by traditional methods of constructing intercontinental reservoirs but also by regulating water in the estuarine areas of rivers. To this end, sea bays, gulfs and sections of sea coasts would be used to accomplish a diversion of water from one slope to the other.

Whether or not any or all of these proposals will solve or at least retard the desiccation and alteration of Russia's great terminal lakes and seas remains to be seen. The realities of modern society mitigate against this possibility. Water diverted from the north to south will cost a great deal and the argument that it should be used for more economically important undertakings than maintaining the level or salinity of a lake or sea will be difficult to counter. The fact that water use in an industrialized world is increasing exponentially does not bode well for esoteric or environmentally motivated endeavors. The requirements for culinary water to feed industry and agriculture at the time such a project might be completed will probably exceed the combined south and diverted north slope water supply total. Under such conditions the decision as to which need will have priority for the use of this water will probably be the same as it is today.

THE GREAT SALT LAKE

During the period of time that white man has been in the Great Basin in Western United States the general trend of the level of the Great Salt Lake has been downward. When the Mormons arrived in the valley of the Great Salt Lake in 1847 the lake's level was approximately what it is today, 4200 ft (1280.16 m) but during the 1860's the level rose to over 4210 ft (1283.21 m). The lake at that time threatened agricultural development along its shores and the inhabitants became so alarmed about their farms and homes being inundated that they appealed to Brigham Young, the Mormon leader, for relief. With his limited knowledge of the basin he contemplated digging a canal to allow the lake to flow into the Snake River. This of course would have been impossible because the level at which Great Salt Lake's predecessor, Lake Bonneville, stabilized (The Provo Level) after it broke through Red Rock Pass in northern Utah into the Snake was 600 ft (182.88 m) above the present level of the lake. Even with modern technology this would be an extremely difficult and expensive undertaking. Luckily for the Mormons the level of the lake declined from its high point 4211.6 ft (1283.7 m) in 1873, rising on occasion to levels approaching 4205 ft (1281.68 m) but falling even further to a low in 1963 of 4191.35 ft (1277.52 m). During the past five decades the lake has reached an average level about five feet (one and one half meters) below that of the previous decades (4195 ft; 1278.64 m).

Because the amount of activity around the lake has grown exponentially, the greatest building and development programs on or near the lake have taken place during the period when the lake was near its lowest level in recorded history. Since WWII a new railway crossing has been built to span the lake replacing the old wooden piling structure with a rock-filled causeway. The highway route around the south end of the lake has changed from a simple two-lane road to a modern multi-laned divided freeway. The development of the Salt Lake airport from a rudimentary landing field to a major international airport has taken place at the southwestern end of the lake. The recognition of the amount and importance of minerals in the lake's brines has led to the construction of several square kilometers of evaporating ponds and refining plants along the lake's shoreline where there was formerly only small salt evaporating works. The construction or enlargement of the bird refuges at the mouths of Weber, Bear and Jordan Rivers has also taken place during the period since the lake declined. At the same time the lake's tourist industry has almost vanished because the facilities, used by tourists during the 1890's when the lake was above 4200 ft, were left many hundreds of meters from the shoreline. The large exposed mudflats around the lake made it difficult and unpleasant to approach the lake from almost any point along the western and southern shorelines.

The declining water level's exposure of several miles of mudflats around the edges of the Great Salt Lake's extremely flat basin hindered tourism but made available a large amount of land which has been used for other purposes. The development of Great Salt Lake's brines for chemicals and minerals made the relicted land which was formerly lake bottom an attractive place to build expanses of evaporating ponds because the land was extremely flat and adjacent to the source of brine. Thus, a large amount of former lake bottom was diked and solar ponding operations for the extraction of magnesium, sodium sulfate, potassium sulfate and other minerals was begun. The utilization of the lake's former bottom created a problem of ownership. Since it had not been formerly utilized nor claimed by anyone the question of title took on great importance in light of the potential income of millions of dollars from the rent, lease or sale of the land to mineral companies. Those private individuals who held title to the land upslope from the lake contended that the newly exposed land belonged to them in pie-shaped wedges extending from their property down to the edge of the lake. The Federal and State governments viewed the problem differently. The former, believed the entire lake and also its bed, belonged to the National Government. The State of Utah made the same claim and surprisingly both based their claims upon the doctrine of navigability and interstate commerce. The Federal brief stated that the lake was the national government's because it fell into the category of navigable waterways that could be regulated only by a higher authority. The State of Utah's claim rested on the belief that title to any navigable body of water totally within a state belonged to the state government. (Fig. 5-2)

The decision, as to who owned the lake bottom, was complicated by the fact that the lake being variable in size would require a second decision defining what constituted a valid shoreline. If a high shoreline were decided upon then the relicted lands above the present level would be non-existent or small and therefore unimportant to the winner. If the shoreline were deemed to be much lower than the present then there would be a large amount of exposed land involved in the decision and it would be important to the winner of a decision. Hence the question of ownership really had two parts: 1) Who owned the Great Salt Lake and its lakebed? and 2) What constituted a valid shoreline? The second part of the decision rested upon several possible precedents. The first was the so-called surveyed meander line which was made by U.S. Government surveyors in 1855. This shoreline determined by actual reconnoissance was somewhat imprecise including a few straight lines drawn across very flat areas surrounding the lake. A second possibility was the level of the lake at the time Utah entered the Union as a sovereign state in 1896. At the instant the government of the state assumed control over its territory the amount

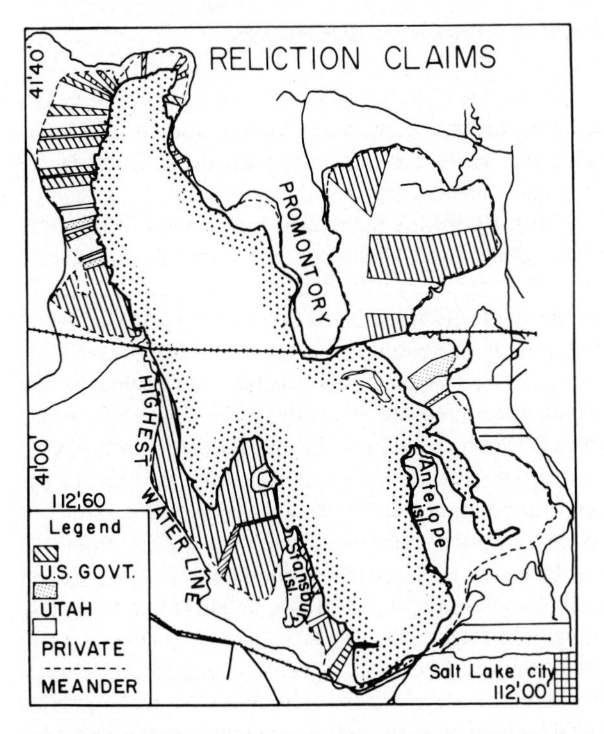

Fig. 5-2. Reliction claims in the Great Salt Lake area. From: Deon Greer, Journal of Geography, March, 1972.

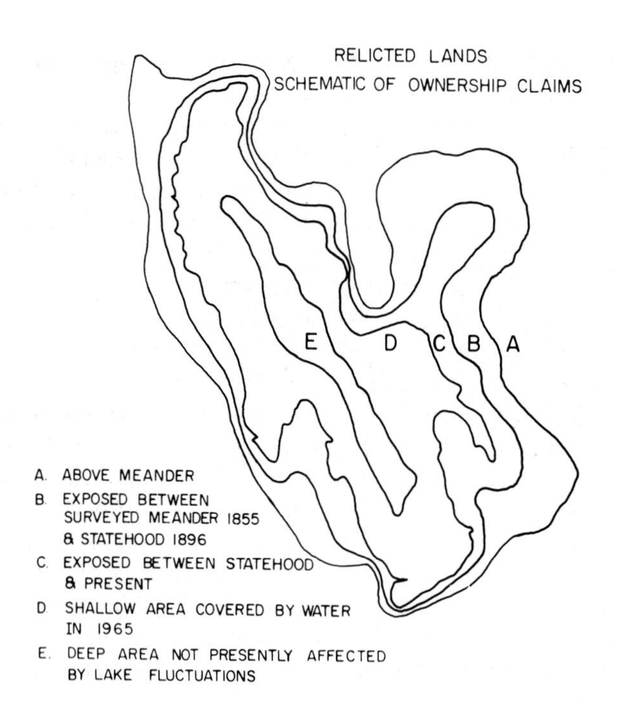

Fig. 5-3. Great Salt Lake area – Relicted lands, schematic of ownership claims.

of land covered by the lake would, it is postulated, be owned by the state. A final possibility was the level of the lake at some modern period, i.e., perhaps at the time the court started its deliberation concerning ownership. (Fig. 5-3)

In a decision rendered during the early 1970's the Supreme Court of the United States decided that the State of Utah held title to the Great Salt Lake and all of the land covered by it. Several years later in 1976 the court held that the title to the lake and its bottom was circumscribed by a line 4200.5 ft (1280.31 m) above sea level, the altitude of the lake's surface at the time Utah entered the Union. All rents, leases and income from the sale of land below this lake level accrued to the State of Utah. Unfortunately the amount of land below this level which was not covered by water in 1977 was nil. In fact the rising lake threatened to destroy development around the lake on the former relicted lands exposed during the previous decades. The principal problem which confronted the state concerning the Great Salt Lake was no longer who owned the lakebed and brine but how to save the millions of dollars in structures which had been erected around the lake during the period of time that the lake was at a relatively low level. The amount of destruction which the lake could conceivably inflict upon man's investments was directly proportional to the height which the lake might rise. By late 1977 the rising lake had concentrated brines. The ponds could support waterfowl, fish and fur animal propagation costs to replace or shore-up dikes eroded by wave action. The following table gives an estimation of the escalating costs which could be expected for the replacement of various structures destroyed by the rising lake stage.

Lake Level	Estimated Cost	Facilities Threatened
4200 ft	$ 5,000,000	Highway causeway to Antelope Island Dykes around ponds raised in height
4205 ft	$ 50,000,000	Freeways, solar ponding operations and bird refuges destroyed
4210 ft	$160,000,000	S. L. Airport, R. R. Causeway, Chemical plants and all other facilities

Faced with the potential destruction of a very large investment by itself and other who had built facilities around the lake, the State of Utah began looking for various strategies to cope with the lake's rising waters. The first and most highly touted strategy was put forward by the salt and chemical companies on the south shore by their "Save Our Shores" organization, easy solution to the problems. This proposal was simply to breach the Southern Pacific Railroad causeway with a 2,000 ft (609.6 m) opening which would cost about three million dollars. This solution would very quickly bring about the leveling of both arms of the lake but since the difference in head is a maximum of about 18 in (½ m) this would only lower the south arm at most 9 in (½ m) and this would only be temporary if the lake continued to rise.

A second strategy proposed by the owners of north arm operations was the construction of a pumping station on the west shore of the lake to pump brine into the relatively flat Great Salt Lake Desert where it would evaporate. The dense brine created by this strategy could then be returned to the lake for utilization by chemical and salt companies. Such a strategy would probably cost over $20 million and should the lake stop rising then as the governor of the state put it "we would have a very expensive white elephant on our hands." Pumping would also flood the U.S. Air Force's bombing and gunnery range west of the lake and, since Hill Air Force Base is the largest employer in the state and the range is important to its mission, it would not appear to be in the best interest of the state to antagonize the source of a $200,000,000 annual payroll.

A third strategy which was presented as a possible alternative to the first two was the construction of a large evaporating pond area on the northeastern shore of the lake. Since these ponds would impound fresh water from the Bear River a greater amount of evaporation would take place, per unit area, from these ponds than from the lake's more concentrated brines. The ponds could support waterfowl, fish and fur animal propagation making them a useful if somewhat expensive alternative. The practicality of this strategy appears to be open to question, however. It is doubtful that sufficient flat land exists in the region to evaporate the amount of inflow which would be required to stabilize the lake. Should the lake begin to fall the ponds would represent an investment which would probably have to be abandoned.

A fourth strategy would require the construction of a considerable number of dams for water storage on the upper reaches of the Bear and other rivers flowing into the lake. This investment would require federal money and could cost in excess of one billion dollars. The argument for investing such a large sum of money would probably have to be based upon the utilization of impounded water by the private sector for agriculture and industry. Since the rights to water for irrigation in the basin have been allocated, storage would have to come from unseasonably high runoff and water could not be guaranteed for use in irrigating new regions on an annual basis. Thus this costly investment could not, therefore, count on amortization through the sale of water which would undercut its reason for being. The amount of time needed to build such a project would probably doom it from the outset anyway, because if the lake continued to rise at its present rate it would destroy all structures around it before the project could be finished.

After debating and studying the lake for more than two years the probability of a decision appears to be no brighter today than it did at the beginning. What has changed in the meantime is the weather. The State of Utah along with the rest of the western United States is undergoing a drought which has apparently halted the rise of the lake at about 4200 ft (1280.16 m). The Great Salt Lake which normally rises to a peak in the spring or early summer of each year peaked during the winter of 1977 and began falling at a faster rate during the summer than in any previous year of its recorded history. Thus perhaps nature has, for the present, solved man's problem of a rising lake stage saving his structures and obviating any need for lake level control, at least for the present.

REFERENCES

Greer, Deon C., 1971. Great Salt Lake, Utah, Map Supplement No. 14, Annals of the Association of American Geographers, 61, 1.

Greer, Deon C., 1972. "The Political Geography of the Relicted Lands of the Great Salt Lake," The Journal of Geography, March.

Marti, Yu. Yu. and Ratkovich, D. Ya., 1976. "Vodokhozyaistvenniy Problemi Azovskovo i Kaspiskovo Moryei" Vodniye Resursi, No. 3, Moskva.

Ratkovich, D. Ya., Zhdanova, I. S. and V. E. Prevalskyy, 1973. "K Probleme Urovennovo Rezhima Kaspiskovo Morya," Vodniye Resursi, No. 3, Moskva.

Ratkovich, D. Ya., 1975. "Ob Upravlennii Rezhimom Kolebanii Urovanya Kaspiiskovo Morya s Pomoshchyu Razdelitelnoi Dambi," Vodniye Resursi, No. 2, Moskva.

Stokes, Wm. Lee, Ed., 1966. Guidebook to the Geology of Utah, No. 20, Utah Geological Society, Salt Lake City.

Voropayev, G. V., 1976. "Yedinaya Bodokhozyaistvennaya Systema Strani," Vodniye Resursi, No. 6, Moskva.

Chapter 6

SALT EFFECT ON THE pH OF HYPERSALINE SOLUTIONS

B. KRUMGALZ

Israel Oceanographic and Limnological Research Ltd., Tel-Shikmona, P.O.B. 8030, Haifa, Israel

INTRODUCTION

The physical chemistry of hypersaline solutions is gaining in importance especially due to recent numerous studies on natural systems such as seawater and specifically the Dead Sea and other hypersaline brines. A number of excellent reviews devoted to the physical chemistry of seawater have been written in recent years (Sillen, 1961; Fofonoff, 1962; Cox, 1965; Goldberg, 1965; Pytkowicz, 1968; Horne, 1965, 1970; Helgeson, 1969; Pytkowicz and Kester, 1971; Stumm, 1973; Whitfield, 1973a; Millero, 1971, 1974a, b, c, d, 1975a, b, 1977). At the same time, several books concerning the chemistry of seawater have been published (Garrels and Christ, 1965; Faust and Hunter, 1967; Horne, 1969; Stumm and Morgan, 1970). However, it should be emphasized that almost all the above-mentioned works deal with different phenomena of the physical chemistry of seawater in which the ionic strength did not exceed 0.8. At these concentrations, the authors could still consider the structure of seawater as that of water disturbed to a varying degree by ions of solvated salts, organic molecules, molecules of gases and suspended and colloidal particles (Horne, 1969). Thus, the understanding of water structure is basic to the understanding of the mechanisms of chemical processes occurring in natural systems. In recent years, many studies have been devoted to the physico-chemical properties and structure of pure water and aqueous solutions (Gurney, 1953; Frank and Wen, 1957; Harned and Owen, 1958; Falkenhagen and Kelg, 1959; Robinson and Stokes, 1965; Samoilov, 1965; Harned and Robinson, 1968; Mishchenko and Poltoratski, 1968; Desnoyers and Jolicoeur, 1969; Eisenberg and Kauzmann, 1969; Nemethy, 1970; Franks, 1972; Horne, 1972; Kell, 1972; Gutmann et al., 1977).

When working with concentrated solutions, the water structure was shown to be completely destroyed (Mishchenko and Poltoratski, 1968). The effects of ionic association and competition between oppositely charged ions for water molecules in their hydration shells are intensified. The so-called "coordination dehydration" of ions caused by water deficiency in their hydration shells can be observed (Mishchenko and Poltoratski, 1968). According to these authors, at the concentrations where all water molecules are used in the ion hydration shells ("the limit of complete hydration"), these solutions can be considered schematically as systems composed of single-layer hydrated ions with their hydration shells touching each other. However, the present theories of electrolyte solutions are only applied to extremely dilute solutions and they cannot describe the behavior of concentrated electrolyte solutions. The physical chemistry of these concentrated solutions is of particular interest for understanding the nature of processes in natural hypersaline brines. In our brief review, we will dwell only on the question of principal interest for natural systems, namely: the salt effect on the activity of hydrogen ions in concentrated aqueous solutions.

Some studies of acid-base equilibrium in concentrated salt solutions have been previously reported (Kilpatrick, 1953; Kilpatrick and Eanes, 1953a, b; Kilpatrick et al., 1953; Paul and Long, 1957; Harned and Owen, 1958, p. 675; Critchfield and Johnson, 1958, 1959, Rosenthal and Dwyer, 1962, 1963) for solutions of bases and mineral or organic acids. The influence of neutral inorganic salts on the pH of buffer solutions has already been observed during the development and establishment of pH standards (Hamer and Acree, 1944; Bates et al., 1951; Bates, 1952, 1964). Bates (1952), at the National Bureau of Standards, noted that the depression of pH was observed in phthalate, phosphate and borax buffer solutions, but it should be emphasized that these results were obtained for diluted solutions with salt concentrations less than 0.05 M.

Recently, Bodenheimer and Neev (1963), Neev and Emery (1967) and Amit and Bentor (1971) have shown that the pH values of Dead Sea water increases progressively on dilution with distilled water. Bodenheimer and Neev (1963) termed this phenomenon "hidden alkalinity", and thought that bacterial reduction of gypsum to calcium sulfide followed by its hydrolysis was responsible for this effect:

$$CaS + 2H_2O = Ca(OH)_2 + H_2S$$

By dilution of Dead Sea water, this reaction will be shifted to the right and consequently the pH value will increase. Later, Neev and Emery (1967) showed that sulphate reduction takes place only within the bottom sediments at depths exceeding 40 m, but Amit and Bentor (1971) emphasized that the effect of "hidden alkalinity" was found in Dead Sea water taken from different depths and sampling areas. These authors also found that this effect was absent in solutions devoid of bicarbonates. Consequently, this phenomenon can be defined by dissociation-association equilibrium in carbonate systems. Recently, Nissenbaum (1969), Ben-Yaakov and Sass (1977) and Sass and Ben-Yaakov (1977) have come to the same conclusion.

SALT EFFECT ON THE pH OF CARBONATE SOLUTIONS

For the carbonate system, it can be shown, on the basis of dissociation-association equilibrium and also conditions of material balance and electroneutrality, that:

$$[H^+] = \sqrt{\frac{K_1 K_2 \cdot [HCO_3^-]}{\gamma_{CO_3^{2-}} \cdot (K_1 + [HCO_3^-] \cdot \gamma_\pm^2)}} \qquad [1]$$

where K_1 and K_2 are the first and the second ionic dissociation constants of H_2CO_3, respectively (K_1 = 4.31 x 10^{-7} (Butler, 1964), 4.32 x 10^{-7} (Read, 1975) or 4.45 x 10^{-7}

(Mishchenko and Ravdel, 1974); $K_2 = 4.69 \times 10^{-11}$ (Mishchenko and Ravdel, 1974); $[H^+]$ is the concentration of H^+ ions; $[HCO_3^-]$ is the concentration of HCO_3^- ions; $\gamma_{CO_3^{2-}}$ is the ionic activity coefficient of CO_3^{2-} ions; and $\gamma_\pm$ is the mean ionic activity coefficient of the H^+ and HCO_3^- ions. For dilute carbonate systems such as sea water, where $[HCO_3^-] = 2.3 \times 10^{-3}$ mol/1000 g H_2O (Kester et al., 1967) and $\gamma_\pm \cong 0.5$ (Garrels and Thompson, 1962; Berner, 1965; Pytkowicz, 1975), one can show that $[HCO_3^-]\ \gamma_\pm^2 \gg K_1$.

Then for this system, equation [1] will take the form:

$$pH = \text{const} + \tfrac{1}{2} \log \gamma_{CO_3^{2-}} \qquad [2]$$

where const $= -\tfrac{1}{2} \log K_1 K_2 = 8.34$.

For dilute carbonate-bicarbonate solutions, where the values of the activity coefficients of CO_3^{2-} ions are close to 1, the pH values according to equation [2] are approximately equal to 8.3. Amit and Bentor (1971) have found that a four-fold dilution of samples of natural saline water, i.e. Mediterranean Sea water (Tel-Aviv) and Red Sea water (Eilat), changed the pH values from 8.11 to 8.33 and from 8.04 to 8.24, respectively.

Pytkowicz (1975) has shown that the activity coefficients of CO_3^{2-} ions decrease very drastically with increasing ionic strength. Therefore in highly concentrated solutions such as Dead Sea water, the value of $\gamma_{CO_3^{2-}}$ will be extremely small and consequently from equation [2], the pH values of the Dead Sea water must decrease very significantly. The pH values of Dead Sea water vary from 5.9 at depths greater than 40 m (Nissenbaum, 1969) to 6.7 at the surface (Amit and Bentor, 1971). Thus the change of Dead Sea water pH under dilution must reflect the change in the $\gamma_{CO_3^{2-}}$ values as a function of the solution ionic strength.

Naturally enough, the question of the accuracy of pH values measured in concentrated solutions occurs. The question is, whether the pH values measured by a glass electrode in concentrated solutions are a good approximation of hydrogen ion activity or whether they are artifacts due to liquid junction error. The results obtained by Ben-Yaakov and Sass (1977) have answered this question very conclusively. They showed that the pH values of Dead Sea water measured by a glass electrode and by an electrochemical cell without a liquid junction are almost identical, and consequently that the low pH values of Dead Sea water are real.

However, it should be taken into consideration that pH measurements in concentrated solutions may depart from the scale established with standard buffer solutions prepared in distilled water. This may be explained by the existence of the liquid junction potential and by the asymmetry potential of glass electrodes when transferring from dilute standard buffers to concentrated solutions. This problem is especially critical when considering pH measurements of seawater. Therefore some pH scales based on buffer solutions prepared in seawater were proposed. Smith and Hood (1964) used a pH scale based on tris(hydroxymethyl)aminomethane buffers prepared in seawater, but Pytkowicz et al. (1966) noted that the data obtained by oceanographic procedures in which dilute buffers have been used were comparable within ±0.006 pH unit.[1]

[1] Not long ago Hansson (1972) again suggested some buffer solutions in synthetic sea water.

Nevertheless, the pH values obtained for concentrated aqueous solutions can only estimate the hydrogen ion activity. MacInnes' (1948) statement: "In possibly all but one case in a thousand it is not necessary to consider the meaning of pH in terms of solution theory at all, but only to accept the numbers as a practical scale of acidity and alkalinity" has been correct up to now.

In our opinion, and also according to preliminary results of Bates (1952), the same phenomenon, namely the change of pH under dilution of solutions reflecting the change in the activity coefficient of bivalent anions as a function of the solution ionic strength, should be observed in other buffer solutions containing some acidic salts. In an attempt to check this assumption, the influence of neutral inorganic salts on the pH of aqueous solutions of NaH_2PO_4 and KH_2PO_4 was examined.

SALT EFFECT ON THE pH OF ORTHOPHOSPHATE SOLUTIONS

The phosphate solutions are not only of academic interest, but they also play a great role in oceanographic problems. Morse (1974) has found that orthophosphate inhibited calcite dissolution significantly.[2] In addition, Avnimelech (1975) has shown the importance of phosphate equilibrium in fish ponds and has investigated some aspects of this equilibrium.

For dihydrogen orthophosphate solutions, the following equation can be obtained on the basis of dissociation-association equilibrium and conditions of material balance and electroneutrality:

$$[H^+] = \sqrt{\frac{K_1 K_2 [H_2PO_4^-]}{\gamma_{HPO_4^{2-}} \cdot (K_1 + [H_2PO_4^-] \cdot \gamma_\pm^2)}} \qquad [3]$$

where K_1 and K_2 are the first and second ionic dissociation constants of H_3PO_4 respectively [K_1 = 7.11 x 10^{-3}, K_2 = 6.34 x 10^{-8} (Mishchenko and Ravdel, 1974)]; $[H_2PO_4^-]$ is the concentration of $H_2PO_4^-$ ions; $\gamma_{HPO_4^{2-}}$ is the ionic activity coefficient of HPO_4^{2-} ions; and $\gamma_\pm$ is the mean ionic activity coefficient of the H^+ and $H_2PO_4^-$ ions. However, for phosphate systems, the magnitudes of K_1 and $[H_2PO_4^-] \cdot \gamma_\pm$ are of the same order (commensurable quantities) in contrast to carbonate solutions. Therefore we cannot neglect the K_1 values in the denominator of equation [3].

Then for these systems:

$$pH = -\log \left[\sqrt{\frac{K_1 K_2 \cdot [H_2PO_4^-]}{\gamma_{HPO_4^{2-}} \cdot (K_1 + [H_2PO_4^-] \cdot \gamma_\pm^2)}} \cdot \gamma_\pm \right] \qquad [4]$$

[2] Sapozhnikov (1977) reported that very high concentrations of the phosphates were found in the upper layer of sea water near the seaboard of Peru.

and finally:

$$\text{pH} = \text{const} + \log \sqrt{\frac{\gamma_{HPO_4^{2-}} \cdot (K_1 + [H_2PO_4^-] \cdot \gamma_\pm^2)}{[H_2PO_4^-] \cdot \gamma_\pm^2}} \qquad [5]$$

where const = $-\frac{1}{2} \log K_1 K_2 = 4.68$.

Equation [5] shows that the pH-concentration relationship in orthophosphate buffer solutions is more complex than the similar relationships in carbonate systems, so the influence of the solution ionic strength on the $\gamma_{HPO_4^{2-}}$ and $\gamma_\pm$ values must be taken into consideration. It must be emphasized that equations similar to equation [4] can be written for every buffer solution containing acidic salts (phthalates, borates, etc.).

Fig. 6-1 shows that the large effect of the addition of neutral inorganic salts (NaCl, NaBr, KCl, KBr) on the pH of buffer solutions containing acidic salts. The ΔpH values in these figures were calculated according to the following equation:

$$\Delta\text{pH} = (\text{pH})_o - (\text{pH})_i \qquad [6]$$

where $(\text{pH})_o$ is the pH of buffer solutions without neutral salts; and $(\text{pH})_i$ is the pH of buffer solutions with added neutral inorganic salts at different concentrations. The effect can be estimated quantitatively by taking into account the fact that the effect of salt additions is expressed primarily as a decrease in the ionic activity coefficients. Let us attempt to estimate this effect using equations [5] and [6]. Substituting equation [5] for buffer solutions without added neutral salts, and with these salts in equation [6] *, one can obtain the following equation:

$$\Delta\text{pH} = \log \left[\left(\frac{\gamma_{HPO_4^{2-},o}}{\gamma_{HPO_4^{2-},i}} \right)^{1/2} \left(\frac{K_1 + [H_2PO_4^-] \;\; \gamma^2_{\pm,o}}{K_1 + [H_2PO_4^-] \;\; \gamma^2_{\pm,i}} \right)^{1/2} \frac{\gamma_{\pm,i}}{\gamma_{\pm,o}} \right] \qquad [7]$$

In the calculations according to equation [7], we should take into consideration that the concentration, expressed in molality units, of $H_2PO_4^-$ ions is equal to the concentration of the salt (NaH_2PO_4 or KH_2PO_4) in these solutions and is constant under addition of other salts. Recently, Childs et al. (1973, 1974), in their investigation of the aqueous system Na^+, K^+, Cl^-, $H_2PO_4^-$, have postulated that the $H_2PO_4^-$ ions associate to dimers

* In all subsequent discussion, subscript o will be applied to values calculated for buffer solutions without added neutral salts, and subscript i to values calculated for buffer solutions with added neutral salts.

with a stoichiometric association constant, $K_A = 0.25 \pm 0.1$ kg mole^{-1} at 25°C. Similar results were obtained for the system $NaH_2PO_4 - NaClO_4$ by Wood and Platford (1975). However, in our studies, we can consider that such a small association constant for anion-anion pairing does not influence the concentration of the $H_2PO_4^-$ species.

The central problem in calculations using equation [7] is the determination of activity coefficients of single ions. At the present time, these values cannot be measured and calculated on the basis of rigorous theoretical criteria.

For the estimation of the values $\gamma_{\pm,o}$ and $\gamma_{HPO_4^{2-},o}$ in the 0.025 m solutions of NaH_2PO_4 or KH_2PO_4, the Debye-Hückel equation was used, and the values $\gamma_{\pm,o} = 0.875$ and $\gamma_{HPO_4^{2-},o} = 0.540$ were obtained.

In mixed electrolyte solutions, the cation activity coefficient is known to be determined primarily by the nature of the anions in the solutions and to be weakly dependent on the nature of the other cations. The same statement can be made for the anion activity coefficient. It is precisely this fact that dictates the use of media with constant ionic strength for determination of the dissociation constants of complex compounds. We can consider that in our experiments, the neutral salts added to buffer solutions create constant ionic strength. By using this assumption, the values of $\gamma_{\pm,i}$ were calculated and are shown in Table 6-1.

Thus all values necessary for the computation of HPO_4^{2-} ions according to equation [7] have been determined. The values of ΔpH were obtained from Fig. 6-1. The calculated $\gamma_{HPO_4^{2-}}$ values are given in the last column of Table 6-1. It is interesting to note that these activity coefficients have the same order of magnitude as the $\gamma_{CO_3^{2-}}$ values in seawater obtained by Pytkowicz (1975) from the ratios of the thermodynamic and the

TABLE 6-1. The values of $\gamma_{\pm}(H^+,H_2PO_4^-)$ and $\gamma_{HPO_4^{2-}}$ for the examined solutions of dihydrogen orthophosphates

Dihydrogen ortho-phosphate (0.025 m)	Added neutral salt	Concentration of added salt	ΔpH	$\gamma_{\pm}(H^+,H_3PO_4^-)$	$\gamma_{HPO_4^{2-}}$
NaH_2PO_4	NaCl	0.5	0.34	0.653	0.092
		1.0	0.50	0.615	0.042
		2.0	0.73	0.612	0.014
		3.0	0.92	0.649	0.006
	NaBr	0.5	0.34	0.666	0.093
		1.0	0.48	0.638	0.046
		2.0	0.69	0.658	0.018
		3.0	0.86	0.732	0.009
KH_2PO_4	KCl	0.5	0.22	0.633	0.156
		1.0	0.32	0.584	0.091
		2.0	0.43	0.568	0.054
	KBr	0.5	0.22	0.646	0.170
		1.0	0.30	0.606	0.104
		2.0	0.40	0.611	0.066

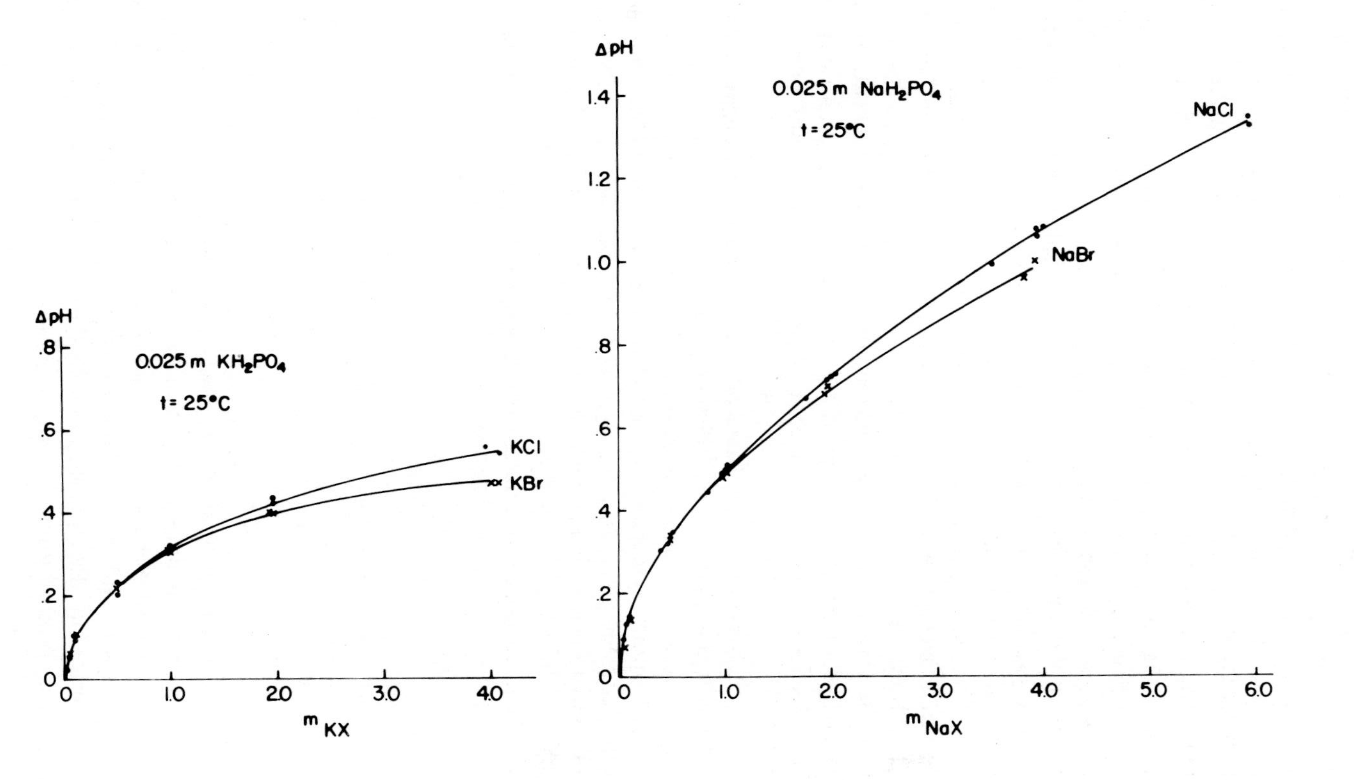

Fig. 6-1. The effect of neutral inorganic mineral salts on the pH of orthophosphate solutions at 25°C. a) Solutions of NaH_2PO_4 (0.025 m); b) Solutions of KH_2PO_4 (0.025 m).

apparent dissociation constants of carbonic acid, and by Leyendekkers (1972), Whitfield (1973b) and Millero (1974a,c; 1975b), who applied the specific interaction model for estimating total activity coefficients. The values of $\gamma_{HPO_4^{2-}}$ for Black Sea water and seawater with S = 33% determined by Skopintsev (1973) at 20°C (0.043 and 0.023, respectively) are of the same order as our results for the system $NaH_2PO_4 - NaCl - H_2O$ at $I_{total} \cong 0.7$. Thus Fig. -1 and Table -1 show, in accordance with equation [7], that the decreasing pH of buffer solutions containing acidic salts (NaH_2PO_4 or KH_2PO_4) with the addition of neutral inorganic mineral salts is defined by drastically decreasing magnitudes of the activity coefficient of the bivalent anion ($\gamma_{HPO_4^{2-}}$).

One can see in Fig. 6-1 that the depression of the pH of buffer solutions is controlled by both the solution ionic strength and also by the nature of the added salts. In our opinion, this phenomenon can be explained by the process of ionic hydration and the resulting change in the water activity, because in these solutions the ionic association is very small. For example, for seawater, Millero (1974a) has shown that most of the sodium and potassium ions are free.

The influence of ion hydration on the activity coefficients in electrolyte solutions was discussed in Stokes and Robinson (1948), Bates et al. (1970) and in Jacobsen and Skou (1977). These authors have shown that in concentrated solutions, the hydration effect exceeds the change in electrostatic interaction energy. This treatment, taking into account the ion hydration but ignoring the change in electrostatic energy, permits the calculation of the activity coefficients in concentrated electrolyte solutions with good approximation.

In conclusion, it has been shown, based on general theoretical assumptions and our experimental results concerning the salt effect on pH values of hypersaline solutions, that concentrating normal seawater show a decrease in pH. Some examples of this phenomenon can be observed during evaporation of seawater in lagoons, in natural or artificial salt-producing ponds or pond used for algae and fish cultivation, etc. Changing the hydrogen ion activity in such systems will affect the solubility of a number of salts and minerals and also the equilibrium and kinetics of many processes. In our opinion, it is necessary to take into consideration the salt effect on the hydrogen ion activity in dealing with natural marine and geochemical systems.

ACKNOWLEDGMENTS

The author wants to thank Prof. J. Gat, Prof. S. Ben-Yaakov, Prof. Y. Avnimelech, Mr. A. Back and Dr. Mary Swenson for their constructive criticism during the preparation of this manuscript. Special thanks are due to Prof. J. Gat for his kindness and generosity and the opportunity to work in his laboratory at the Weizmann Institute where this work was partly carried out. Research supported by the Center for Absorption in Science, the Ministry for Immigrant Absorption, State of Israel.

REFERENCES

Amit, O. and Y. K. Bentor, 1971. pH-dilution curves of saline waters. Chem. Geol., 7, 307.
Avnimelech, Y., 1975. Phosphate equilibrium in fish ponds. Verh. Int. Verein. Limnol., 19, 2305.
Bates, R. G., 1952. A fundamental approach to the establishment of pH standards. Analyst, 77, 653.
Bates, R. G., 1964. Determination of pH. Theory and practice. Wiley, New York, 435 pp.
Bates, R. G., W. E. Bower, R. G. Miller and E. R. Smith, 1951. pH of solutions of potassium hydrogen d-tartrate from 0° to 60°C. J. Res. Nat. Bur. Stand., 47, 433.
Bates, R. G., B. R. Staples and R. A. Robinson, 1970. Ionic hydration and single ion activities in unassociated chlorides at high ionic strengths. Anal. Chem., 42, 867.
Ben-Yaakov, S. and E. Sass, 1977. Independent estimate of the pH of Dead Sea brine. Limnol. Oceanogr., 22, 374.
Berner, R. M., 1965. Activity coefficients of bicarbonate, carbonate and calcium ions in sea water. Geochim. Cosmochim. Acta, 29, 947.
Bodenheimer, W. and D. Neev., 1963. On the change of pH in Dead Sea brine on dilution with distilled water. Bull. Res. Counc. Israel, 11S, 150.
Butler, J. N., 1964. Ionic equilibrium (a mathematical approach). Reading, Mass., Addison–Wesley Pub. Co., 547 pp.
Childs, C. W., C. J. Downes and R. F. Platford, 1973. Thermodynamics of aqueous sodium and potassium dihydrogen orthophosphate solutions at 25°C. Aust. J. Chem., 26, 863.
Childs, C. W., C. J. Downes and R. F. Platford, 1974. Thermodynamics of multicomponent electrolyte solutions: aqueous mixtures of two salts from among NaCl, KCl, NaH_2PO_4 and KH_2PO_4 at 25°C. J. Solut. Chem., 3, 139.
Cox, R. A., 1965. Physical properties of seawater. In: Chemical oceanography, J. P. Riley and G. Skirrow, eds., Vol. 1, p. 73-120. Academic Press, New York.
Critchfield, F. E. and J. B. Johnson, 1958. Titration of weak bases in strong salt solutions. Anal. Chem., 30, 1247.
Critchfield, F. E. and J. B. Johnson, 1959. Effect of neutral salts on the pH of acid solutions. Anal. Chem., 31, 570.
Desnoyers, J. E. and C. Jolicoeur, 1969. Hydration and thermodynamical properties of ions. In: Modern aspects of electrochemistry, J. O'M. Bockris and B. E. Conway, eds., Plenum Press, New York.
Eisenberg, D. and W. Kauzmann, 1969. The structure and properties of liquid water. Oxford Univ. Press, Oxford.
Falkenhagen, G. and G. Kelg, 1959. Modern condition of electrolyte solutions theory. In: Modern aspects of electrochemistry, J. O'M. Bockris, ed., Butterworths Scientific Publications, London.
Faust, S. D. and J. V. Hunter, 1967. Principles and applications of water chemistry. Wiley, New York, 643 pp.
Fofonoff, N. P., 1962. Physical properties of sea water. In: Sea water, M. N. Hill, ed., Vol. 1, p. 3–30. Interscience, New York.
Frank, H. S. and W. Y. Wen, 1957. Structural aspects of ion-solvent interaction in aqueous solutions: a suggested picture of water structure. Disc. Farad. Soc., 24, 113.
Franks, F., 1972. The physics and physical chemistry of water. In: Water, a comprehensive treatise, F. Franks, ed., Vol. 1. Plenum Press, New York.
Garrels, R. M. and C. L. Christ, 1965. Solutions, minerals and equilibria. Harper & Rowe, New York, 450 pp.
Garrels, R. M. and M. E. Thompson, 1962. A chemical model for sea water at 25°C and one atmosphere total pressure. Am. J. Sci., 260, 57.
Goldberg, E. D., 1965. Minor elements in sea water. In: Chemical oceanography, J. P. Riley and G. Skirrow, eds., Vol. 1, p. 163–196. Academic Press, New York.
Gurney, R. W., 1953. Ionic processes in solution. Dover, New York.

Gutmann, V., E. Platner and G. Resch, 1977. Structural consideration about liquid water and aqueous solutions. Chimia, 31, 431.

Hamer, W. J. and S. F. Acree, 1944. A method for the determination of the pH of 0.05 molal solutions of acid potassium phthalate with or without potassium chloride. J. Res. Nat. Bur. Stand., 32. 215.

Hansson, I., 1972. Thesis, Anal. Chem. Univ. Göteborg, quoted according to E. Hogfeldt, On the construction of single ion activity functions and some comments on the formulation of conventions to describe sea water equilibria, in "The Nature of Seawater", Ed. E. D. Goldberg, Dahlem Konferenzen, Berlin, 1975, p. 281–312.

Harned, H. S. and B. B. Owen, 1958. The physical chemistry of electrolyte solutions. 3rd edition. Reinhold Publ. Corp., New York, 803 pp.

Harned, H. S. and R. A. Robinson, 1968. Multicomponent electrolyte solutions. Pergamon, Oxford.

Helgeson, H. C., 1969. Thermodynamics of hydrothermal systems at elevated temperatures and pressures. Am. J. Sci., 267, 729.

Horne, R. A., 1965. The physical chemistry and structure of sea water. Wat. Resour. Res., 1, 263.

Horne, R. A., 1969. Marine chemistry. Wiley-Interscience, New York, 568 pp.

Horne, R. A., 1970. Sea water. In: Advances in hydroscience, Vol. 6, p. 107–140. Academic Press, New York.

Horne, R. A., 1972. Water and aqueous solutions. Wiley Interscience, New York, 837 pp.

Jacobsen, T. and E. Skou, 1977. A simple hydration treatment on activity coefficients in concentrated electrolyte solutions. Electrochim. Acta, 22, 161.

Kell, G. S., 1972. Continuum theories of liquid water. In: Water and aqueous solutions, R. A. Horne, ed., p. 119. Wiley, New York.

Kester, D. R., I. W. Duedall, D. N. Connors and R. M. Pytkowicz. 1967. Preparation of artificial sea water. Limnol. Oceanogr., 12, 176.

Kilpatrick, M., 1953. The dissociation constants of acids in salt solutions. I. Benzoic acid. J. Am. Chem. Soc., 75, 584.

Kilpatrick, M. and R. D. Eanes, 1953a. The dissociation constants of acids in salt solutions. II. Acetic acid. J. Am. Chem. Soc., 75, 586.

Kilpatrick, M. and R. D. Eanes, 1953b. The dissociation constants of acids in salt solutions. III. Glycolic acid. J. Am. Chem. Soc., 75, 587.

Kilpatrick, M., R. D. Eanes and J. G. Morse, 1953. The dissociation constants of acids in salt solutions. IV. Cyclohexanecarboxylic acid. J. Am. Chem. Soc., 75, 588.

Leyendekkers, J. V., 1972. The chemical potentials of seawater components. Mar. Chem., 1, 75.

MacInnes, D. A., 1948. Criticism of a definition of pH. Science, 108, 693.

Millero, F. J., 1971. The physical chemistry of multicompound salt solutions. In: Biophysical properties of skin, H. R. Elden, ed., p. 329–376. Wiley, New York.

Millero, F. J., 1974a. The physical chemistry of seawater. Ann. Rev. Earth Planet. Sci., 2, 101.

Millero, F. J., 1974b. Seawater as a multicomponent electrolyte solution. In: The Sea; ideas and observations, E. D. Goldberg, ed., Vol. 5, p. 3–80. Wiley-Interscience, New York.

Millero, F. J., 1974c. The physical chemistry and structure of seawater. In: Structure of water and aqueous solutions, W. Luck, ed., p. 513–522. Verlag Chemie, Weinheim.

Millero, F. J., 1974d. Equation of state of seawater. Naval Res. Rev., 27, 40.

Millero, F. J., 1975a. The physical chemistry of estuaries. In: Marine chemistry in the coastal environment, T. M. Church, ed., p. 25–55. ACS Symp. Ser. 18, Washington, D. C.

Millero, F. J., 1975b. The state of metal ions in seawater. Thalassia jugosl., 11, 53.

Millero, F. J., 1975c. The physical chemistry of estuaries, Chapter 2, in "Marine Chemistry in Coastal Environment", Ed. T. M. Church, ACS Symp. Ser. 18, Washington, D.C., p. 25–55.

Millero, F. J., 1977. Thermodynamic models for the state of metal ions in seawater. In: The Sea; ideas and observations, E. D. Goldberg, ed., Vol. 6, p. 653–693. Wiley-Interscience, New York.

Mishchenko, K. P. and G. M. Poltoratsky, 1968. The problem of thermodynamics and structure of aqueous and nonaqueous electrolyte solutions. (in Russian). Khimiya, Leningrad.

Mishchenko, K. P. and A. A. Ravdel, 1974. A brief reference book of physico-chemical quantities. (in Russian). Khimiya, Leningrad.

Morse, J. W., 1974. Dissolution of calcium carbonate in sea water. V. Effects of natural inhibitors and the position of the chemical lysocline. Am. J. Sci., 274, 638.

Neev, D. and K. O. Emery, 1967. The Dead Sea; depositional processes and environment of evaporates. Bull. Geol. Surv. Israel, 41, 147 pp.

Nemethy, G., 1970. The structure of water and the thermodynamic properties of aqueous solutions. Annali d'Ist. Superiore di Sanita, 6 (Special Issue), 487.

Nissenbaum, A., 1969. Studies in geochemistry of the Jordan River – Dead Sea system. Unpublished Ph.D. Thesis, University of California, Los Angeles.

Paul, M. A. and F. A. Long, 1957. H_0 and related indicator acidity functions. Chem. Rev., 57, 1.

Pytkowicz, R. M., 1968. The carbon dioxide-carbonate system at high pressures in the oceans. Oceanogr. Mar. Biol. Ann. Rev., 6, 83.

Pytkowicz, R. M., 1975. Activity coefficients of bicarbonates in seawater. Limnol. Oceanogr., 20, 971.

Pytkowicz, R. M. and D. R. Kester, 1971. The physical chemistry of seawater. Oceanogr. Mar. Biol. Ann. Rev., 9, 11.

Pytkowicz, R. M., D. R. Kester and B. C. Burgener, 1966. Reproducibility of pH measurements in seawater. Limnol. Oceanogr., 11, 417.

Read, H. J., 1975. The first ionization constant of carbonic acid from 25° to 250°C and to 2000 bar. J. Solut. Chem., 4, 53.

Robinson, R. A. and R. H. Stokes, 1965. Electrolyte solutions. 2nd edition, revised. Butterworths Scientific Publications, London, 571 pp.

Rosenthal, D. and J. S. Dwyer, 1962. Acid-base equilibria in concentrated salt solutions. I. Potentiometric measurements, indicator measurements and uncharged bases in dilute acid solutions. J. Phys. Chem., 66, 2687.

Rosenthal, D. and J. S. Dwyer, 1963. Acid-base equilibria in concentrated salt solutions. II. Acid-base titrations and measurements of pH and H_0. Anal. Chem., 35, 161.

Samoilov, O. Ya., 1965. Structure of aqueous electrolyte solutions and the hydration of ions. Authorized translation from the Russian by D. J. C. Ives, Consultants Bureau, New York, 185 pp.

Sapozhnikov, S. S., 1977. The increase of phosphates concentration owing to the work of "biofilter" in high productive regions of ocean. in "Chemicooceanographic Researches", Ed. B. A. Skopintsev and V. N. Ivanenkov. In Russian. Nauka, Moscow, p. 57–59.

Sass, E. and S. Ben-Yaakov, 1977. The carbonate system in hypersaline solutions: Dead Sea brines. Mar. Chem., 5, 183.

Sillen, L. G., 1961. The physical chemistry of sea water. In: Oceanography, M. Sears, ed., p. 549–581. Am. Assoc. Advanc. Sci., Washington, D. C.

Skopintsev, B. A., 1973. Activity coefficients of some ions in Black Sea water. In: Researcher in sea chemistry, p. 140–147. Acad. Sci. USSR Trans. P. P. Shirshov Inst. Oceanol., Vol. 63, Nauka, Moscow.

Smith, W. H., Jr. and D. W. Hood, 1964. pH measurement in the ocean: a seawater secondary buffer system. In: Recent researches in the fields of hydrosphere, atmosphere and nuclear geochemistry, Y. Miyake and T. Koyama, eds., p. 185–202. Ken Sugawara Festiva Volume, Maruzen Co., Tokyo.

Stokes, R. H. and R. A. Robinson, 1948. Ionic hydration and activity in electrolyte solutions. J. Am. Chem. Soc., 70, 1870.

Stumm, W., 1973. Chemical speciation. In: Chemical oceanography, J. P. Riley and G. Skirrow, eds., Vol. 1. Academic Press, New York.

Stumm, W. and J. J. Morgan, 1970. Aquatic chemistry, an introduction emphasizing chemical equilibria in natural waters. Wiley-Interscience, New York, 583 pp.

Whitfield, M., 1973a. Sea water as an electrolyte solution. In: Chemical oceanography, J. P. Riley and G. Skirrow, eds., Vol. 1. Academic Press, New York.

Whitfield, M., 1973b. A chemical model for the major electrolyte component of seawater based on the Bronsted-Guggenheim hypothesis. Mar. Chem., 1, 251.

Wood, R. H. and R. F. Platford, 1975. Free energies of aqueous mixtures of NaH_2PO_4 and $NaClO_4$: evidence for the species $(H_2PO_4)_2^{-2}$. J. Solut. Chem., 4, 977.

Chapter 7

OXYGEN ISOTOPE FRACTIONATION IN SALINE AQUEOUS SOLUTIONS

K. HEINZINGER

Max-Planck–Institut für Chemie (Otto-Hahn-Institut), Mainz, Germany

The fractionation of the oxygen isotopes between the hydration water of the ions and bulk water can be determined with the CO_2-equilibration technique. If CO_2 is equilibrated with an aqueous electrolyte solution, the ions not having exchangeable oxygen, (CO_2^*) and with pure water of the same oxygen isotope composition (CO_2) as the solvent, we can define a quantity δ by:

$$\delta = (1 - \alpha_{CO_2^*-CO_2}) \cdot 10^3 \qquad [1]$$

with

$$\alpha_{CO_2^* - CO_2} = (^{18}O/^{16}O)_{CO_2^*}/(^{18}O/^{16}O)_{CO_2} \qquad [2]$$

δ is in general different from zero, a consequence of the change of the binding of the water molecules if an electrolyte is added to pure water. In a very good approximation, at least as far as oxygen isotope effects are concerned, it can be distinguished between three kinds of water molecules in the solution. Water molecules in the hydration shell of cations, in the hydration shell of anions and bulk water in a binding situation similar to pure water. Therefore δ can be expressed as:

$$\delta = r(n_c\delta_c + n_A\delta_A) \qquad [3]$$

where r is the concentration in moles of salt per mole of water. $\delta_C(\delta_A)$ is the fractionation factor between hydration water of the cations (anions) and pure water and n is the corresponding hydration number. With this definition positive δ values indicate enrichment of ^{18}O in the hydration shells of the ions, negative ones depletion.

Measurements of this kind have first been performed by Taube (1954). He found in the limits of error of his measurements that δ depends linearily on the salt concentration and is independent of the anion, indicating a constant separation factor between hydration water of cations and bulk water and no isotope effect in the hydration shells of anions. Sofer and Gat (1972) extended these measurements to other salt solutions which are of special interest in geochemistry. They found a linear relationship between δ/r and the charge over Pauling radius of the cations. In connection with the investigations of oxygen isotope effects in $CuSO_4 \cdot 5H_2O$, aqueous solutions of $CuCl_2$ and $CuSO_4$ were measured and it was found contrary to the results of Taube that δ/r was different for the two solutions (Maiwald and Heinzinger, 1972). Subsequent investigations of potassium halide solutions confirmed the anion dependence (Götz and Heinzinger, 1973). In addition it was shown that δ/r changes significantly when H_2O is replaced by D_2O as solvent.

The availability of calculations of energy surfaces of a water molecule in the field of alkali ions by Clementi and Popkie (1972) and by Kistenmacher et al (1973a,b) provided the basis for calculations of gas phase separation factors for the oxygen isotopes between water molecules hydrated to a cation and free water molecules (Bopp et al, 1974). From the force constants derived from these energy surfaces the fractionation factors were calculated from the normal frequences of vibration on the basis of the Bigeleisen formalism using the Wolfsberg and Stern version of the Schachtschneider program (Wolfsberg and Stern, 1964). (Dashed lines if Fig. 7-1). In order to relate these gas phase separation factors to the measured fractionation in solution, information on the transfer of free water to bulk water and of the hydration complex from the gas phase to the solution is necessary. The transfer of free water to bulk water is accomplished by multiplying the gas phase separation factor with the vapor pressure isotope effect (Szapiro and Steckel, 1967). The resulting full lines in Fig. 7-1 thus give the separation factor between the gas phase hydration complex and pure liquid water. For the transfer of the hydration complex there exists no similar obvious relationship. It might be expected that by forming the hydrogen bonds between the hydration water and the bulk water in the solution the ion-oxygen bond becomes weaker while vibrational and wagging frequencies increase. In respect to the oxygen isotope effect both changes tend to cancel, at least partly. Therefore it seems to be justified in a first approximation to use the same partition function ratios for the hydration complex in the gas phase as well as in the solution. Thus the full lines will be used for comparison with experimental results.

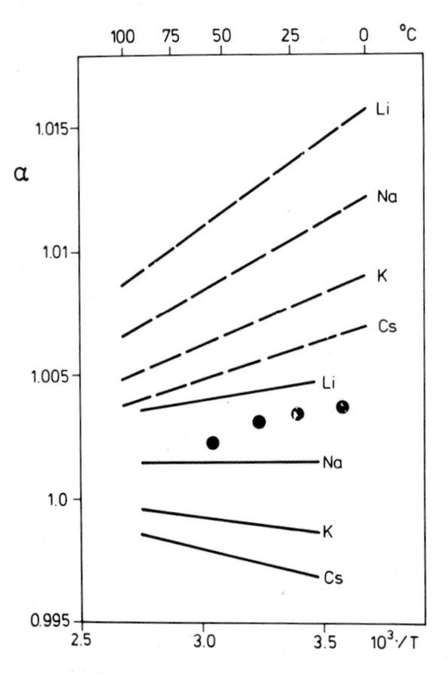

Fig. 7-1. Calculated oxygen isotope fractionation factors between hydration water of alkali ions and free water in the gas phase (dashed lines) and in solution (full lines) in the temperature range 0°–100°C.

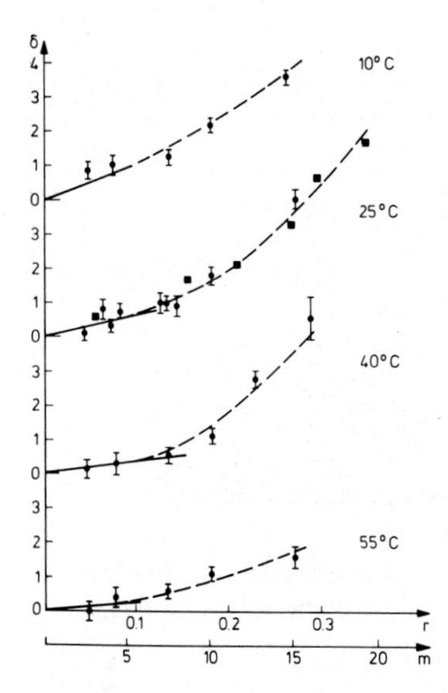

Fig. 7-2. The concentration dependence of δ for LiCl solutions at four different temperatures. The ■ are measurements by Taube.

In principle it would be possible to calculate also the influence of the halide ions on the oxygen isotope fractionation in aqueous solutions on the basis of ab initio calculations (Kirstenmacher et al, 1973c). But the uncertainty in respect to the transfer of the anion hydration complex from the gas phase to the solution is appreciably larger because the oxygen atoms become directly involved in hydrogen bonds in the solution. As in addition the anion effects are significantly smaller, it would be unjustified to draw any quantitative conclusions from such calculations.

In order to compare the calculated separation factors with experimental ones it is necessary to separate cationic and anionic contributions. It is also interesting to check the calculated temperature dependence.

As the available experimental evidence was not sufficient for this comparison, additional measurements were necessary. From the alkali ions (only for these spherically symmetric ions energy surfaces are available) the small Li^+ shows the by far largest effect. As in addition the solubility of LiCl is very high, such solutions have been investigated over the whole concentration range up to 15 molal in the temperature range 10°–55°C (Bopp et al, 1977). The results of these measurements are shown in Fig. 7-2. In the concentration range up to about 5 molal the measurements are fitted to a straight line because here enough water molecules are available to have constant hydration numbers and a linear relationship between δ and r can be expected. For higher concentrations no information is available on the change of cation and anion hydration numbers with concentration. From x-ray and neutron diffraction studies it is concluded that in a 15 molal solution all water molecules are tetrahedrally coordinated with Li^+ and it is expected that these $Li^+ \cdot 4H_2O$ groups do not share water molecules (Narten et al, 1973). These investigations as well as NMR measurements find that the Li-O distance varies only little with concentra-

TABLE 7-1.

δ/r values for various alkali halide solutions at 25°C.

Solute	δ/r	Ref.
LiCl	6.4 ± 0.5	Taube (1954) and Bopp et al (1977);
LiI	5.4 ± 0.9	Bopp et al (1977)
NaCl	0.0	Taube (1954) and Götz and Heinzinger (1972)
NaI	−4.7 ± 0.5	Bopp et al (1977)
	−1.5	Taube (1954)
KCl	−7.7 ± 0.3	Sofer and Gat (1972) and Götz and Heinzinger (1972)
KBr	−9.5 ± 0.8	Götz and Heinzinger (1972)
KI	−11.3 ± 0.8	Götz and Heinzinger (1972)
CsCl	−11.9 ± 1.0	Bopp et al (1977)

tion(Langer and Hertz, 1977). Thus it can be expected that the influence of the chloride ions on these $Li^+ \cdot 4H_2O$ groups is similar to the influence of the water molecules at lower concentrations and therefore

$$\delta(r = 0.25) = \delta_{Li} \qquad [4]$$

For the other cations and the anions the δ's have to be deduced in a different way by using the slopes δ/r at low concentration and the hydration numbers.

In Table 7-1 the δ/r values at 25°C for alkali halide solutions available in the literature have been collected.

A set of consistent hydration numbers for various alkali and halide ions is available from molecular dynamics simulations (Heinzinger and Vogel, 1976). They are defined as the integral over the ion-oxygen radial pair correlation function up to its first minimum. The problem here is that some of the water molecules in the hydration shell so defined are disoriented as can be seen from the angular distribution of the water dipoles (Heinzinger, 1976). This means that different force constants apply for the water molecules, resulting in different δ values. To prevent an unreasonable averaging in analysing the experimental results only the water molecules with an energetically favourable orientation were considered. In the limits of error it can be deduced that all four cations have 4 water molecules with favorable orientation in spite of quite different Pauling radii. In the case of the anions no obvious dependence of the relevant force constant from the water molecule orientation can be seen. Therefore as an average value of the hydration number of the anions in 2.2 molal solutions $n_A = 7$ was chosen. Counter-ion and concentration dependences were neglected.

With these hydration numbers and the value for δ_{Li} given above the remaining δ_C and δ_A can be determined from the δ/r values given in Table 7-1. The result of the fitting procedure is given in Table 7-2. The errors are estimated to be ± 0.2 not considering the uncertainties in the hydration numbers.

In Fig. 7-3 theoretical and experimental δ values are compared. Considering the uncertainties in the determination of the experimental results and the approximations used

TABLE 7-2.

δ_C and δ_A values for various alkali and halide ions at 25°C.

Ion	Li	Na	K	Cs	Cl	Br	I
δ	3.1	1.6	−0.4	−1.4	−0.9	−1.1	−1.3

in evaluating the theoretical values the agreement can be called fair. In both cases δ_{Li} and δ_{Na} are positive as expected for positively hydrating ions and K^+ as well as Cs^+, known as negatively hydrating ions, show negative δ values. The interpolation leads to a Pauling radius of 1.25 Å for the change of the sign of δ in agreement with conclusion by Samoilov (1972) who found that the transition from positive to negative hydration at 21.5°C occurs at a Pauling radius of 1.18Å .

For the temperatures 10°, 25°, 40° and 50°C the values for δ_{Li} and δ_{Cl} can be determined from the measurements of the LiCl solutions as shown in Fig. 7-2 by the use of Equation [4]. The results are given in Table 7-3. The error for the δ_{Li} is again estimated to be ±0.2 while for the anions a larger error has to be assumed because of the additional error in the slope at low concentrations and the uncertainties in the hydration numbers. While the δ_{Li} show the expected decrease with increasing temperature no trend is visible for the δ_{Cl}, most probably because of the relatively much larger errors.

For comparison with the theoretical values the δ_{Li} from Table 7-3 are shown in Fig. 7-1 as points. Considering the approximations used in the calculations and the uncertainties in the experimental results the agreement might be called fair. Especially the temperature dependence is very well reproduced. In conclusion it can be stated that the oxygen isotope fractionation in aqueous solutions can, at least qualitatively, be understood on the basis of ab initio calculations.

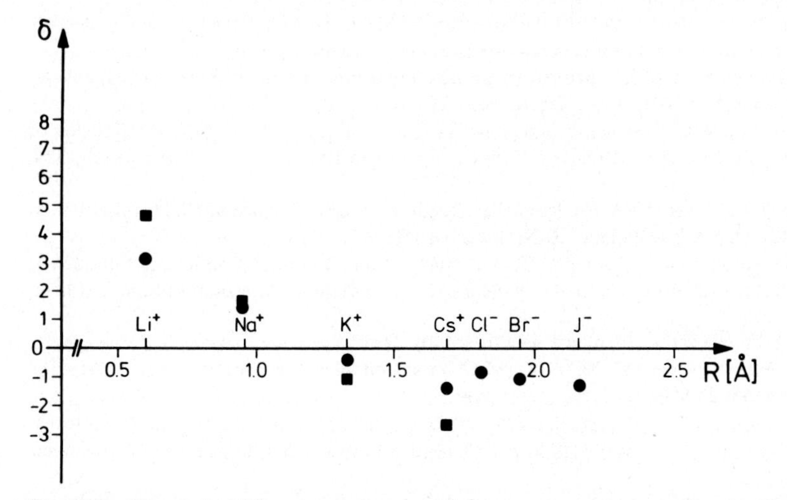

Fig. 7-3. Theoretical (■) and experimental (●) δ values for various alkali and halide ions as a function of Pauling radii.

TABLE 7-3.

The temperature dependence of δ_{Li} and δ_{Cl} as determined from LiCl solutions.

Temperature °C	δ_{Li}	δ_{Cl}
10	3.5	−0.5
25	3.1	−0.9
40	2.9	−1.1
55	1.7	−0.6

SUMMARY

The fractionation of the oxygen isotopes between hydration water and bulk water in alkali halide solutions has been investigated by the CO_2 equilibration technique. It is shown that contrary to previous conclusions the anion effect cannot be neglected. The single ion contributions are determined with the help of hydration numbers deduced from molecular dynamics simulations. The results are in fair agreement with fractionation factors calculated on the basis of ion-water pair potentials derived by Clementi and co-workers from ab initio calculations for the gas phase.

REFERENCES

Bopp, P., K. Heinzinger and P. C. Vogel, 1974. Calculations of the Oxygen Isotope Fractionation between Hydration Water of Cations and Free Water. Z. Naturforsch. 29a, 1608–1613.

Bopp, P., K. Heinzinger and A. Klemm, 1977. Oxygen Isotope Fractionation and the Structure of Aqueous Alkali Halide Solutions. Z. Naturforsch. 32a, 1419–1425.

Clementi, E., and H. Popkie, 1972. Study of the Structure of Molecular Complexes. I. Energy Surface of a Water Molecule in the Field of a Lithium Positive Ion. J. Chem. Phys. 57, 1077–1094.

Götz, D., and K. Heinzinger, 1973. Sauerstoffisotopieeffekte und Hydratstruktur von Alkalihalogenid–Lösungen in H_2O und D_2O. Z. Naturforsch. 28a, 137–141.

Heinzinger, K., 1976. Molecular Dynamics Study of Aqueous Solutions. V. Angular Distribution of Water Dipoles in the Hydration Shells of Various Alkali- and Halide Ions. Z. Naturforsch, 31a, 1073–1076.

Heinzinger, K., and P. C. Vogel, 1976. A Molecular Dynamics Study of Aqueous Solutions. III. A Comparison of Selected Alkali Halides. Z. Naturforsch. 31a, 463–475.

Kistenmacher, H., H. Popkie and E. Clementi, 1973a. Study of the Structure of Molecular Complexes. II. Energy Surfaces for a Water Molecule in the Field of a Sodium or Potassium Cation. J. Chem. Phys. 58, 1689–1699.

Kistenmacher, H., H. Popkie and E. Clementi, 1973b. Study of the Structure of Molecular Complexes. V. Heat of Formation for the Li^+, Na^+, F^- and Cl^- Ion Complexes with a Single Water Molecule. J. Chem. Phys. 59, 5842–5848.

Kistenmacher, H., H. Popkie and E. Clementi, 1973c. Study of the Structure of Molecular Complexes. III. Energy Surface of a Water Molecule in the Field of a Fluorine or Chlorine Anion. J. Chem. Phys. 58, 5627–5638.

Langer, H., and H. G. Hertz, 1977. The Structure of the First Hydration Sphere in Electrolyte Solutions. A Nuclear Magnetic Relaxation Study. Ber. Bunsenges. 81, 478–490.

Maiwald, B., and K. Heinzinger, 1972. Isotopieeffekte in den Hydratstrukturen von festem und gelöstem Kupfersulfat. Z. Naturforsch. 27a, 819–826.

Narten, A. H., F. Vaslow and H. A. Levy, 1973. Diffraction Pattern and Structure of Aqueous Lithium Chloride Solutions. J. Chem. Phys. 58, 5017–5023.

Samoilov, O. Ya., 1972. Residence Times of Ionic Hydration. In: Water and Aqueous Solutions (R. A. Horne ed.), pp. 597–612, Wiley Interscience.

Sofer, Z., and J. R. Gat, 1972. Activities and Concentrations of Oxygen-18 in Concentrated Aqueous Salt Solutions: Analytical and Geophysical Implications. Earth Planet. Sci. Lett. 15, 232–238.

Szapiro, S., and F. Steckel, 1967. Physical Properties of Heavy-Oxygen Water. 2. Vapour Pressure. Trans. Far. Soc. 63, 883–894.

Taube, H., 1954. Use of Oxygen Isotope Effects in the Study of Hydration of Ions. J. Phys. Chem. 58, 523–528.

Wolfsberg, M., and M. J. Stern, 1964. Validity of Some Approximation Procedures Used in the Theoretical Calculation of Isotope Effects. Pure Appl. Chem. 8, 225–242.

Chapter 8

HELIOTHERMAL LAKES

P. SONNENFELD AND P. P. HUDEC

University of Windsor, Windsor, Ontario, N9B 3P4, Canada

INTRODUCTION

Heliothermal lakes develop a hot water lens due to absorption of solar radiation. No other energy source is involved. The solar origin of the heat was first suggested by Ziegler (1898) and first demonstrated both by observation and experimentally by Von Kalecsinszky (1901). The term "heliothermal" was first used by Maxim (1930, 1936). A number of natural and man-made lakes have since been described, and sporadic research has been conducted on the physical requirements for solar heating, chemistry, and biology of the lakes. Although the study of solar lakes is a relatively recent undertaking, heliothermal lakes have been used for balneological purposes since the dawn of history.

The physical basis of heliothermal energy entrapment is the presence of density stratification. A moderately concentrated brine is overlain by a layer of less dense water and the interface between the two layers acts as a one-way mirror for incoming radiation. Findenegg (1935, 1937) had referred to density stratified lakes as "meromictic lakes", recognizing thereby a static meromixis derived from geologically determined presence of saltwater, and a dynamic meromixis, derived from the decomposition of organic matter leading to an increasing enrichment of bottom waters in bicarbonate or sulfate ions. Meromictic lakes do not develop convection in their lower layers, become stagnant, with anaerobic bottom layers that are enriched in hydrogen sulfide. However, not all meromictic lakes are also heliothermal ones.

For heliothermal energy entrapment a density difference of 15 g/l appears to be a minimum value. Only density stratified chloride and sulfate lakes attain this value, carbonate and bicarbonate lakes do not. The greater the density difference (Maxim, 1936), the more stable the heat storage in heliothermal lakes. The nomenclature of stratification in heliothermal lakes is given in Table 8-1. Lakes with a lesser density difference may develop meromictic stratification and an anoxic hypolimnion, but do not develop a heliothermal mesolimnion. The upper part of the anaerobic hypolimnion becomes the habitat of a large assemblage of pink bacteria which absorb incoming light leaving the lower hypolimnion in the dark. The summer thermocline is then between warm epilimnion and cold hypolimnion (Anagnostidis and Overbeck, 1966).

ORIGINS OF HELIOTHERMAL LAKES

Many heliothermal lakes have formed in man-made reservoirs or in abandoned salt and sulfate mines; thus a classification is possible either by predominant anion content or by the mode of origin, natural or artificial.

TABLE 8-1.

Nomenclature of stratification in heliothermal lakes.

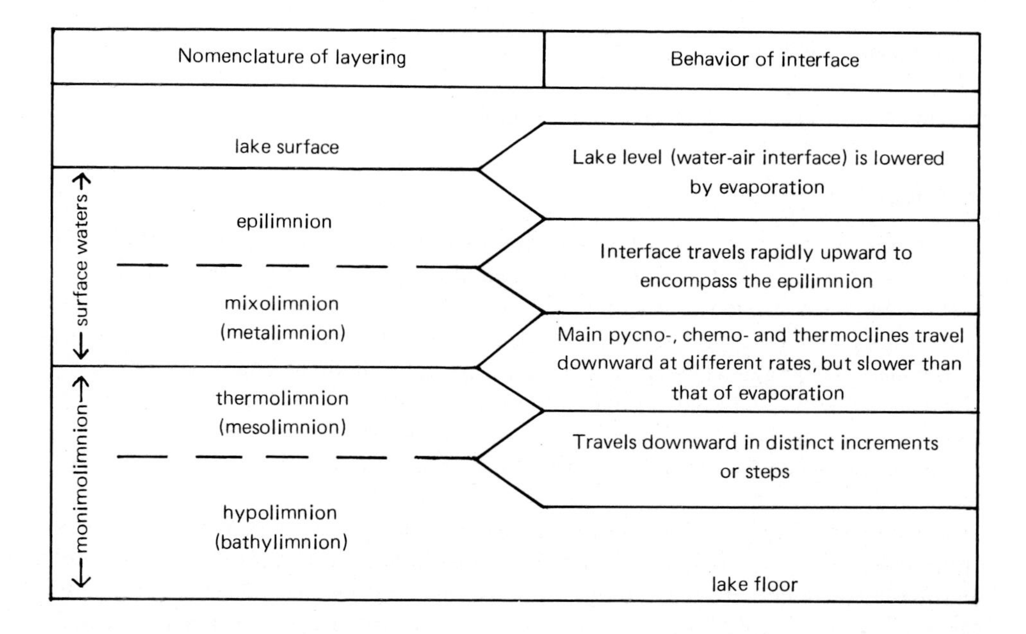

Type of interface	determined by
thermocline	temperature differential
chemocline	chemical concentration
pycnocline	density differential

Under natural origin conditions, the hypersaline brine is initially produced by one or more of the following processes:

a. evaporation of sea water in restricted inflow basins.

b. evaporation of fresh water in land-locked basins.

c. seepage of saline brines into basins, depressions or lakes. The natural hypersaline lakes are usually, but not always, found in semi-arid and arid climates with significant deficits in their annual water balance.

In artificial bodies of water the hypersaline brine is produced by leaching of exposed rock salt or sulfate salts in abandoned mines, or by industrial brine waste disposal or storage in ponds or lakes. Artificial brine ponds are commonly found where salt caverns are used for storage of liquified gas.

A hot lens develops in the brine whenever a surface layer of fresh water or lower density sea water covers the brine. The source of the lower density surface water can be rain water, runoff, groundwater, streams, sea water influx, etc. Density stratified fjords normally do not attain a sufficient density difference to produce significant heliothermal effects and neither does the catastrophic entry of seawater into coastal freshwater lakes. Lake Hemmelsdorf near Lübeck, West Germany (Griesel, 1935) is an example in which storms pushed Baltic seawater into the lake nearly a century ago. It is thus a meromictic, but not a heliothermal lake.

CONDITION OF FORMATION

Any sufficiently concentrated sulfate or chloride brine if overlain by a significantly lower density layer becomes a heliothermal energy trap. The thickness of the surface layer of few centimeters is sufficient to begin the heating. The maximum depth of the surface layer allowing heating is determined by the light transmissibility of the layer. Some energy must reach the lower dense brine. The supply of surface layer waters must be sufficient to maintain the low density surface layer and thus to sustain the heliothermal effect. If this supply is not sufficient throughout the year, epi- and mixolimnion evaporate and heating of the brine no longer takes place. Such lakes are occasionally or seasonally heliothermal. If the surface layer becomes too thick, not enough energy penetrates the interface to produce significant entrapment. A careful balance between rates of inflow and rates of water loss through evaporation is needed to maintain the system, to perpetuate the capacity to store the incoming radiation.

It is not so much the air temperature or the amount of available insolation but rain and running water which regulate stratification and with it heat accumulation in heliothermal lakes. Lake temperatures increase most, when clear skies follow several days of moderately strong rain (Rozsa, 1911). Sonnenfeld et al. (1976b) reported 53°C in the thermolimnion of Jan Thiel Bay, Curaçao, on a bright sunny day, but 59°C on an overcast day a week later, after some more rains. Diffuse sunlight is thus as good a source of energy as direct solar irradiation.

Saline waters are practically opaque to infrared rays; radiation of cold bodies made up of wave lengths greater than 2,000 nm cannot pass at all through waters of even a few centimeters thickness. The uppermost 100 cm of water strip incoming radiation of all wave lengths above 900 nm, leaving only 1.2 percent of the thermal effect available at the surface (Dzens-Litovskii, 1953). When air temperature exceeds surface water temperature in late afternoon, even the epilimnion subdivides into thermally stratified layers producing poikilothermy (Rozsa, 1913; Hutchinson, 1957); only the immediate surface layer records significant temperature fluctuations in a 24-hour period.

If an exposed dense brine is overrun by less dense waters, the stratification re-establishes itself. Even after severe stirring of hypersaline brines a temperature layering forms in as little as thirty minutes (Herrmann et al., 1973). Once established, the thermal stratification can last for centuries (Maxim, 1936), i.e. as long as a surface layer of less saline waters can be maintained in the face of high rates of evaporative losses.

Wind contributes little to the mixing, as the agitation is limited to the less dense epilimnion (Dzens-Litovskii, 1953). Indeed, heliothermal lakes occur both in sheltered and open locations. In our laboratory and in nature in Gran Roque, Venezuela, it could be observed that continuously blowing strong winds do not by themselves induce mixing of the layers, but merely a tilting of the interfaces. The winds accelerate the removal of moist air, increase evaporation, and in that way speed up the disappearance of the epilimnion. The mixing front or lower interface of the mixed layer advances downward more slowly than the water surface is lowered by evaporation until the latter overtakes the mixing front: the hypolimnion becomes exposed to the atmosphere without any mandatory overturn of the lake.

In most heliothermal lakes, a Secchi disc remains visible only to a depth of about 45–50 cm; the transparency of the frequently opalescent lakes is reduced by incipient precipitation of salts, by colloidal matter, dust, and by suspended organic remains unable to rot due to lack of oxygen. Agitation only further reduces the transparency (Maxim, 1936), mainly by stirring up settled fines.

Most heliothermal lakes have a finite mesothermal layer, sandwiched between cooler epi- and hypolimnion. The thickness of this thermolimnion is a function of radiation intake and of the lake bottom morphology. The smaller the angle of the lake surface with the near-shore lake bottom, the more radiation is absorbed by the lake floor and either dissipated into the ground or reflected and absorbed within the monimolimnion (Rozsa, 1915). Only a small fraction of this reflected radiation can escape by penetrating the interface with the mixolimnion (Hudec and Sonnenfeld, 1974).

The lake bottom, particularly when covered with a highly reflective gypsum, trona or mirabilite layer, reflects incoming solar radiation, focussing its absorption onto the region below the reflective low density–high density interface. The travel path of rays prior to their complete absorption is thus compressed into a thin layer of water. A black muddy bottom radiates long-wave heat which is absorbed by the brine almost immediately in a very short travel path. Experimentally, Liesegang (1916) proved that heating from below introduces and stabilizes multiple density stratification.

Many heliothermal lakes have a broad area of shallow water and a relatively small deeper pool (Smith, 1947; Hudec and Sonnenfeld, 1974). Thermophile molecules (such as NaCl) increase the salinity of the warm layer during the day. Very little cooling by conduction is needed at night before this still hot, but now denser brine begins to seep into the deep, displacing the cooler, less dense bottom waters. Tropical lakes with a relatively shallow deeper pool (about 5 m deep) thus often lack cooler bottom waters.

The more stratified the bottom waters, the slower the nightly cooling of lower layers, as equalizing currents are restricted in circulation. A warm homogeneous saturated solution cools down faster at night than a less concentrated solution composed of several layers of distinct density (Rozsa, 1911).

DISTRIBUTION OF HELIOTHERMAL LAKES

Most heliothermal lakes have hitherto gone unobserved, were then discovered accidentally, as few field scientists carry remote-sensing thermometers. Thus an attempt to plot reported occurrences (Fig. 8-1) is really more a measure of the intensity of search effort in various regions, than a true indication of the distribution pattern of such lakes.

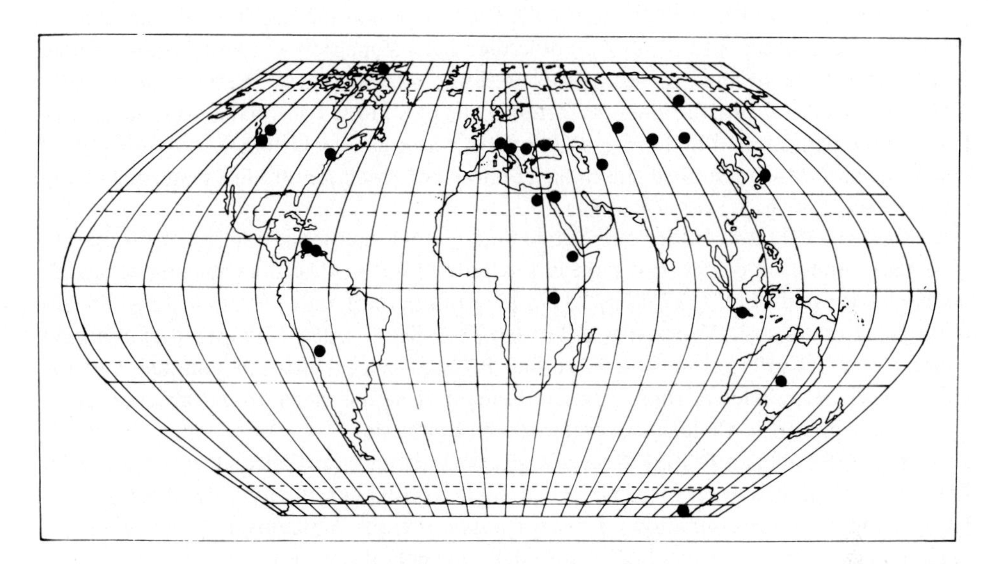

Fig. 8-1. Occurrence of heliothermal lakes.

Most of the heliothermal lakes have been described from Siberia, such as Lake Kuchuk and another small sodium-chloride lake next to Lake Burla in the middle of the Kulunda steppe, Lake Sultan-Sandzhar in the Kara-Kum region, Lake Chapchachi on the salt dome of the same name along the lower Volga River, Lake Dus-Khol in the Tuva region, Lake Doronino east of the Yablon Mountains, Lake Selenga east of Lake Baikal, Lake Inder in the Aralo-Caspian depression, a nameless lake near the Semiluk Canal south of Lake Balkhash, lakes Molla-Kar and Goluboye in Turkmenia, lakes Malyi Rassol, Abalakh, Tus-Kel', Moskogolookh (the Ṣuntara and Kempenda lakes) in Yakutia, Lake Shunet in Kazakhstan, as well as several lakes in the Crimea (Dravert, 1908; Maksheev, 1915; Frishenfel'd, 1930; Litvinova, 1938; Tychino, 1953, Dinkuletsku, 1959; Dzens-Litovskii, 1953, 1966, 1968, Patrichiu, 1961; Prosetskii, 1964; Rusin, 1967; Pinneker, 1972).

Dzens-Litovskii (1953) described in detail a series of lakes in abandoned mine workings of the Ilek salt dome in the southern Urals: Only 150 m separate Lake Tuzluk, a shallow pool with bottom waters reaching 67°C, from Lake Razval, a deeper lake with steep sides formed in part by mine timber. The latter lake, formed in 1906, has no shallow shelf area, is perennially cold, precipitates hydrohalite, and attains summer temperatures above 0°C only in the surface waters. Another nearby lake, Dunino, which formed in 1896 by the flooding of several sinkholes, also remains cold in its bottom waters. However, shallow pools near the shore heat up to 60°C and more, scalding the unsuspecting bathers.

Lake Sovata and adjacent lakes on a Romanian salt dome had been originally investigated by Von Kalecsinszky (1901) and Rozsa (1911, 1913, 1915). A rather comprehensive study was then undertaken by Maxim (1929, 1930, 1936). Artificial salt pans in Yugoslavia likewise produce heliothermal heating (Herrmann et al., 1973). Collet (1925) had mentioned a warm water lens in Lake Ritom in central Switzerland, which may not be entirely heliothermal in origin, but a case of biogenic meromixis.

In Africa, Fulda (1928) referred to Lake Bahr-el-Assal in the Republic of Djibouti as a solar lake, Melack and Kilham (1972) described Lake Mahega in western Uganda as such. Smith (1947) found that in salt pools of Siwas oasis in Egypt the salt is seasonally redissolved and is then overridden by artesian spring waters. U.S. navy divers operating in the Suez Canal found in 1975 a scalding hot water lens in Bittern Lake. Por (1972) and Eckstein (1970) described the Solar Pond on the east coast of Suez Peninsula fed by underground seepage from the Red Sea.

In the United States, Hudec and Sonnenfeld (1974) mentioned a reservoir in upstate New York, Anderson (1958) Hot Lake in the State of Washington in an abandoned sulfate mine. Sonnenfeld et al. (1976a, b, 1977, 1978) found hot water lenses in Lago Pueblo in the Los Roques atoll, Venezuela, and in Jan Thiel Bay, Curaçao, composed of stratified concentrated seawater, as well as in sulfate lakes of central British Columbia, Canada. Other localities are known from Ellesmere Island, northern Canada, from California, from the Atacama Desert of Chile, and a crater lake in Indonesia.

Wilson and Wellman (1962) described Lake Vanda and Shirtcliffe (1964) the nearby Lake Bonney, both in Antarctica, Yoshimua (1936, 1937) Lake Sinmiyo, Miyake, Ryu-Kyu Islands, Japan. The latter shows a case of poikilothermy (Hutchinson, 1957) with one maximum beneath the chemocline, a subsidiary one near the surface.

CONCLUSIONS

Heliothermal lakes are not restricted to areas of high angle of incidence of solar radiation, i.e. low latitudes. They occur within 80°of latitude on either side of the equator. They are the product of solar energy entrapment in the shallows of a hypersaline lake covered by low-salinity waters. Both continental (sulfate and chloride) and paralic (chloride) natural heliothermal lakes are known in semi-arid and arid regions. Artificial heliothermal lakes are either derived from flooding of abandoned rock sale or sulfate mines or from discharge of industrial brines, and are independent of the climatic regime.

SUMMARY

Heliothermal lakes, or lakes with a hot water lens produced by absorption of solar energy occur in all latitudes. They are produced when a low density layer overlies a higher density, brine layer. Density difference requirement is 15 g/l or greater. The natural brines capable of such high densities are sulfate or chloride brines. The low density layer acts as a one-way miror to solar radiation, and prevents convection of heated lower brine to the surface, thus limiting evaporative cooling.

Early work on heliothermal lakes is reviewed and locations of such lakes are given.

REFERENCES

Anagnostidis, K. and Overbeck, J., 1966. Methanoxidierende und hypolimnische Schwefelbakterien. Studien zur ökologischen Biocönotik der Gewässermikroorganismen. Dtsch. bot. Ges., Ber., 79(3): 163–174.

Anderson, G.C., 1958. Some limnological features of a shallow saline meromictic lake.. Limnol. Oceanogr., 3(3): 259–270.
Collet, L. W., 1925. Les Lacs, leur mode de formation, leurs eaux, leur destin. Eléments d'hydrogéologie. Paris: Librairie, Octave Dion, 320 p.
Dinkuletsku, T., 1959. Solenye ozera reguliruyemoy temperatury geliotermy. Vopr. kurortol., fizioterap. i lech. fiz. kul't., No. 2.
Dravert, P., 1908. Ekspeditsiya v Suntarskii solenosnyi raion Yakutii. Trudy Yakutskogo oblastnogo statisticheskogo komiteta, issue No. 1.
Dzens-Litovskii, A. I., 1953. Mineral'noye ozera Iletskogo solyanogo kupola i ikh termicheskii rezhim (Mineral lakes of the Ilek salt dome and their thermal regime). Akad, Nauk SSSR, Tr. Labor. Ozeroved., 2: 108–138.
Dzens-Litovskii, A. I., 1966. Solyanoi karst SSSR (the salt karst in USSR). Leningrad: Nedra.
Dzens-Litovskii, A. I., 1968. Solyanye ozera SSSR i ikh mineral'nye bogatstva. Leningrad: Nedra, 119 p.
Eckstein, Y., 1970. Physicochemical limnology and geology of a meromictic pond on the Red Sea shore. Limnology and Oceanography 15(3): 363–372.
Findenegg, I., 1935. Limnologische Untersuchungen im Kärntner Seengebiete. Ein Beitrag zur Kenntnis des Stoffhaushaltes in Alpenseen. Intern. Rev. Hydrobiol., 32: 369–423.
Findenegg, I., 1937. Holomiktische und meromiktische Seen. Intern. Rev. Hydrobiol., 35: 586–610.
Frishenfel'd, G. E., 1930. O geologicheskom stroyenii Leno-Vilyuiskogo vodorazdela i o genezise Kempendyaiskikh mestorozhdenii kamennoi soli. Moskov. obshch. ispyt. prir., Byul., Ser. Geol., 8(3–4): 1–30.
Fulda, E., 1928. Der Assalsee in Somaliland und seine Bedentung für die Erklärung der Entstehung mächtiger Salzlager. Zeitschr. deutsch. geol. Ges., 79(1): 70–75.
Griesel, R., 1935. Die Aussüssung des Hemmelsdorfer Sees. Mitteil, Geogr. Ges., Lübeck, 2. Reihe, 38: 77–83.
Herrmann, A. G., et al., 1973. Geochemistry of modern seawater and brines from salt pans: main components and bromine distribution. Contrib. Mineral. Petrol., 40(1): 1–24.
Hirschmann, J., 1965. Suppression of natural contraction in open ponds by concentration gradient. Proc. First Intern. Sympos. Water Desalinization, Washington, p. 478–487.
Hirschmann, J. 1970. Salt flats as solar-heat collectors for industrial purposes. Solar Energy, 13: 83–97.
Hudec, P. P. and Sonnenfeld, P., 1974. Hot brines on Los Roques, Venezuela. Science, 185(4149): 440–442 and 186(4169): 1074-1075.
Hudec, P. P. and Sonnenfeld, P., 1978. Comparison of Carribean solar ponds with inland solar lakes of British Columbia. This volume p. 000.
Hutchinson, G. E., 1957. A Treatise on Limnology, New York: John Wiley & Sons. 3 vols.
Liesegang, R. E., 1916. Zur Theorie der heissen ungarischen Salzseen. Intern. Rev. ges. Hydrobiol., 7(6): 469–471.
Litvinova, N. N., 1938. Kurorty i lechebnye mestnosti Kazakhstana. Izdat. Akad. Nauk Kazakh. S.S.R., Alma, Ata.
Maksheev, N. N., 1915. Celebnye istochniki v Zakaspiiskoy oblasti. Materialy po izucheniyu Zakaspiiskoi oblasti, Zakaspiiskoi oblasnoi statisticheskii kometet (B.M.)
Maxim, I. A., 1929. Contributiuni la explicarea fenomenului de incalzire al apelor lacurilor sarate din Transilvania. Kontribution zur Erklärung des Erwärmungsprozesses des Wassers der Salzteiche von Transilvanien. I. Lacaurile Sovata: Die heissen Seen von Sovata. Univ. Cluj, Rev. Muz. Geol. Mineral., 3(1): 49–86.
Maxim, I. A., 1930. Contribu'iuni la explicarea fenomenului de incalzire al apelor lacurilor sarate din Transilvania. Kontribution zur Erlärung des Erwärmungsprozesses des Wassers der Salzteiche von Transilvanien. II. Lacurile de la Ocna-Sibiului: Die Teiche von Ocna-Sibiului. Univ. Cluj, Rev. Muz. Geol. Mineral., 4(1): 47–111.
Maxim, I. A., 1936. Contributiuni la explicarea fenomenului de incalzire al apelor lacurilor sarate din Transilvania. Kontribution zur Erklärung des Erwärmungsprozesses des Wassers der Salzteiche von Transilvanien. III. Lacurile Sarate dela Turda: Die Teiche von Turda. Univ. Cluj, Rev. Muz. Geol. Mineral., 6(1–2): 209–320.

Melack, J. M., and Kilham, P., 1972. Lake Mahega: a mesothermic, sulphato-chloride lake in western Uganda. African Jour. Tropical Hydrobiol. and Fisheries, 2(2): 141–150.

Patrichiu, V., 1961. Kak ustroit' geliotermy. Vopr. kurortol., fizioterapii i lechebnoi fizkul'tury, No. 2.

Pinneker, E. V., 1972. Geliotermy ozera Dus-Khol' or Svatikovo) v Tuve. (Heliotherms of Lake Dus-Khol' or Svatkovo in Tuva). In: Geologiya i Gidrogeologiya solyanykh mestorozhdenii, 56: 104–109.

Por, F. D., 1972. Limnology of the heliothermal Solar Lake on the coast of Sinai (Gulf of Elat). Rapp. Comm. Int. Mer Medit., 20(4): 511–513.

Prosetskii, E. P., 1964. Estestvennye geliotermy i perspektivy ikh lechebnogo ispol'zovaniya. Materialy po izucheniya lechebnykh mineral'nykh vod i gryazei v bal'neoteckhnike. Moscow: Izdat. Tsentral. Instituta Kurortologii i Fizioterapii. pp. 73–79.

Rozsa, M., 1911. Neuere Daten zur Kenntnis der warmen Salzseen. Berlin: R. Friedlander & Son. 32 pp.

Rozsa, M., 1913. Uber die periodische Entstehung doppelter Temperaturmaxima in den warmen Salzseen. Ann. Hydrogr. u. Marit. Meteorol., 41: 511–513.

Rozsa, M., 1915. Die physikalischen Bedingungen der Akkumulation von Sonnenwärme in den Salzseen. Physik. Zeitschr., 16(6): 108–111.

Rusin, N. P., 1967. Teplovoi balans nashei planety. Novoe v zhizni, nauke i tekhnike, Ser. 13, No. 1. Moscow: Snanie.

Shirtcliffe, T. G. L., 1964. Lake Bonney, Antarctica: Cause of elevated temperatures. Jour. Geophys. Res., 69(24): 5257–5268.

Smith, C. L., 1947. Hydrography of the salt pools (Armstrong College expedition to Siwa, 1935). L'Inst. Fr. d'Archéol. Orient., Inst. d'Egypte, Cairo, Bull., 28: 139–159.

Sonnenfeld, P., et al., 1976a. Stratified heliothermal brines as metal concentrators. Acta Cient. Venezolana, 27: 190–195.

Sonnenfeld, P., et al., 1976b. The chemistry of precipitates and brines in restricted, density stratified lagoons. IIIrd. Latinamerican Geol. Congr., Acapulco, Mex., Memoria. In press.

Sonnenfeld, P., and Hudec, P. P., 1977a. Stratified brines in restricted basins as sources of oil and oil-field brines. IInd. Intern. Sympos. Water-Rock Interaction, Proc., 2: 42–49.

Sonnenfeld, P., et al., 1977b. Base-metal concentration in a density-stratified evaporite pan. In: J. H. Fisher (editor): Reefs and Evaporites – Concepts and Depositional Models. Studies in Geology No. 5, p. 181–187. Tulsa: Amer. Assoc. Petrol. Geol.

Tychino, Ya. I., 1953. Nekotorye cherty termicheskogo rezhima mezhkristal'noi rapy ozera Inder. (Some features of the thermal regime in intercrystalline brine of Lake Inder. Akad. Nauk SSSR, Tr. Labor. Ozeroved., 2: 139–147.

von Kalecsinszky, A., 1901. Uber die ungarischen warmen und heissen Kochsalzseen als natürliche Wärme-Accumulatoren. Földtani Közlöny, 31: 409–431.

Weinberger, H., 1964. The physics of the solar pond. Solar Energy, 8(2): 45–56.

Wilson, A. T., and Wellman, H. W., 1962. Lake Vanda: an Antarctic lake. Nature, 196(4860): 1171–1173.

Yoshimura, S., 1937. Abnormal thermal stratifications of inland lakes. Japan. Imper. Acad. Sci., Proc., 13: 316–319.

Yoshimura, S., and Miyadi, D., 1936. Limnological observations of two crater lakes of Miyaki Island, western North Pacific. Japan, Jour. Geol. Geogr., 13: 339–352.

Ziegler, G., 1898. An den Herausgeber des Prometheus: Absonderliche Temperaturverhältnisse in einem Sölbenhalter. Prometheus, 9:79 and discussion 9:325.

Chapter 9

COMPARISON OF CARIBBEAN SOLAR PONDS WITH INLAND SOLAR LAKES OF BRITISH COLUMBIA

P. P. HUDEC AND P. SONNENFELD

Department of Geology, University of Windsor, Windsor, Ont. N9B 3P4, Canada

INTRODUCTION

The principal requirements of solar ponds are that they be density stratified, with sufficient density differences between the upper and lower layers to restrict convective cooling of the lower layer and allow it to be heated by solar radiation. Thus any body of water, be it derived from ocean or continent can be called a solar pond providing that the conditions are right for solar heating to take place.

The authors visited and sampled solar ponds that are derived from sea water (in the Caribbean), and those that are fresh water fed (in British Columbia, Canada). This paper discusses the origin of the temperature and density stratification of these ponds, and their chemistry. It suggests basic requirements for solar pond development, and poses some questions regarding the relationship of the chemistry of the feed waters and the final concentrated brine.

CARIBBEAN PONDS

The authors visited several localities in the Caribbean that have both natural and artificial solar ponds. These are located on Los Roques atoll, Venezuela, and in Jan Thiel Bay, Curaçao. The Los Roques atoll has been studied in greater detail, and the results of these studies [Sonnenfeld et al. (1976a, b, 1977, 1978)], and Hudec and Sonnenfeld (1974) will be summarized in this paper. For details, the reader is referred to the above publications.

The location of the Los Roques atoll is given in Fig. 9-1. The island of El Gran Roque on which the Lago Pueblo is located is on the northern fringe of the atoll.

Lago Pueblo, a heliothermal pond.

The Lago Pueblo is a seasonally stratified pond. Fig. 9-2 gives the temperature and density stratification with depth typically obtained during the November–December rainy season. At other times, the temperature and density profile equalized at approximately 33°C and 1.17 g/cm^3 density.

Table 9-1 gives the temperature readings taken within one hour's time at 45 cm depth and at the bottom at various points in the pond shown in Fig. 9-3. As can be seen, there is a significant lateral variation in the 45 cm and bottom temperatures.

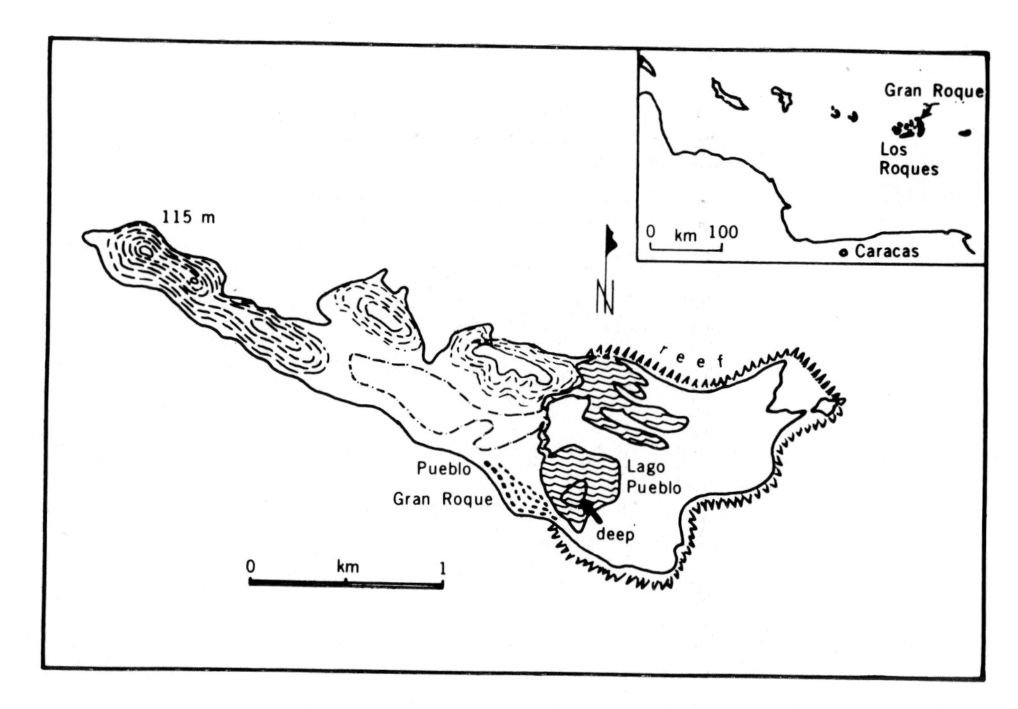

Fig. 9-1. Location map of Los Roques; Aruba Curacao, Bonaire, and Las Aves islands are to the west of Los Roques.

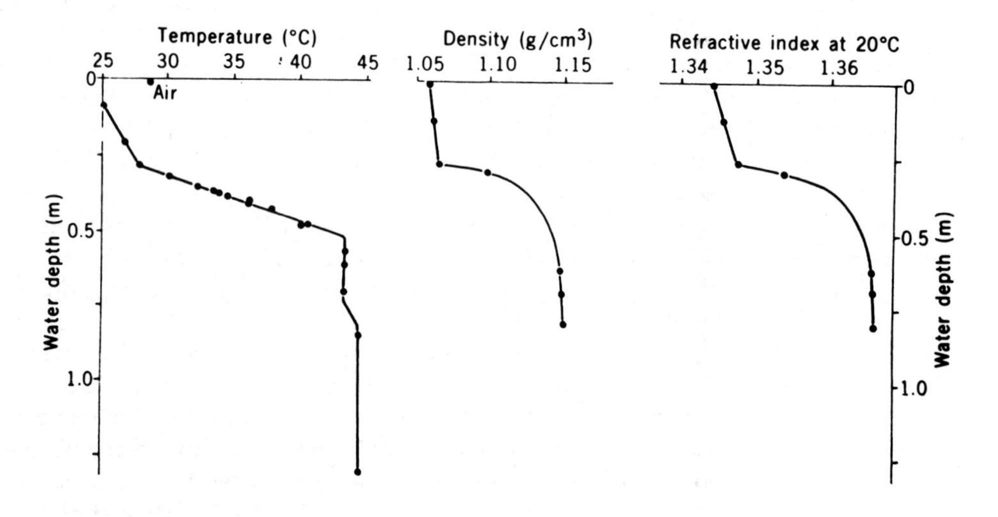

Fig. 9-2. Density stratification in Lago Pueblo, Los Roques, December 10, 1973. Temperature, density and refractive index indicate the position of epi-, mixo- and hypolimnion.

TABLE 9-1.

Temperature readings at various points of Lago Pueblo, showing the different temperatures obtained at the same depth level (45 cm). The lowest depth given for each station represents pond bottom. The location of the temperature reading stations is shown in Fig. 9-3.

Location	Depth cm	Temperature °C	Location	Depth cm	Temperature °C
1	45	33.5	6	45	33.0
	55	36.5		60	36.0
2	45	32.5	7	45	32.0
	60	35.5		52	34.0
3	45	33.5	8	45	34.0
	55	34.5		60	37.0
4	30	29.5	9	45	32.0
	45	31.5		60	38.0
	60	36.0			
5	45	31.5	10	45	37.0
	60	34.5		52	38.5

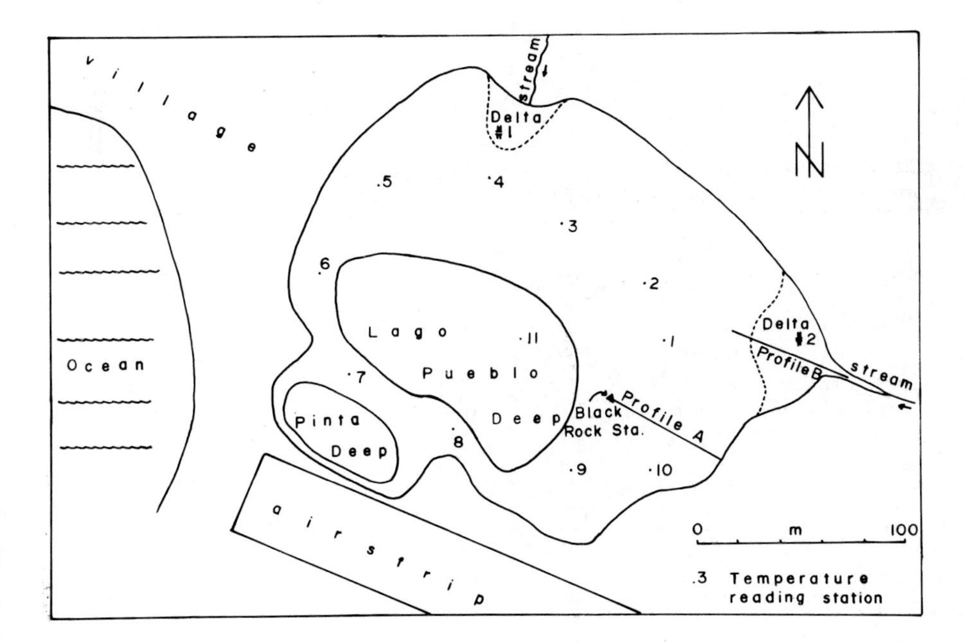

Fig. 9-3. Detailed plan of Lago Pueblo, Los Roques, showing location of continuous temperature profiles of individual temperature reading stations.

To determine the lateral temperature variation from the central deep to the shore of the pond, temperature readings were taken along a line from the Lago Pueblo deep to the eastern shore. Temperatures were obtained at 45 cm and bottom at various points along the line. They were also recorded for different levels of the deep (Fig. 9-4).

During the rainy season, water flows into the pond through two small streams which carry sea water from an adjacent lagoon. The streams form small deltas of calcite sand in the pond. The water was 15 cm deep in most of delta 2 and in the adjacent stream; incoming water volume varied from 10 m³ in the morning and evening to about 20 m³ in the early afternoon. The water velocity was approximately 3 m/sec in the stream entering the delta.

The temperature profile of delta 2 is shown in Fig. 9-5. Over most of the delta, the incoming water spread to about 3 cm thickness over the hot pond water. The thickness of cooler surface waters increased to about 15 cm just off the delta. The hot water extended upstream for several meters despite the rapid flow of water. A kind of "hump" of hot water formed near the crest of the delta, where stream velocities were falling off.

Table 9-1 and Figs. 9-4 and 9-5 indicate that the lateral hot water layering across the pond is not uniform. The depth to the hot water layer is dictated by the bottom topography of the pond, i.e., the shallower the water, the closer the hot water layer approaches the surface. This is particularly true in the near shore environment.

Our results suggest that the hot water layering is to a degree independent of density layering. It is hydrologically impossible to maintain the lateral density differences com-

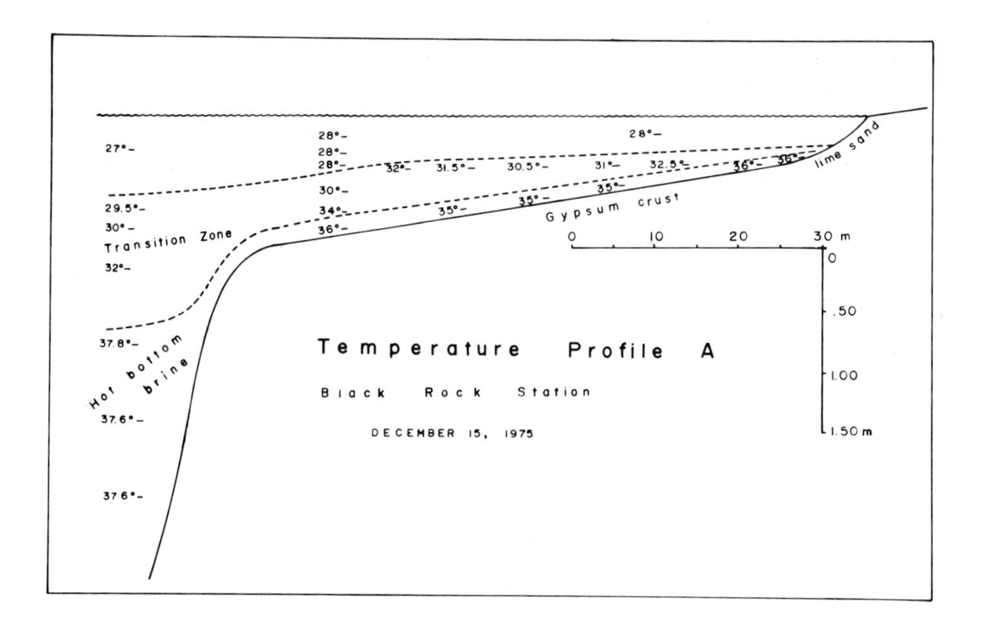

Fig. 9-4. Temperature profile from Lago Pueblo Deep to shore, December 15, 1975.

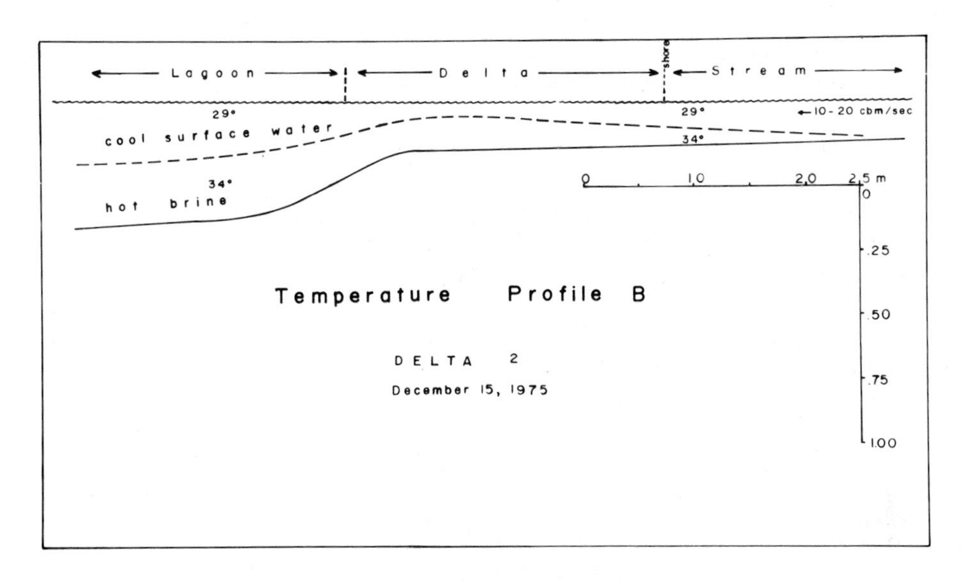

Fig. 9-5. Temperature profile of Lago Pueblo and its eastern delta 2 on December 15, 1975.

parable to the temperature differences. The shallower parts of the pond are hotter, and the hot water layer is closer to the surface because more solar energy is concentrated in a smaller water column, resulting in a steep thermal gradient.

The temperature stratification in the Lago Pueblo Deep is given in Fig. 9-6. Temperature readings in the epilimnion do not change very much from day to day because of energy removal through evaporation. Temperatures in the thermolimnion depend on daily solar irradiation, cloudiness and amount of suspended matter in the pond. As the temperature readings were taken in the morning, the increase in bottom temperatures is not due to direct irradiation, but to hotter, more saline waters sliding down at night from shallow parts of the pond.

Standard nomenclature for meromictic lakes is applied. However, it follows from the fore-going that the nomenclature within the mixing layer reflects more the temperature stratification than a density (concentration) stratification, since a layer of same density within the lake can have either high or normal temperature, depending on the depth of the water.

Jan Thiel Bay, a heliothermal lagoon.

In Jan Thiel Bay of Curaçao, Netherlands Antilles, a similar thermal regime prevails when a high water level allows sea water to establish a common sea level across several otherwise separated lagoons and creates local epilimnia. Temperature recorded in the thermolimnion reached 53°C on a sunny day, 59°C a week later on an overcast day following some rains.

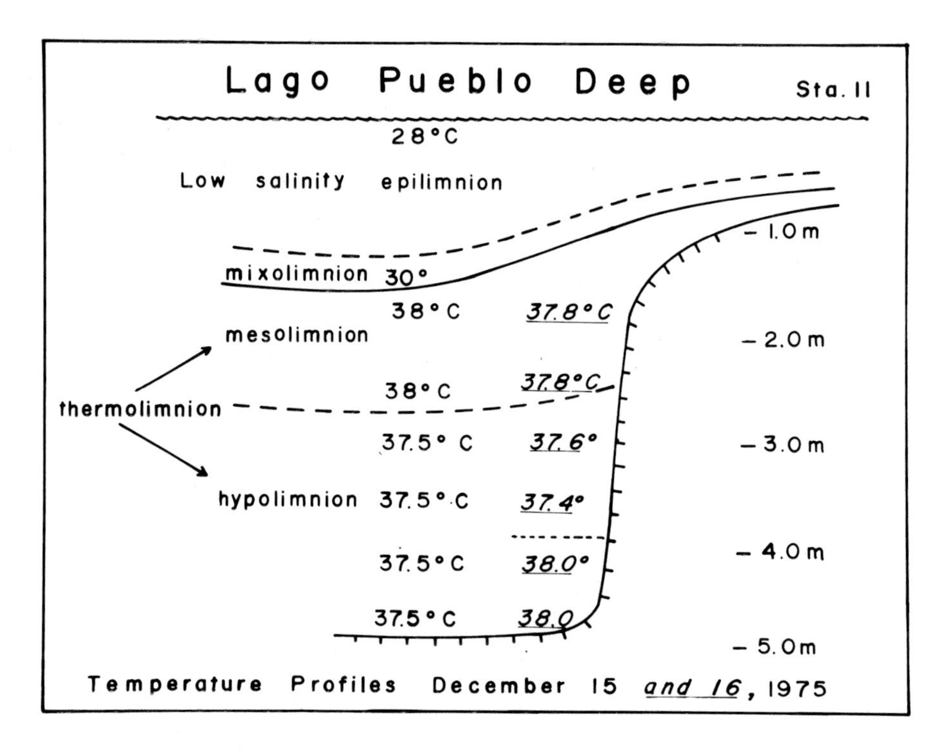

Fig. 9-6. Thermal stratification in Lago Pueblo Deep, Los Roques.

Chemistry of Lago Pueblo waters

Table 9-2 gives the chemistry of the Lago Pueblo deep. Samples were taken at intervals from the surface to the bottom on six different times from December 1975 to December 1976 by M. Lew, a graduate student (Lew, 1977). A statistical analysis of the results were made, comparing the concentration of the various elements by the linear regression method.

The analysis (Table 9-2) showed that K, Mg, and Li are the conservative elements giving the highest correlation, i.e., they are not removed from the brine by precipitation, adsorption, or reaction. Na and Cl are removed annually to an unknown, but probably slight degree by precipitation of halite during the dry season, which is then harvested by local inhabitants. Since Na and Cl occur in different proportion in the seawater than they do in the mineral halite, the ratios of these elements are continually altered by precipitation.

Calcium (not shown in Table 9-2) gave no correlation, attesting to its removal as gypsum which lines the bottom of the lagoon. However, water samples from the hypolimnion gave an instant precipitate of gypsum when sulfuric acid was added. Gypsum precipitation is thus controlled more by the scarcity of SO_4^{--} anions than by the availability of Ca^{++} cations.

TABLE 9-2.

Variation of the Lago Pueblo water from normal Los Roques sea water.

Variables Y	X	Equation (line of best fit)	Correl. Coeff.	Los Roques Sea Water (var. X) mg/l	Calculated Equivalent (var. Y) mg/l	Los Roques Sea Water (Var. Y) mg/l	Enrichment of var. Y percent
Na	Cl	Na = 1.1668Cl + 9567	.90	20332	14006	11326	24
K	Cl	K = .0423Cl + 165	.98	20332	1004	544	85
Mg	Cl	Mg = .0824Cl – 638	.97	20332	1674	1521	10
Li	Cl	Li = 9.09×10^{-6}Cl + .483	.95	20332	.67	.20	235
Li	Mg	Li = 1.01×10^{-4}Mg + .647	.90	1521	.70	.20	250
Li	K	Li = 2.12×10^{-4}K + .494	.93	544	.61	.20	205
Li	Na	Li = 6.02×10^{-4}Na + .812	.81	11326	.88	.20	340

Lago Pueblo waters are basically concentrated sea water. If the equation of the line of best fit is used to calculate the probable original composition of the pond water (using Los Roques sea water as standard), it becomes evident that they are enriched in certain elements. The last column of Table 9-2 gives percent enrichment of the various elements. Lithium is enriched about 3 times relative to sea water, potassium less than 2 times, sodium about 24%, and magnesium about 10%. The precipitation and removal of halite should result in sodium depletion rather than enrichment; an enrichment in sodium and other metals is due to groundwater seepage and fresh surface water runoff, carrying the weathering products of the metadiabase and metalamprophyres exposed on the rocky hills to the north of the pond.

BRITISH COLUMBIA SALINE LAKES

Twelve lakes in south-central British Columbia were visited in May, 1975. All lakes are located in the vicinity of Kamloops and Clinton, an area of semi-arid climate. The lakes occupy closed basins, and are fed by small streams, groundwater, surface run-off and snow melt. Fig. 9-7 shows the location of the examined sulfate lakes. The carbonate lakes are small unnamed lakes NW of Clinton.

The surface and bottom water samples were analyzed for their major cation and anion content, as well as for some trace metals. The results of the analysis are given in Table 9-3. Correlation matrix was determined for the above results, and for certain combinations of cations and anions. The correlations of 95% significance or better are presented in Table 9-4.

The correlations have been ranked on basis of correlation coefficient which indicates the quality of fit of points to the common line. The equation of the line, if traced to some initial unit value, can be used to determine the initial composition of the water (provided certain assumptions are made). Thus, for instance, if most dissolved components

TABLE 9-3.

Chemistry of selected lakes in central British Columbia. Samples 1–9 from the Kamloops area, 10–19 from the Clinton area.

Names		Fe	Mg	Ca	Pb	Na	K	Cl	SO_4	Density
Inks Lake	top	0.45	1588.6	296.3	0.77	3325.0	334.2	11.35	11267.50	1.01494
Inks Lake	bot	1.62	3788.2	369.9	1.21	611.1	648.2	26.95	21150.60	1.03097
Wallender Lake	top	1.36	2615.1	241.9	0.91	3600.0	290.2	56.68	14558.40	1.01981
Wallender Lake	bot	2.61	6183.3	546.8	1.46	8333.3	707.8	72.34	34024.00	1.04733
Ironmask Lake		0.16	160.1	4.3	0.00	658.7	117.2	1.52	82.90	1.00125
West Long Lake	top	0.24	3580.4	87.4	0.66	1666.7	432.5	4.26	14589.70	1.01846
West Long Lake	bot	2.91	6611.0	114.8	1.02	3250.0	794.3	7.09	27735.60	1.03497
Batchelor Lake	top	0.35	540.1	144.3	0.44	104.0	129.8	0.77	2045.40	1.00164
Batchelor Lake	bot	0.40	613.4	110.8	0.18	79.4	13.4	0.00	2292.50	1.00215
Last Chance Lake		1.56	37.9	5.3	1.23	17857.1	621.3	127.30	1783.20	1.04565
Telegraph Road Lake No. 1	top	1.25	85.5	4.7	0.13	1198.4	12.7	1.06	0.86	1.00218
Telegraph Road Lake No. 2	bot	0.52	18.2	5.4	0.27	5396.8	91.2	32.98	9.30	1.01230
Three Mile Lake	top	1.10	1869.7	230.7	0.90	2030.0	184.0	20.00	17715.80	1.02204
Three Mile Lake	mid	3.13	14175.2	456.3	1.72	5714.3	526.9	28.77	55280.0	1.06652
Three Mile Lake	bot	3.92	14663.9	442.1	1.17	5952.4	471.8	28.42	50518.4	1.06682
Long Lake	top	0.56	213.8	7.5	0.58	3860.0	397.1	21.28	3449.6	1.01008
Long Lake	bot	0.19	263.9	7.2	0.31	3730.2	424.7	27.61	4791.1	1.01064
Unnamed Lake	top	0.00	130.7	6.4	0.24	2630.0	155.7	9.54	413.8	1.00371
Unnamed Lake	bot	0.16	130.7	6.2	0.28	1452.4	146.3	9.72	416.4	1.00371

TABLE 9-4.

Correlation of variables in sulfate and carbonate lakes of central British Columbia.

Variables		Line of Best Fit Equation		Correlation Coefficient	
X	Y	Slope, Y/X	Y-intercept	r	Rank
Mg + Ca	SO_4	3.619	2197.6	.978	1
Mg	SO_4	3.719	2456.0	.974	2
ΣFe, Mg, Ca, Pb, Na, K, Cl, SO_4	Density	$8.917 + 10^{-7}$	1.003	.956	3
Na + K	Cl	$6.83 + 10^{-3}$	−2.25	.923	4
Na	Cl	$7.057 + 10^{-3}$	−.815	.923	4a
Pb	Density	$3.899 + 10^{-2}$	.9941	.914	5
Cl + SO_4	Density	$1.124 + 10^{-6}$	1.006	.846	6
SO_4	Density	$1.123 + 10^{-6}$	1.006	.895	7
Mg	Density	$4.179 + 10^{-6}$	1.009	.872	8
Ca	SO_4	83.296	562.6	.869	9
Pb	SO_4	28202.0	−6153.0	.830	10
Pb	K	398.0	55.91	.813	11
Ca	Pb	$2.282 + 10^{-3}$	.339	.809	12
Mg	Ca	$3.128 + 10^{-2}$	61.01	.785	13
Ca	Density	$9.115 + 10^{-5}$	1.007	.758	14
K	Density	$6.530 + 10^{-5}$	.999	.749	15
Mg	Pb	$8.301 + 10^{-5}$	.445	.739	16
Na + K + Mg	Cl	$2.918 + 10^{-3}$	4.78	.636	17
Na + K	Density	$3.22 + 10^{-6}$	1.08	.629	18
Na	Density	$3.191 + 10^{-6}$	1.009	.604	19
Cl	Density	$3.722 + 10^{-4}$	1.012	.537	20

in the water are equated to 0, i.e., taken out of the water, the density of water should approach that of distilled water. This is indeed the case. The correlation of all the analyzed dissolved components with density yields the initial density of 1.003. The difference of density from distilled water equivalent at 20°C is attributed to additional dissolved metals and anions not analyzed for, especially the carbonate anion.

What is perhaps most surprising about the correlations is that there are any. The lakes sampled are found in diverse geologic terrains, underlain by basalts, granites, limestones, quartzites, etc. The waters draining the terrains, although not sampled, must have significantly different initial composition. As the waters become concentrated in the lakes by evaporation, the ratios of non-precipitating cations and anions should remain constant. Thus, little or no correlation should be expected among lakes. Yet, as Table 9-4 clearly demonstrates, correlations are significant, especially in some metal and cation combinations.

The precipitating salts at all of these lakes are Na, Ca carbonates and bicarbonates, Na, Mg sulfates, and minor NaCl. Significant accumulations of these salts have been reported in the bottom sediments of the lakes (Cummings, 1940; Goudge, 1924; Reinecke, 1920, Jenkins, 1918). Since some lakes precipitate carbonates, other sulfates,

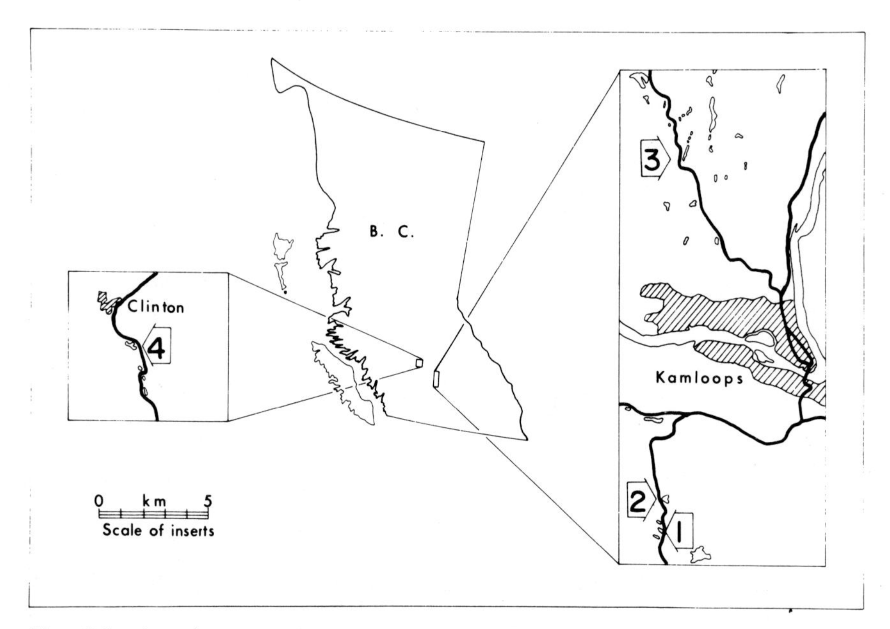

Fig. 9-7. Location map of density stratified sulfate lakes in British Columbia. 1. Inks Lake; 2. Wallender Lake; 3. West Long Lake; 4. Three Mile Lake.

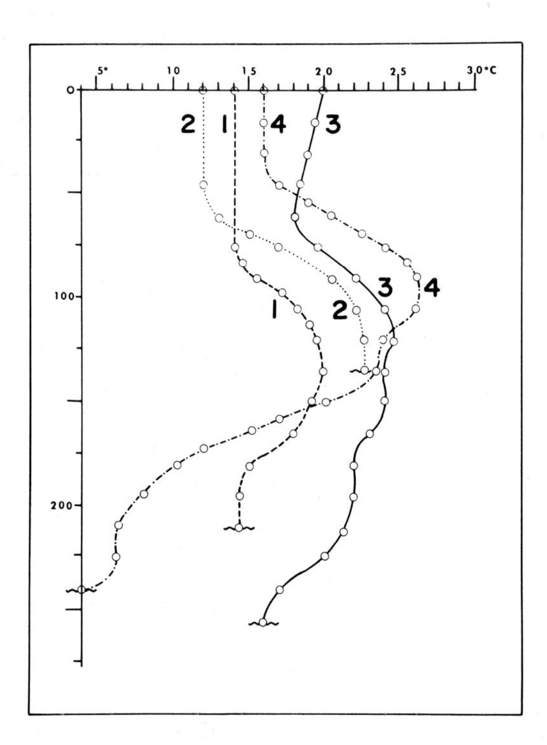

Fig. 9-8. Temperature profiles of four sulfate lakes in British Columbia. 1. Inks Lake 2. Wallender Lake 3. West Long Lake 4. Three Mile Lake.

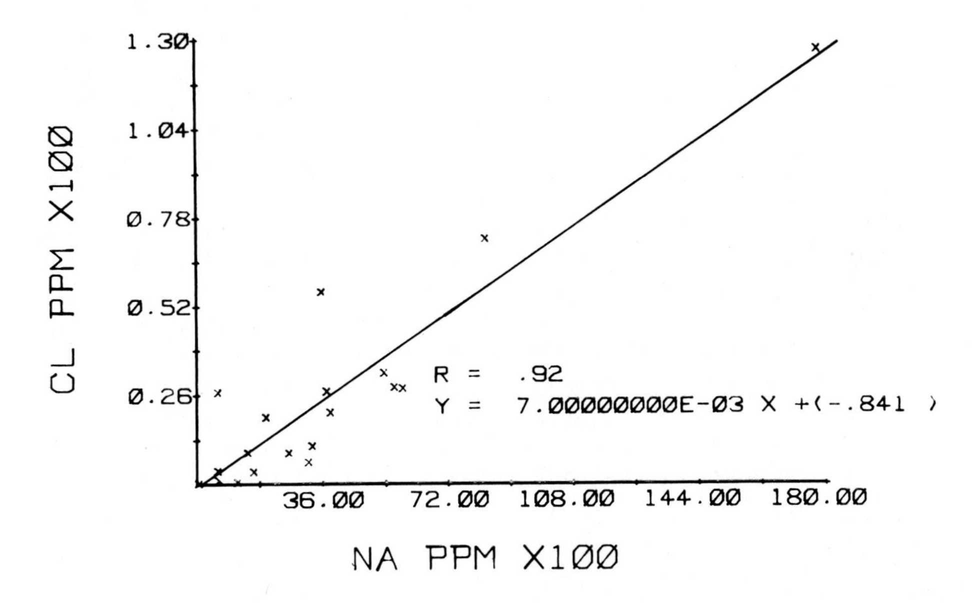

Fig. 9-9. Correlation between sodium and chlorine in carbonate and sulfate lakes of central British Columbia.

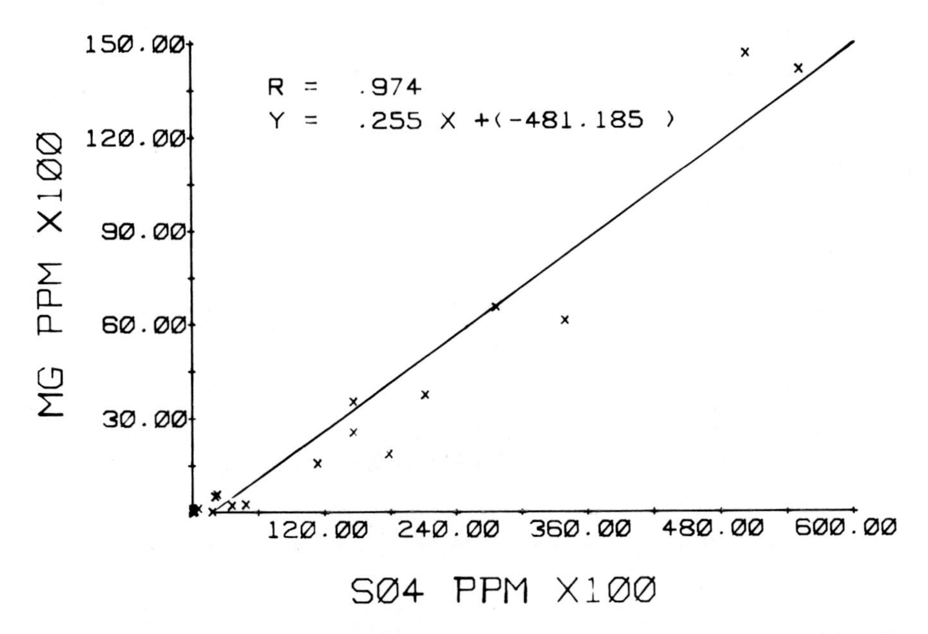

Fig. 9-10. Correlation between Mg^{++} and SO_4^{--} ions in sulfate and carbonate lakes of central British Columbia.

and still others combinations of various salts, it is puzzling to note the good correlations between different combinations of elements dissolved in the waters in different lakes: Na and K correlate well with Cl; (Fig. 9-9), Ca and Mg correlate well with SO_4 (Fig. 9-10).

The sulfate radical with the associated cations is the main determinant of the water density (Fig. 9-11). The carbonate lakes also contain dissolved sulfates, but to a lesser degree, and thus have lower densities. If the densities of most of the carbonate lakes are extrapolated to those of sulfate lakes, the calculated sulfate contents become comparable to that of sulfate lakes.

It can therefore be concluded that the saline lakes in British Columbia have a similar history: as the water evaporates, the lakes become first carbonate-rich, precipitating carbonate salts. Further evaporation leads to progressive concentration of sulfates. The next stage is precipitation of sodium sulfate, followed by precipitation of magnesium sulfate. Minor chloride is concentrated, and comes out in trace amounts with the Na and Mg sulfates. The original composition of the ground water becomes less important the more concentrated the lake waters, because precipitation of the various salts tends to homogenize the remaining water. The only difference between carbonate lakes and sulfate lakes is the degree of concentration through evaporation that the lake has undergone.

Trace Metals

Whereas high metal values have previously been reported for Lago Pueblo, anomalously low concentrations of base metals have been found in the waters of saline lakes of British Columbia. This is surprising in view of the presence of base metal deposits in the area and of basic rocks underlying much of the terrain. No valid explanation can be offered for this phenomenon at this time.

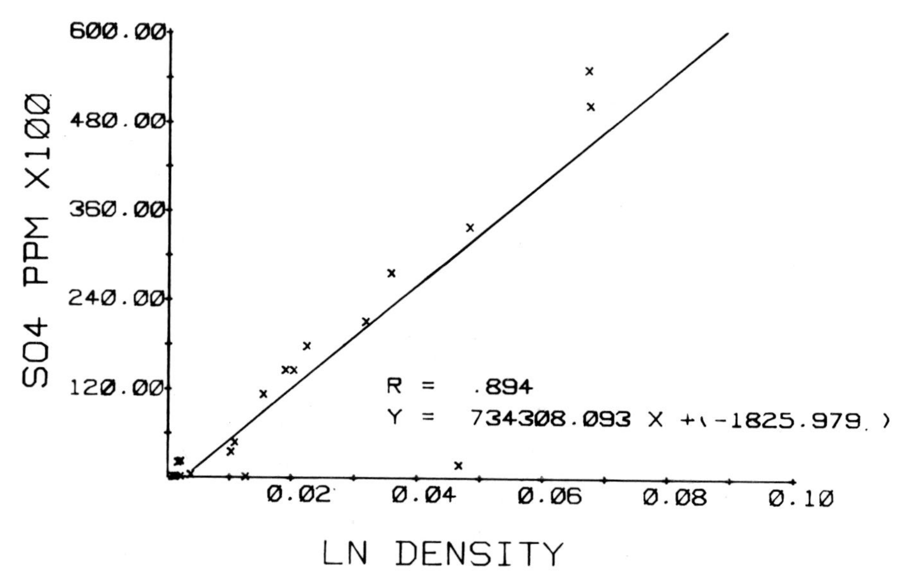

Fig. 9-11. Correlation between sulfate ion and density of waters in sulfate and carbonate lakes of central British Columbia.

Solar heating and water chemistry

Temperature profiles were taken of all the lakes, as well as surface and bottom water samples, and the bottom mud samples. Four lakes were found to be temperature stratified, having a warm water lens from 50–70 cm below the surface. The temperature profiles of the lakes are given in Fig. 9-8.

The warm lens extends to between 200 and 250 cm below the surface; beyond this depth, normal cool waters are found. If the bottom of the lake lies within 50 to 200 cm depth, warm waters reach the bottom. Closer examination of profile 3 and 4 (Fig. 9-8) reveals step-like progression in the cooling curve. This is interpreted as successive warm fronts penetrating the deeper horizons.

Only sulfate lakes exhibit solar heating; carbonate lakes do not. This can be attributed to the density differences between surface and bottom waters. Based on the data presented, it would appear that a density difference of 0.015 is required before a lake can be solarly heated.

This value is not attained in carbonate lakes. Since the sulfate content was shown to be the principal determinant of density, the density difference translates to 13.36 mg/g of sulfate content difference. Heating decreases the density of the thermolimnion, but the stability of thermal stratification increases with the density difference to the epilimnion.

Comparison of Caribbean and British Columbia lakes

The major difference between the saline lakes of these two regions is the obvious one, the nature of the dominant anion. The Caribbean lakes, specifically Lago Pueblo waters, being of marine origin, are chlorine brines. The significant addition of salts through the groundwater system does not alter the principal nature of the brines. Density differences between hypersaline hypolimnion and surface inflow of sea water is sufficient to maintain a thermal stratification until the surface layer evaporates. The lakes in British Columbia are in continental land-locked basins and the dominant anions are carbonate and sulfate, with chlorine only a subsidiary element. The dominant cations is in both cases sodium, followed by magnesium and potassium. Calcium is not a conservative element in either case but is depleted by early precipitation.

The temperature difference between epi- and thermolimnion is only 6–8°C in the tropical chlorine lakes, but absolute temperatures of the thermolimnion are ranging from 38–59°C and more. In sulfate lakes the temperature differences may be double that of chlorine lakes, but absolute temperatures of the thermolimnion do not reach half the values obtained in chlorine lakes. Moreover, the thickness of the hot water lens is much smaller due to the lesser intake of solar energy. Consequently the hot water lens heats up underlying water levels by conduction in step-like fashion.

SUMMARY

Caribbean solar ponds are seawater fed, seasonally density stratified, hypersaline lakes or lagoons, which act as solar heat collectors. The chlorine brine is somewhat enriched in certain cations by weathering products of nearby hills, but is primarily derived from the concentration of sea water.

Saline lakes in British Columbia are either carbonate or sulfate lakes, the latter containing the more concentrated brines that have previously precipitated much of their carbonate content. Only sulfate lakes have a sufficient density difference between surface an bottom layers to act as solar heat collectors. The hot water lens is smaller than that found in the chloride lakes, probably due to lower solar energy input at the higher latitude.

The chemistry of the B.C. lakes is remarkably homogeneous considering the diverse bedrock geology through which the groundwaters feeding the lakes pass.

REFERENCES

Cummings, J. M., 1940. Saline and hydromagnesite deposits of B. C. B. C. Dept. Mines, Bull. (4) 1–60.

Goudge, M. F., 1924. Magnesium sulphate in B. C. Investigations of Mineral Resources and the Mining Industry. Canada Dept. Mines, Mines Branch Bull., 642: 1–189.

Reinecke, L., 1920. Mineral deposits between Lillooet and Prince George, B. C. G.S.C. Mem. 118: 1–129.

Jenkins, O., 1918. Spotted lakes of epsomite in Washington and B.C. Am. J. Sci. 46: 638–644.

Hudec, P. P. and Sonnenfeld, P., 1974. Hot brines on Los Roques, Venezuela. Science, 185 (4149): 440–442 and 186 (4169): 1074–1075.

Lew, M. M., 1977. Seasonal variations in the physical and chemical properties of Lago Pueblo solar pond. Los Roques Archipelago, Venezuela. Unpubl. M.Sc. thesis, University of Windsor.

Sonnenfeld, P., et al., 1976a. Stratified heliothermal brines as metal concentrators. Acta Cient. Venezolana, 27: 190–195.

Sonnenfeld, P., et al, 1976b. The chemistry of precipitates and brines in restricted, density stratified lagoons. IIIrd Latinamerican Geol. Congr., Acapulco, Mex., Memoria. In press.

Sonnenfeld, P., and Hudec, P. P., 1977a. Stratified brines in restricted basins as sources of oil and oilfield brines. IInd Intern. Sympos. Water-Rock Interaction, Strasbourg, Proc.

Sonnenfeld, P., et al., 1777b. Base-metal concentration in a density-stratified evaporite pan. In: J. H. Fisher (editor): Reefs and Evaporites – Concepts and depositional Models. Studies in Geology No. 5, p. 181–187. Tulsa: Amer. Assoc. Petrol. Geol.

Sonnenfeld, P. and Hudec, P. P., 1979. Comparison of Caribbean solar ponds with inland solar lakes of British Columbia. (This volume, pp. 93–100).

Chapter 10

STUDIES OF MARINE SOLUTION BASINS -- ISOTOPIC AND COMPOSITIONAL CHANGES DURING EVAPORATION

A. NADLER[1] and M. MAGARITZ

Isotope Department, The Weizmann Institute of Science, Rehovot, Israel

The question of chemical and isotopic composition of saline solutions had been the subject of intensive studies in recent years both in the field and in the laboratory. The systems studied were saline lakes (Lotze, 1957; Fontes and Gonfiantini, 1967; Irion, 1970; Friedman, et al., 1976) and coastal sabkhas (Lloyd, 1966; Butler, 1969; Levy, 1977). Tracing of the chemical and isotopic changes in the evaporating solutions had been studied in the laboratory (Braitsch, 1971) and in artificial salt pans (Lloyd, 1966; Herrmann et al., 1973). The advantage of the latter type of study is the possibility of studying all evaporation stages simultaneously on a natural surface without experimental interferences. Sampling takes place within a few hours and under known meteorological conditions.

Our study was undertaken in small solution basins on the Mediterranean coast of Israel near Haifa. These basins were up to 2.0 m wide and 0.5 m deep. Twenty four such solution basins were sampled on October 26, 1972, on a small island (100 m^2) 5 m from the shore. A few of the basins were in the spray zone at the time of sampling. The basins were formed on eolianite rocks (Michelson, 1970). On the bottom of some of the basins halite and small quantities of gypsum and aragonite were found.

Evaporation and precipitation from sea water can be described mathematically by a formula similar to that of Rayleigh distillation law. Allegre et al. (1977) show that using a trace component with a very low bulk partition coefficient (b.c.p.) as a reference, bulk partition coefficient for other elements can be derived.

In this paper we shall try to evaluate chemical and isotopic changes during evaporation using the Mg concentration as a reference. Mg concentration indicates the degree of the evaporation of the solution. Mg was chosen because: (i) it can accurately be measured; (ii) it remains in the solution without precipitating as a solid phase. Mg is removed from solution only at the stage of concentration which was not reached by any of our samples.

The change of isotopic composition of oxygen and hydrogen in water during the process of evaporation had been studied by Craig and Gordon (1963). In that paper, however, the salinity effect on the isotopic fractionation factor was not discussed. Craig et al. (1965) showed that in experiments conducted on fresh water the vapor removal

1 Present address: Institute of Soil and Water, Agricultural Research Organization, Beit Dagan, Israel

from an evaporating solution is characterized by a kinetic isotopic effect. A stationary isotopic composition was reached by the liquid phase and atmospheric vapor after the liquid had been reduced to 15–25% of its original volume. Lloyd (1966), using experimental and natural data concluded that isotopic enrichment during sea water evaporation is caused by a kinetic effect coupled with a significant contribution by an exchange reaction.

The important factors controlling the isotopic enrichment in the solution are the salinity, humidity and the isotopic differences between the vapor and the water. In natural brines the atmospheric exchange is the predominant factor over kinetic evaporation effects. Friedman et al. (1976) in study of brines from Owens Lake, California, calculated using the Rayleigh equation and Li^+ concentration as an index for the degree of evaporation. Herrmann et al. (1973) studied the partitioning of bromide between halite in the sediments and the NaCl saturated brines and determined the partition factor between the mineral and solution. We will attempt to use the chemical and isotopic data to evaluate (using Rayleigh distillation mathematics) their behaviour in a fractional crystallization and evaporation sequence.

ANALYTICAL TECHNIQUES

The samples were collected in both polyethylene and glass bottles directly from the pools. pH and temperature measurements were performed during sampling. The pH was determined by a pH-meter manufactured by the Biological Services Electronic Laboratory of the Weizmann Institute.

Hg_2I_2 was added to the solution in order to prevent biological activity. Although no acidification was carried out to prevent Ca precipitation no evidence was found for its taking place.

The chemical composition was determined a few days later: Ca^{+2}, Mg^{+2}, Sr^{+2}, K^+, Li^+, Na^+ – by atomic absorption spectrophotometry with Perkin Elmer instrument model 306; Cl^- – by chloride titrator manufactured by Aminco Cotlove American Inst. Co.

Br^- was determined by Van der Meulen method which is based on oxidation of bromides into bromates by NaOCl at pH = 8, and determination of the resulting bromates by the iodine method.

Based on duplicates, reruns and analysis of references solutions we estimate the analytical error to be better than ± 2.5%.

The oxygen isotopic composition of the water samples were determined using the technique of Epstein and Mayeda (1953). The only change to that technique was the increase of equilibration time to at least 48 h due to the high salinities. Deuterium abundance was determined by the technique described by Friedman (1953). Extremely saline samples were analysed through the method of equilibration with water vapor as described by Sofer and Gat (1975). Back calculation from activities to concentrations were carried out using the fractionation factors given by Sofer and Gat (1975). The mean analytical reproducibility was of the order of 0.2‰ for $\delta^{18}O$ and 2‰ for δD.

RESULTS AND DATA EVALUATION

The chemical and isotopic data are given in Table 10-1. In the following discussion we shall also use the data of Herrmann et al. (1973) from Secovlje artificial salt pans which we shall refer to as the H.K. samples.

For the interpretation of the chemical changes we shall follow the formulation presented by Allegre et al. (1977); for the discussion of the isotopic data we shall use the Friedman et al. (1976) formulation, without any mechanistic implications. Using the Newman et al. (1953) formula:

$$C^{i} = C^{i}_{o,\ell} \; f^{(\bar{D}_i - 1)} \qquad [1]$$

where (i) = index of element i; C^{i}_{ℓ} = element concentration in liquid; $C^{i}_{o,\ell}$ = element concentration in the initial liquid; f = fraction of residual liquid; $\bar{D}_i$ = Bulk partition coefficient for the i element.

For two elements, one of them having $\bar{D} \ll 1$, the equation will have the form:

$$C^{i}_{\ell} = C^{i}_{o,\ell} \left(\frac{C^{*}_{o,\ell}}{C^{*}_{\ell}} \right)^{(\bar{D}_i - 1)} \qquad [2]$$

where $C^{*}_{o,\ell}$ and C^{*}_{ℓ} are the concentration of the element with very low b.p.c. In a logarithmic form:

$$\log (C^{i}_{\ell}) = \log (C^{i}_{o,\ell}) + (\bar{D}_i - 1) \log (C^{*}_{o,\ell}) - (\bar{D}_i - 1) \log C^{*}_{\ell} \qquad [3]$$

By plotting log C^{i}_{ℓ} vs. log C^{*}_{ℓ} assuming $\bar{D}i$ is constant throughout the process (or for part of it), the data points fall on a straight line with the slope of $(1 - \bar{D}_i)$.

In the case of mineral precipitation at some stage, $\bar{D}_i$ can be calculated from the point where precipitation starts by deviation of element concentration from the initially straight line. In cases that Di values change systematically during evaporation they can be derived from the tangent of the curvature on the Log C^{d}, log C^{i}_{ℓ} plot (Albareda, 1976).

In the following section we shall discuss the analytical data and the theoretical model.

In the case of natural pools where water is replenished several times, and mineral formed at the pool bottom may dissolve, deviations from a straight line (concentration vs. residual volume) are encountered. These mixtures of solutions are shown in the figures by a separate symbol and will be discussed in the isotopic data section.

Li^{+}: Following the change of log (Li^{+}) as a function of log [Mg^{+2}] eq. 3 suggests that Li^{+}, like Mg^{+2}, is retained in solution. [Mg^{++}] was chosen as a fraction monitor because of the relatively larger analytical error caused by the low concentrations of Li^{+}.

K^{+}: Potassium data for our and H.K. samples are shown in Fig. -1a. If no K^{+} were lost from solution then its concentration in the most saline solution would have to be $[K^{+}]_{S.W.} \times R^{*}_{Mg}$ = 12.290 mg/l.

$$\left[\frac{[Mg^{*+2}]\ \text{sample}}{[Mg^{*+2}]_{S.W.}} = R_{Mg} \right]^{*} \qquad [4]$$

TABLE 10-1.

Chemical and Isotopic data.

Sample No.	Mg*	Na	K	Li	Ca	Sr	Cl*	Br	pH	TMP(°C)	$\delta^{18}O$(‰)	δD(‰)
1	55750	50830	8070	4.85	177	6.0	178100	2050	7.22	29.0	.08	3.4
2	49400	59300	7750	4.26	234	9.6	179400	1560	7.32	28.5	.51	7.3
3	41800	70590	6940	3.38	245	27.0	185100	1240	7.32	32.5	1.79	17.3
4	24300	97890	4680	2.35	396	34.3	186900	859	7.68	32.0	2.79	20.0
5	20100	94130	4520	1.90	495	24.0	183400	757	7.82	29.0	1.07	16.8
6	18700	94126	4120	2.20	464	51.8	188100	706	7.62	28.5	1.91	14.1
7	15200	83770	4500	1.17	760	20.5	155400	483	7.70	32.5	2.22	18.9
8	14900	101660	3150	1.62	542	39.0	182900	525	7.58	28.0	2.15	22.6
9	11000	101660	2500	1.03	693	31.3	183400	490	7.84	27.0	1.59	15.3
10	10640	18825	2180	1.17	906	27.1	158200	392	7.73	32.0	3.96	22.5
11	7430	63380	1450	0.74	1560	33.7	113200	323	7.64	27.0	5.59	24.9
12	6500	44430	1130	0.88	1630	28.9	84100	280	8.00	27.0	3.99	26.2
13	5370	43130	890	0.59	760	20.5	74100	203	8.02	25.0	2.42	20.3
14	5240	46400	890	0.35	1690	25.3	75400	185	8.00	29.5	2.92	16.2
15	5070	113890	1290	0.44	1690	40.0	186400	202	8.02	27.0	2.07	16.2
16	2900	19580	705	0.23	781	15.1	39700	120	7.92	24.0	3.01	16.4
17	2330	18070	581	0.19	740	13.3	31800	102	8.70	29.0	2.88	21.7
18	2260	16190	510	0.17	714	13.2	30600	97	8.50	29.0	2.55	20.8
19	1980	13180	446	0.19	536	9.6	25000	80	8.70	25.0	2.22	17.9
20	1780	12420	426	0.15	516	10.2	24300	77	8.70	25.0	2.09	20.9
21	1774	12050	414	0.17	500	8.4	23600	78	8.50	26.0	1.85	10.6
22	1740	12800	414	0.19	531	9.6	22800	85	8.20	25.0	1.87	21.2
23**	1756	11670	387	0.16	500	9.0	22300	74	8.10	25.0	1.62	10.7

* All chemical analysis are in mg/l

** Surface seawater from the location

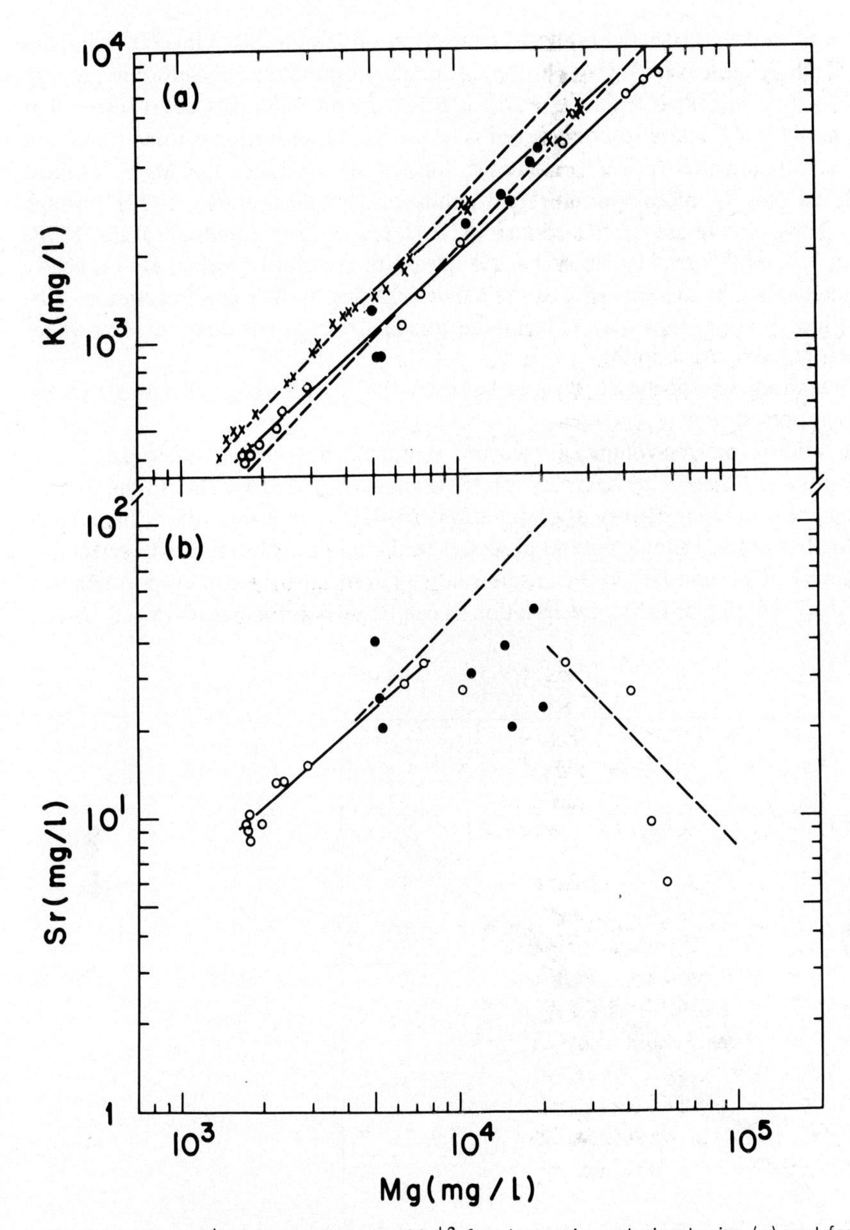

Fig. 10-1a. Log K^+ plotted against log Mg^{+2} for the marine solution basins (o) and for the Secovlje sample (Herrmann et al., 1973) (X). The full circles (●) represent recently flooded solution basins (see text). A dashed line represents a slope of +1 or -1. , A full line represents the linear best fit through the samples (not including a flooded solution basins). A small depletion of K^+ for both sets of data can be noticed. $\bar{D}_K$ values are 0.12 (and 0.16 for H.K. samples).

10-1b. Log Sr^{+2} plotted against log Mg^{+2} for marine solution basins. Note a two state removal of Sr; an early one with a $\bar{D}$ = 0.132; a latter one, starting at the stage of halite precipitation, where most of the Sr^{+2} is removed from the solution.

As can be seen in Table 10-1 the observed concentration is only 8070 mg/ℓ; this implies that 35% of the K^+ is lost from the solution during the evaporation-precipitation process. The b.p.c. for K^+ calculated from Fig. 10-1a is 0.12. In the H.K. data set it is seen that the point in which K^+ starts to be removed is where its concentration is three times that of sea water. A. Carpenter (private communication, 1978) concluded that Mg^{+2} is gained from residuals in the more concentrated ponds of Hermann et al. (1973) through historically long-term re-use of the ponds. B. F. Jones (private communication, 1978) suggest that this Mg^{+2} gain is caused by the release of previously sorbed Mg on nearly all of the likely solid phases exclusive of the halides existing in solar pond operation. The above mentioned suggestions may explain the reason for apparent early removal of K^+ and Br^- before halite precipitation.

There are three significant differences between the marine solution basins and the artificial solar ponds:

(1) The ratio of solution volume/surface area is much higher in the former case.

(2) The flux of brines in the later case is large against a non existing flux in the former.

(3) A complicated long history of evaporation, artificial man intervention characterize the solar ponds, against a simple natural processes in the marine solution basins system.

The removal of K^+ and Br^- in the m.s.b. is much later in the stage of evaporation, not until after the beginning of halite precipitation as can be seen in Figures 10-1a and 10-2.

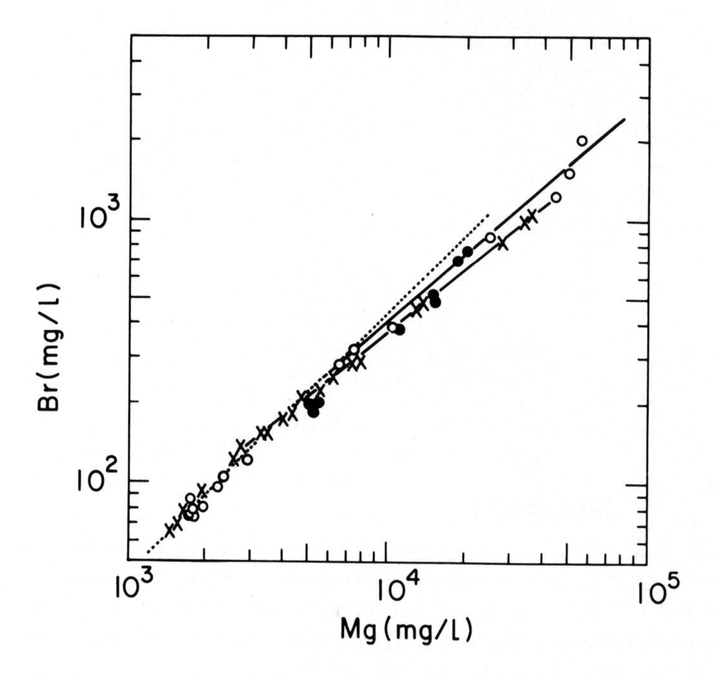

Fig. 10-2. Log Br^- vs. log Mg^{+2} for the marine solution basin and H.K. data. For symbols see Fig. 10-1a. A small depletion of Br^- for both sets of data can be noticed; $\bar{D}_{Br^-}$ = 0.14 (and 0.188 for H.K. samples).

Br^-: Bromide, similarly to potassium, shows in both sets of data about 20% defficiency in the most concentrated samples (Fig. 10-2). Also in the case of bromide the H.K. set of data show the depletion in Br^- starts before halite precipitation. The explanation is similar to the one discussed for the K^+. The calculated b.p.c. for our data is 0.14 and for the H.K. set is 0.188. Herrmann et al. (1973) measured the Br partition coefficient for the halite found in the salt pans and their D values range from 0.13 to 0.15. Our data agrees with their measurements within the limits of the data scatter.

Na^+: Although sodium is a major component it can be seen from Fig. 10-3a that the sodium data can be treated in a similar way to the rest of the elements. As can be seen from our data (Fig. 10-3a) excess of Na^+ is found in the stage before halite precipitates. This excess is related to dissolution of halite from a previous evaporation cycle found at the bottom of the pools. In the H.K. case the pool lacked halite precipitation and no enrichment by dissolution occurred. As can be seen in Fig. 10-3a the removal of Na^+ from the solution saturated for halite does not have a constant b.p.c. value. The values change from 1.13 at the beginning of precipitation (which is similar to the $\bar{D}_{Cl}$) to a value 2.14. The same pattern is found for H.K. data set (the range being 1.13–2.26).

This behaviour indicates that at each point along the evaporation process more Na^+ is removed from the solution compared to the rate of raise of Mg^{+2} concentration. During halite precipitation Na^+ and Cl^- are removed in equal amount from the solution. Since the initial Na/Cl (eq/eq) ratio of sea water is smaller than 1 (0.86) relatively more Na^+ is precipitated compared to Cl^-. In such a case, shown also by Levy (1977) the line of Cl^- concentration as a function of Mg^{++} increase stays almost parallel to the abcissa (Fig. 10-3b), while the line for Na^+ gets gradually curved (Fig. 10-3a). As a result $\bar{D}_{Na}$ changes during halite precipitation.

Cl^-: The chloride data (Fig. 10-3b) are similar to that of sodium; excess chloride is found before NaCl saturation and depletion occurs as halite precipitates, b.p.c. ($\bar{D}$) equals 1.17

Ca^{+2}: The calcium, which is removed almost totally from evaporated sea water by $CaSO_4 \cdot 2H_2O$ precipitation (Braitsch, 1971), has a high $\bar{D}$ value (2.00 and 2.44 for ours and H.K. data set respectively).

As can be seen from Fig. 10-4a, a few samples (No. 6, 8, 9, 13) appear below the normal evaporation line. We suggest, based on isotopic data (see discussion below) that these samples represent recent floodings of the basins by sea water. Another group of samples (No. 5, 7, 14, 15) are following the normal evaporation line although they too are the result of flooded brines – based on isotopic data.

The fact that this group of samples represent mixtures can only be deduced from the isotopic data, but not from the chemical information. This is due to dissolution and re-equilibration with the bottom sediments (See Na^+ section).

SO_4^{-2}: The beginning of SO_4^{-2} precipitation is clearly shown by the graphic method. In Fig. 10-4b, H.K. data set show the point where gypsum precipitates with a b.p.c. of 0.28.

Sr^{+2}: The data shown in Fig. -1b indicate Sr^{+2} removal in two stages: (1) an early one in the region of gypsum precipitation ($\bar{D}$ = 0.132), (2) a latter one, in which most of the Sr^{+2} is removed, starting at the point where only 0.1 of the original volume remains. The sharp decline of Sr^{+2} concentration (resembling that of Ca^{+2}) suggests Sr^{+2} precipitation as a major phase and not as a trace element. Muller and Puchelt (1961) believe that celestite ($SrSO_4$) precipitates in the transition region between carbonate and gypsum;

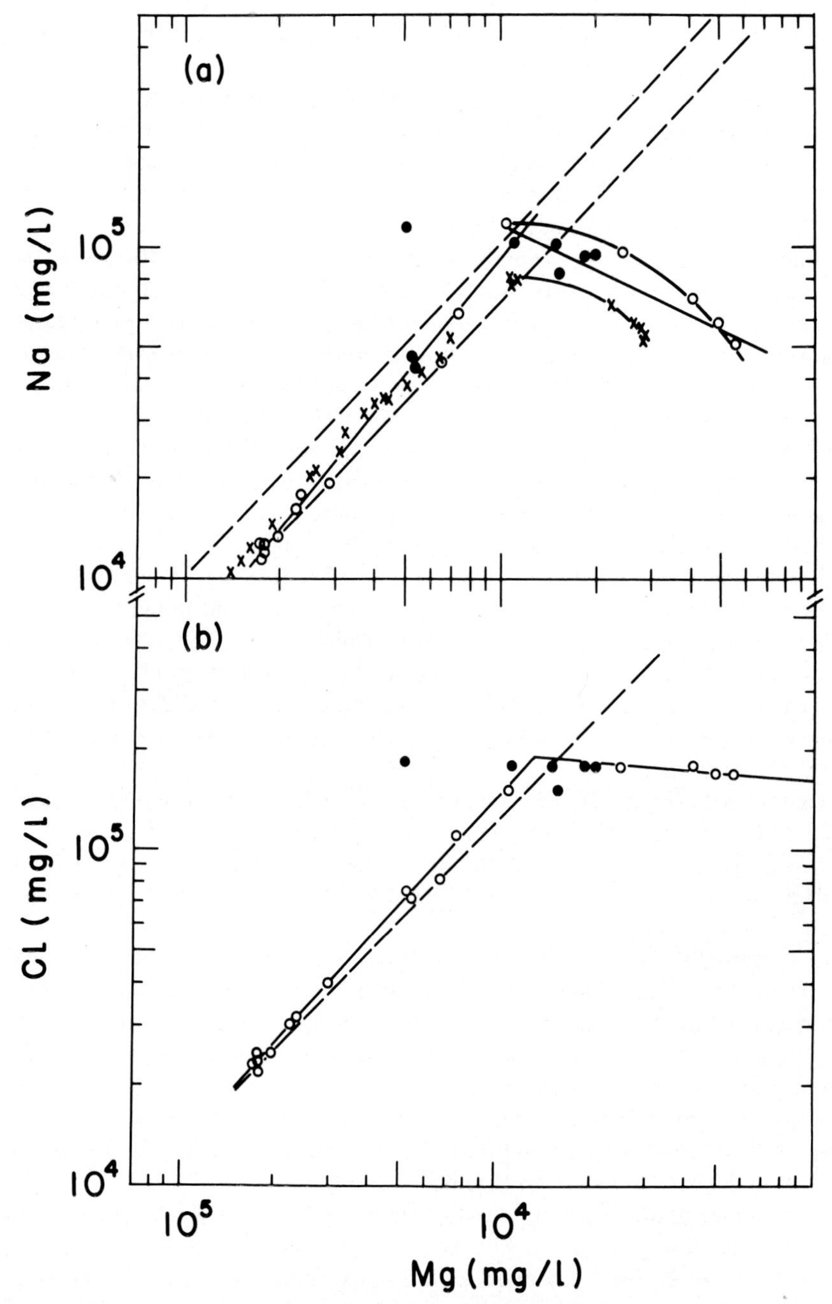

Fig. 10-3a. Log Na^+ vs. log Mg^{+2} for the marine solution basins and H.K. data. For symbols see Fig. -1a. Note the excess of Na^+ found in the stage before halite precipitation. $\bar{D}_{Na^+}$ change during precipitation from 1.13 to 2.14 (1.13–2.26 for H.K. samples).

10-3b. Log Cl^- vs. log Mg^{+2} for the marine solution basins and H.K. data. For symbols see Fig. -1a.

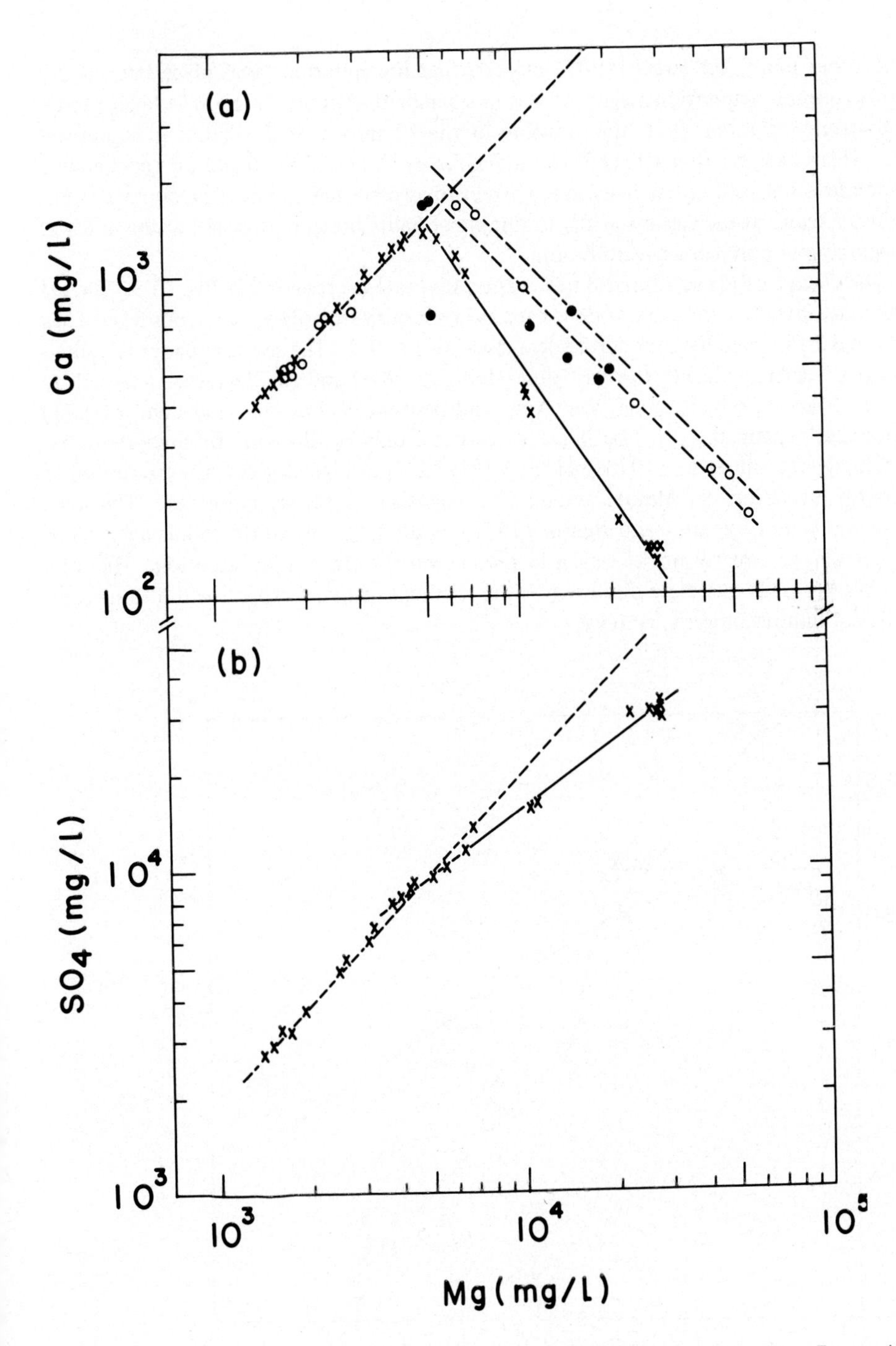

Fig. 10-4a. Log Ca^{+2} vs. log Mg^{+2} for the marine solution basins and H.K. data. For symbols see Fig. -1a.

10-4b. Log SO_4 vs. log Mg^{+2} for H.K. data. For symbols see Fig. 10-1a.

on the other hand, Usdowski (1967) suggests that precipitation takes place later at the end of gypsum precipitation stage which is in agreement with our data. Sr^{+2} begins to be quantitatively removed from the solution at the beginning of the halite precipitation stage. This confirms Braitsch (1971) model for Sr^{+2} behaviour made on geochemical consideration suggesting that it is more correct to assume that celestine precipitates from normal sea water at the earliest at the beginning of halite precipitation and at the latest at the beginning of polyhalite precipitation.

pH: The change of pH as a function of evaporation rate is presented in Fig. 10-5. The pH is controlled by the carbonate system causing a sharp rise of pH values from 8.0 (of sea water) to 8.7 followed by a moderate decrease down to 7.2. The rise is apparently caused by biogenic activity during the day time (Emery, 1946) and the decrease is caused by increase of salinity which agrees with Amit and Bentor (1971). Amit and Bentor (1971) describe and explain the pH raise in Dead Sea water only by dilution. Bodine (1976) explains the Herrmann et al. (1973) pH drop (Fig. 10-5) at high degrees of evaporation as caused by formation of "Minute amounts of magnesium hydroxy chloride". The same pH rise was found by Amit and Bentor (1971) by dilution only of the saline water. The phenomenon was clearly noticed even in the absence of Mg^{+2} in the solution. The drop of pH in our case is moderate because it is the resultant of two factors; the dilution factor and the continuous biogenic activity.

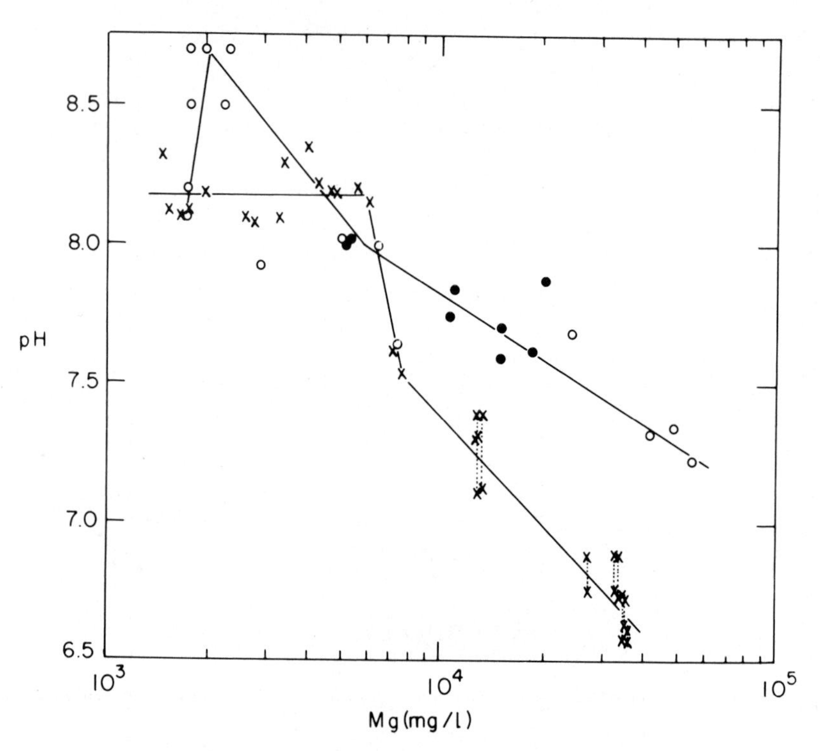

Fig. 10-5. pH plotted vs. log Mg^{+2} for marine solution basins and H.K. data. For symbols see Fig. 10-1a.

OXYGEN AND HYDROGEN ISOTOPIC VARIATION

The oxygen and hydrogen isotopic composition of the samples varied systematically as a function of Mg^{+2} concentration (Table 10-1). The solution became enriched in both O^{18} and D during the earlier stages of evaporation and then reached a maximum; additional increase of salinity was followed by a depletion in these two isotopes.

One can use the formulation of a Rayleigh law for the evaluation of the δO^{18} and δD isotopic data in exactly the same manner as the chemical species were treated. Such a treatment of isotopic data was applied by Lloyd (1966) for his experimental data and by Friedman et al. (1976) for the Ownes Lake samples. In the following discussion we shall use Friedman's et al. (1976) formulation:

$$R / R_o = f^{D-1} \qquad [5]$$

This can be expressed in a logarithmic form similarly to the chemical species formulation.

$$\log (1000 + \delta) / (1000 + \delta_o) = (D - 1) (\log Mg_o - \log Mg) \qquad [6]$$

The plot of $\log (1000 + \delta)/(1000 + \delta_o)$ against log Mg for oxygen and hydrogen isotopic measurements are shown in Fig. -6a and 6b.

In Fig. 10-6a we show both $\delta^{18}O$ values on activity scale and on the concentration scale as corrected for water activity after Sofer and Gat (1972). However, this transformation from scale to scale does not change the general trend of the curve as seen in Fig. 10-6a. Flooding of the solution basins by sea water can be identified best by the isotopic data (Figs. 10-6a and 6b). Due to dissolution of bottom sediments, as was discussed earlier, this could not be clearly seen using the chemical data.

During evaporation the solutions became enriched in Oxygen-18 and Deuterium up to the stage in which the solution volume is reduced to 25% of the initial volume. From this point on it gets depleted in both heavy isotopes. Measured values for oxygen (on activity scale) and deuterium became even lower than the original sea water for the most saline solution.

Earlier workers also noted that the isotopic data points do not fit a straight line on plots similar to our Figs. 10-6a and 6b. They suggested that the fractionation factor changes during evaporation due to salinity effect (Lloyd 1966, Fontes and Gonfiantini 1967, Friedman et al. 1976).

We have calculated, using eq. 10-6, the bulk partition coefficient $(\bar{D})$* for different segments along the evolution line of the brines. The calculated $(\bar{D})$ values are shown in Fig. 10-6c as a function of Mg concentration. The rate of change of D values, for both isotopes, can be divided into three categories:

* We use this notation $(\bar{D})$ for reasons of consistency with the chemical data. We suggest to keep on using such a notation whenever calculating effects which sum up several processes between the solution and the environment. α (isotopic fractionation) is being used in cases where the main mechanism operating is fractionation between vapor and liquid.

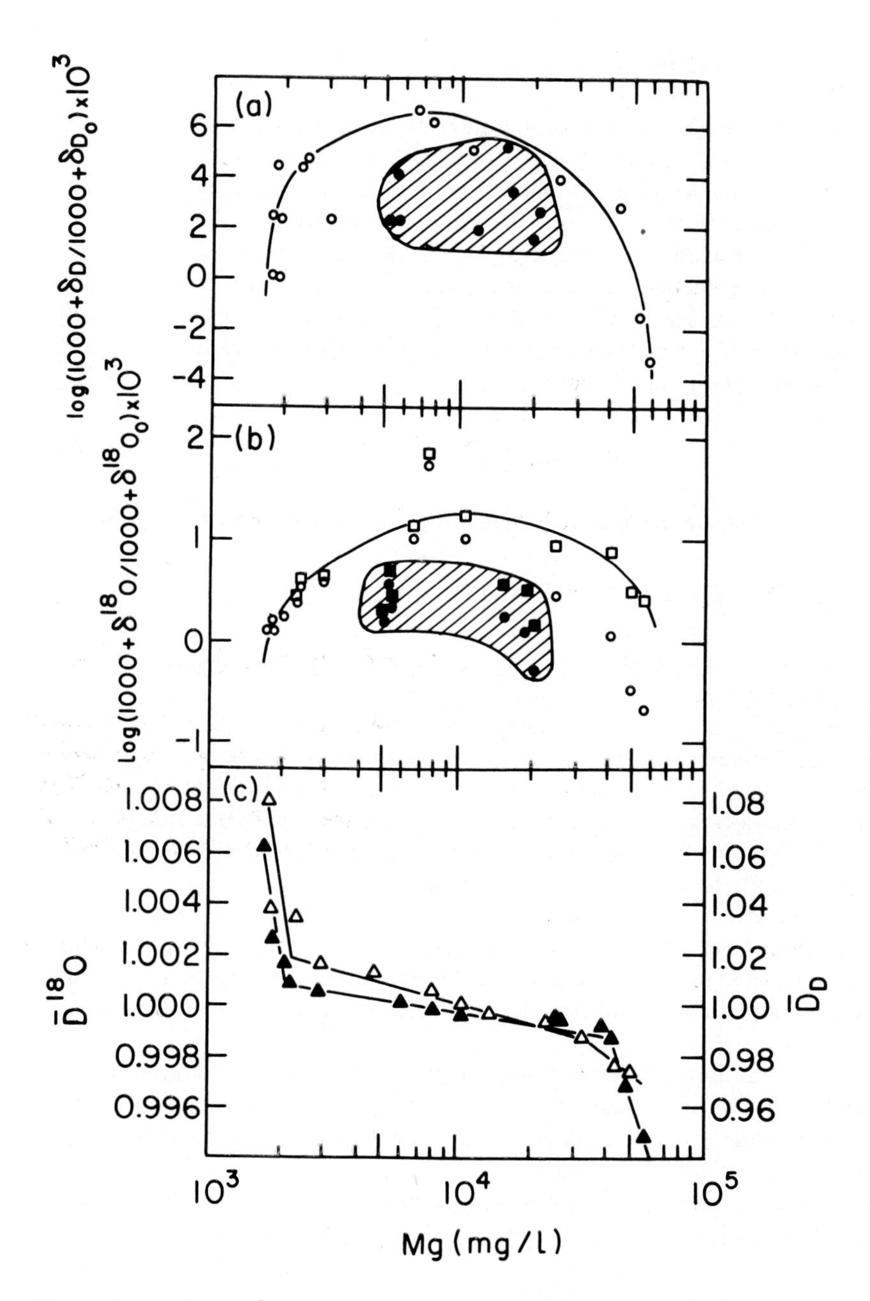

Fig. 10-6a. Log $(1000 + \delta D)/(1000 + \delta D_O)$ vs. log Mg^{+2} for the marine solution basins data. For symbols see Fig. 10-1a. The shaded area represents the flooded samples.

10-6b. Log $(1000 + \delta^{18}O)/(1000 + \delta^{18}O_O)$ vs. log Mg^{+2} for the marine solution basins. For symbols see Fig. 10-1a. The squares (open and closed) are corrected values after Softer and Gat (1972).

10-6c. $\bar{D}_{18O} + \bar{D}_O$ vs. log Mg^{+2} for the above data. △ – D_{18O} ▲ – $\bar{D}_O$. The triangles (open and closed) represent $\bar{D}_{18O}$ respectively.

(1) A drastic change from nearly equilibrium value (Craig et al. 1963) down to 1.000. This drop occurs along a narrow range of salinities. Lowenstam and Epstein (1957) had already noted that $\bar{D}$ values for the earlier stage of evaporation are close to the theoretical fractionation between liquid and vapor.
(2) During the larger part of the evaporation process, the change in $\bar{D}$ takes place along a moderate slope. Values change from slightly above 1.000 (1.002 and 1.010 for δO^{18} and δD respectively) to values slightly below one (0.999–0.990 respectively).
(3) In the last segment (where the residual volume is ∿5% of the initial volume) we have found a drastic drop in $\bar{D}$ values. In comparison to "effective alpha" of Friedman et al. (1976), our data extends the range of the partition coefficient (which goes below 1.000 for Deuterium). This extension of the range may be related to the difference of the atmosphere prevailing over the sampling sites: a desert, low humidity climate in Friedman's case and a marine humid climate in our case.

The bulk partition coefficient includes at least two components:
(1) The fractionation factor between the liquid and vapor which was discussed by Craig and Gordon (1963).
(2) The second component is an exchange reaction of H_2O molecules between the atmosphere and the water. The rate of evaporation drops with increase of salinity (Harbeck, 1955) thus lowering the contribution of the first component. On the other hand, the rate of exchange depends on the availability of water molecules in the atmosphere (and in high humidities the rate is higher).

The mean isotopic composition of atmospheric vapour in Israel's coastal plain has been given by Tzur (1971) as $\delta^{18}O = -11.5‰$ and $\delta D = -80‰$; vapour in isotopic equilibrium with Mediterranean surface waters are (at 25°C) $\delta^{18}O = -7.4‰$ and $\delta D = -65‰$. The humidity "seen" by the brine pools can be expected to be a mixture between these values.

CONCLUSIONS

One can use the mathematics of Rayleigh distillation law to evaluate simultaneously chemical and isotopic sets of data from an evaporation process. The results of such procedures are: (1) For trace elements like bromine the calculated b.p.c. agrees with the measured one. (2) For elements which form solid phases such as Na, Ca, Cl, and SO_4 the b.p.c. is related to the rate of dissolution of previously precipitated minerals from the bottom of the brine pool and is proportional to the anion/cation ratio in the solution. (3) Sr^{++} is removed from solution (as celestite) when the degree of evaporation is about 10, in agreement with Braitsch's conclusions (1971). (4) The b.p.c. for the hydrogen and oxygen isotopic compositon changes systematically from near the equilibrium value down to values lower than 1.000. Solutions with δD and $\delta^{18}O$ values even lower than the starting solution (sea water) are thus formed.

ACKNOWLEDGMENTS

We thank Dr. Z. Sofer for part of the deuterium analyses and are grateful to Drs. Blair F. Jones, T. B. Coplen, I. Friedman and J. R. Gat for their helpful notes on earlier versions.

REFERENCES

Albareda, F., 1976. Some trace element relationships among liquid and solid phases in the course of the fractional crystallization of magma. Geochim. Cosmochim. Acta, 40, 667–673.

Allegre, C. J., Treuil, M., Minster, J. F., Minster, B. and Albarede, F., 1977. Systematic use of trace element in igneous process. Part I: fractional crystallization processes in volcanic suites. Contrib. Miner. Petrol. 60, 57-75.

Amit. O., and Y. K. Bentor, 1971. pH dilution curves of saline waters (from the Dead Sea, Red Sea, the Mediterranean and salt springs from the Tiberias area. Chem. Geol., 7, 307–313.

Bodine, M. W., Jr., 1976. Magnesium hydroxychloride: a possible pH buffer in marine evaporite brines, Geol. 3, 76-80.

Braitsch, O., 1971. Salt deposits: their origin and composition, Springer-Verlag. New York, Heidelberg, Berlin, 197 pp.

Butler, G. P., 1969. Modern evaporite deposition and geochemistry of coexisting brines, the sabkha, Tracial Coast, Arabian Gulf, J. Sed. Petrol., 39, 70-89.

Craig, H. and Gordon, L. I., 1965. Deuterium and Oxygen-18 variations in the ocean and the marine atmosphere. In: Stables Isotopes in Oceanographic Studies and Paleotemperatures, Consiglio Nazionale delle Ricerche, Piza, p. 9–130.

Craig, H., Gordon, L. I. and Horibe, Y., 1963. Isotope exchange effects in the evaporation of water: I. Low temperature experimental results, J. Geophys. Res., 68, 5079-5087.

Emery, K. O., 1946. Marine solution basins, J. of Geol., 54, 209-228.

Epstein, S. and Mayeda, T., 1953. Variation of O^{18} content of waters from natural sources, Geochim. Cosmochim. Acta, 4, 213-224.

Fontes, J. C. and Gonfiantini, R., 1967. Comportement isotopique au cours de l'evaporation de deux bassins sahariens, Earth Plant. Sci. Lett., 3, 258-266.

Friedman, I., 1953. Deuterium content of natural water and other substances, Geochim. Cosmochim. Acta, 4, 89-103.

Friedman, I., Smith, G. I. and Hardcastle, G., 1976. Studies of quaternary saline lake. II. Isotopic and compositional changes during dessication of the brine in Owens Lake, Calif. 1969-1971. Geochim. Cosmochim. Acta, 40, 501-511.

Harbeck, G. E., Jr., 1955. The effect of salinity on evaporation, U.S. Geol. Survey, Prof. Pap., 272-A, 1-6.

Herrmann, A. G., Knake, D., Schneider, J., and Peters, H., 1973. Geochemistry of modern seawater and brines from salt pans: main components and bromine distribution. Contrib. Mineral. Petrol, 40, 1-24.

Irion, G., 1970. Mineraligisch-sedimentpetrographische une Geochemische Untersuchungen am Tuz Golu (Salzsee), Turkei, Chem. Erde, 29, 163-226.

Levy, Y., 1977. The origin and evolution of brine in coastal sabkhas, Northern Sinai, J. Sed. Petrol. 47, 451-462.

Lloyd, R. M., 1966. Oxygen isotope enrichment of sea water by evaporation. Geochem. Cosmochim. Acta, 130, 801-814.

Lotze, F., 1957. Steinsalz und Kalisalze. I. Teirl. Berlin – Nikolassee. Brontraeger.

Lowenstam, H. and Epstein, S., 1957. On the origin of sedimentary aragonite needles of the Great Bahama Bank, J. Geol. 65, 364-375.

Michelson, H., 1970. The geology of Mt. Carmel shore line. M. Sc. thesis, the Hebrew University, Jerusalem, Israel, 60 pp.

Muller, G. and Puchelt, H., 1961. Die Bildung von Coelestine ($SrSO_4$) aus Meerwasser, Naturewiss., 48, 301-302.

Neumann, H., Mead, J. and Vitalino, C. J., 1954. Trace element variation during fractional crystallization as calculated from the distribution law. Geochim. Cosmochim. Acta, 6, 90-99.

Sofer, Z. and Gat. J. R., 1972. Activities and concentrations of oxygen-18 in concentrated aqueous salt solutions: analytical and geophysical implication. Earth Plan. Sci. Lett., 15, 232-238.

Sofer, Z., and Gat. J. R., 1975. The isotope composition of evaporation brines: effect of the isotopic activity ratio in saline solutions. Earth Plan. Sci. Lett., 26, 179–186.

Tzur, Y., 1971. Isotope effects in the evaporation of water into air. Ph.D. Thesis, Weizmann Institute of Science, Rehovot, Israel, 109 pp.

Usdowski, E., 1967. Der Einbau von Sr in Gips une Anhydrit. Ann. Meet. Deutsche Mineral. Gesell., Berlin.

Chapter 11

EVAPORITIC ENVIRONMENTS IN NORTHERN SINAI

YITZHAK LEVY

Geological Survey of Israel, 30 Malkei Israel Street, Jerusalem, Israel

INTRODUCTION

The northern coast of the Sinai peninsula (Fig. 11-1) is bordered along about two–thirds of its length by the Bardawil lagoon, a body of water about 600 km^2, it is about 80 km long and its maximal width is 18 km. The maximum depth of the lagoon water is about 3 m. The lagoon is separated from the Mediterranean Sea by a long curving sand barrier. Mediterranean water enters the lagoon through two main inlets in the barrier, each about 300 m wide and 2–3 m deep. The arid climate in this area (daily temperature in the summer ranges from 30–35°C, average rainfall 10 cm/yr and the evaporation rate is 2 m/yr), and the restricted connection between the lagoon and the Mediterranean lead to the formation of highly saline waters in the lagoon. Lagoon waters adjacent to the coast show chlorosities up to 60 g/l Cl^-. The chemistry and sedimentology of the lagoon was described by Levy (1974).

At the southern reaches of the lagoon a series of small sand barriers interrupted by very shallow straits separate a restricted, very shallow-water environment to the south from the rest of the lagoon. In this southern area lagoon waters concentrate to salinities

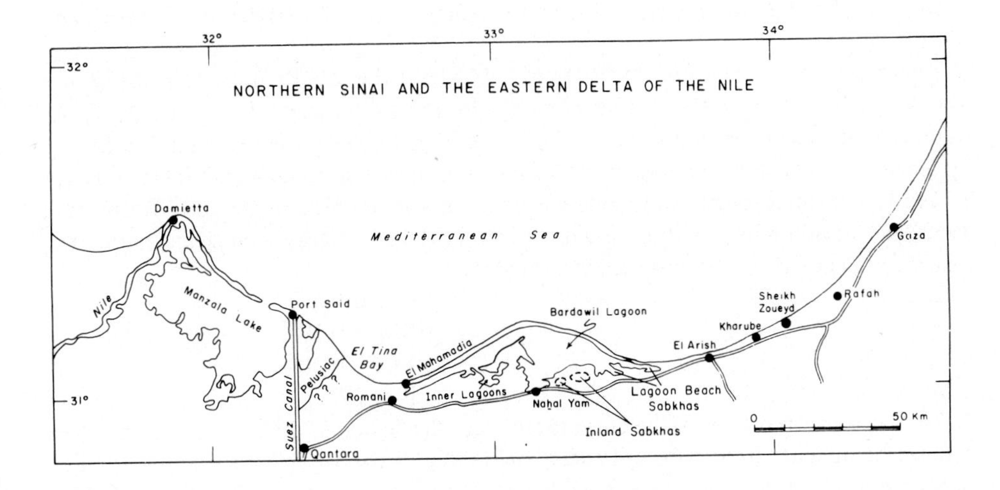

Fig. 11-1. Northern Sinai and the eastern delta of the Nile.

which cause precipitation of gypsum and halite, thus forming extensive salt pans, marine beach sabkhas, in which the water depth is only a few centimeters.

To the south of these lagoon beach sabkhas, sand dunes extend inland. In the coastal portion of the dune field, small sabkhas occupy the depressions between the sand dunes. These inland sabkhas are completely disconnected from the lagoon at present (Fig. 11-1).

The groundwater table in the area is as little as 1 cm to as much as 30 cm below the sabkha surface during most of the year. After rainy periods, which last no longer than a few days, the sabkhas are flooded by groundwater flowing from the neighboring dunes. These flood waters which may attain a maximum depth of 40 cm evaporate within a few weeks. The chemical composition of the sabkha brines varies considerably probably due to evaporation and dilution of the concentrated brines and to interactions between the brines and the sediments (Levy, 1977a and b). In contrast to the immense sabkha of Abu Dhabi which occupies a broad coastal plain, the Bardawil coastal plain consists of many small sabkhas, separated by sand dunes.

METHODS

The study of the area was performed through sampling of brines and sediments from the Bardawil lagoon and the adjacent sabkhas. Lagoon water samples from different depths throughout the water column were taken, during different seasons. Sediment samples were taken using a grab sampler; core samples were obtained at a number of stations by driving transparent plastic tubes into the sediment with a hammer. The sediment was kept in the core by a tight pneumatic seal at the top of the tube (Sanders, 1968).

Interstitial waters were extracted from the sediment at different depths (0–60 cm) by centrifugation.

Brines were sampled from 24 different sabkhas at the sabkha surface, in the winter months, or from trenches or shallow boreholes during the summer. The chlorosity of the water samples was measured using the Mohr method, sulphate was determined as $BaSO_4$ by gravimetry, calcium and magnesium by E.D.T.A. titration, sodium and potassium with an E.E.L. flame-photometer, bromide by modified Van der Meulen method (Bloch et al., 1952) and bicarbonate by titration with 0.1 N. HC1 using methyl orange indicator. The precision of analytical methods used is as follows:

Na ± 2%	Cl ± 0.5%
K ± 2%	SO_4 ± 0.5%
Ca ± 2%	HCO_3 ± 10%
Mg ± 2%	Br ± 2%

The bottom samples from the cores taken from the lagoon at different depths down to 1 m and sediments from outcrops and trenches dug in the sabkhas to a depth of half a meter were subjected to grain size analysis and the mineralogy of each fraction was determined by X-ray diffraction.

THE LAGOON

Lagoon Water

The spatial changes in chlorosity in June 1970 may be summarized as follows: There is an increase in chlorosity from inlet A to the east up to Mat Iblis and from there to the south up to Nahal Yam. Again the chlorosity increases from Nahal Yam to the east (Fig. 11-2).

The spatial distribution of chlorosities described above is valid only if Inlet B is blocked and Inlet A is partially open, the situation at the time of study. Any change in the state of the inlets will bring a change in the hydrology of the lagoon and hence a different spatial distribution of the chlorosity of the waters.

The differences in chlorosity at different depths of the water column are more pronounced in the western arm than in the main lagoon, especially in the summer months (May–September). While the increase in chlorosity with depth is gradual in the main lagoon, in the western arm it is gradual up to a height of 50 cm from the bottom and increases abruptly below. In May 1969, the surface water showed a chlorosity of 26.6 g/l, which increased gradually with depth attaining 28.9 g/l at depth of 2 m; below this it increased considerably and attained 41.5 g/l at a depth of 2.5 m. These results suggest that the water which evaporated at the surface of the lagoon sank to near-bottom depths and due to their higher specific gravity formed a near-bottom returning current of saline waters. This water migrates from the main lagoon to the western arm replacing less saline waters. If Inlet A was completely opened, these saline waters would flow from the lagoon to the Mediterranean, but since Inlet A was blocked by sand to a depth of about 1 m at the period of sampling, the saline waters were entrapped in the western arm.

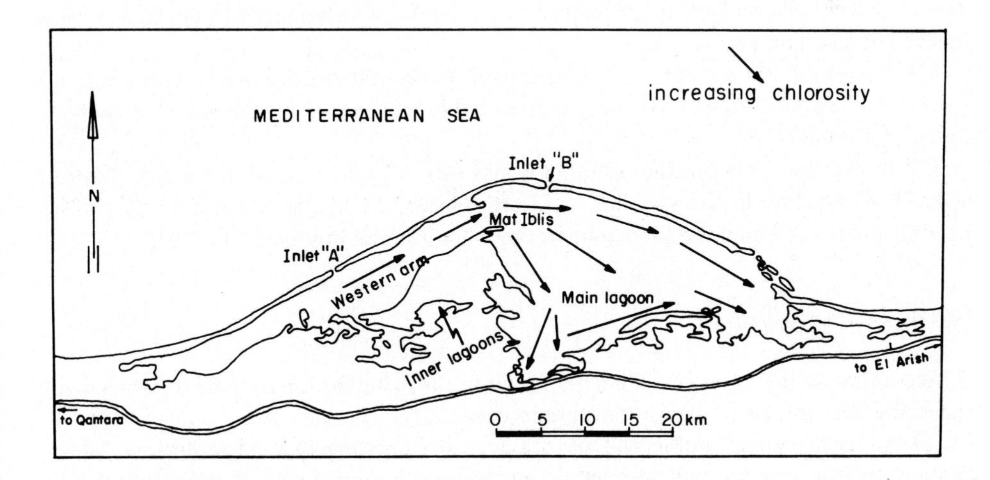

Fig. 11-2. Areal changes in chlorosity of the Bardawil Lagoon

Scruton (1953) has suggested the mechanism of returning near-bottom saline water counter currents from lagoons to the sea as an explanation for the maintenance of lagoon water at a certain salinity, thus preventing transformation into salt pans.

Seasonal changes in lagoon water chlorosity are probably influenced by several factors. The most evident are the following:

(1) In the winter the lagoon water is diluted to a certain minimum salinity, and is almost completely mixed. The dilution is due to both rain water (100 mm/yr) and the high rate of exchange between the lagoon and the Mediterranean.

(2) In the summer, the high rates of evaporation and the lower rate of exchange of lagoon and Mediterranean waters cause high chlorosities in different regions of the lagoon as well as the stratification in the water column.

The ratios of Na/Cl, K/Cl, Mg/Cl, Ca/Cl and SO_4/Cl for waters in the main lagoon and western arm show that these brines resulted from evaporation of Mediterranean Sea water. Water of the inner lagoons (Fig. 11-2) shows a positive anomaly in both Ca/Cl and SO_4/Cl ratios reflecting the dissolution of gypsum from the bottom sediments of the inner lagoons in the lagoon waters (Levy, 1971).

The Interstitial Water

The results of chemical analyses of the interstitial waters from the Bardawil lagoon show that they differ from normal marine interstitial waters in the following aspects:

(1) They show an increase in salinity with increasing depth.

(2) The ionic ratios differ from those obtained for normal marine interstitial waters and for brines obtained by evaporation of normal sea water to the same chlorosity (Levy, 1974b).

The processes responsible for the formation of the interstitial waters are the following:

(1) Evaporation of normal sea water in the lagoon to chlorosities attaining 200 g/l Cl^-. This may have taken place, if the inlets connecting the Mediterranean to the lagoon were blocked at a certain period.

(2) Mixing of the concentrated brines with Mediterranean Sea water which entered the lagoon at a later stage. The mixing is probably responsible for the observed gradient in salinity in the cores.

(3) Interaction between the resulting brines (80–90 g/l Cl^-) and the clayey bottom sediments, resulting in the peculiar ionic ratios observed in the interstitial water. The possible processes leading to these ionic ratios were discussed by Levy (1974b).

Lagoon Sediments

According to the types of sediments found in the different lagoon portions, two main types of environments of deposition were recognized (Levy, 1974a):

(1) An environment influenced by sea water, consisting of sand sediments, which contain marine diatoms and pelecypods but no gypsum crystals. These sediments are prevalent in the western arm (Fig. 11-3).

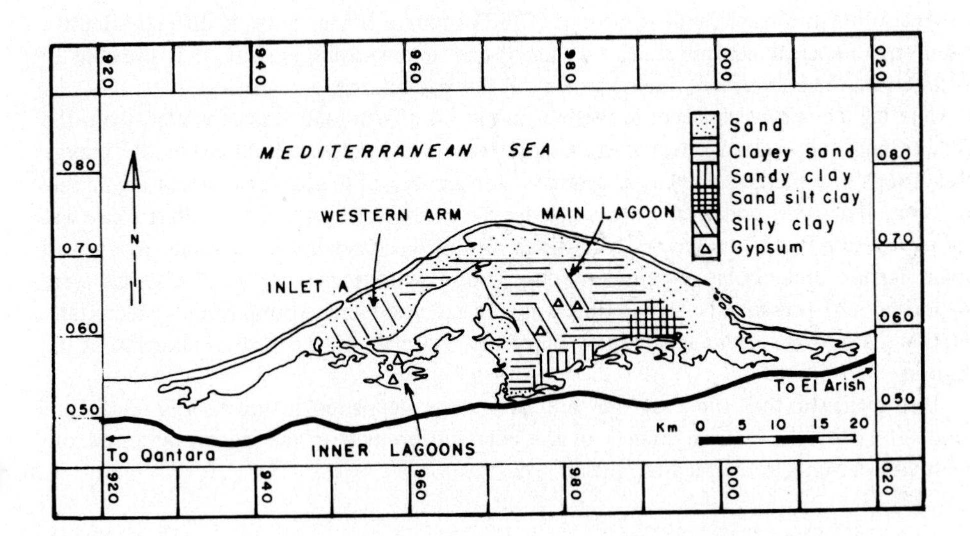

Fig. 11-3. Distribution of various types of sediments in the Bardawil Lagoon

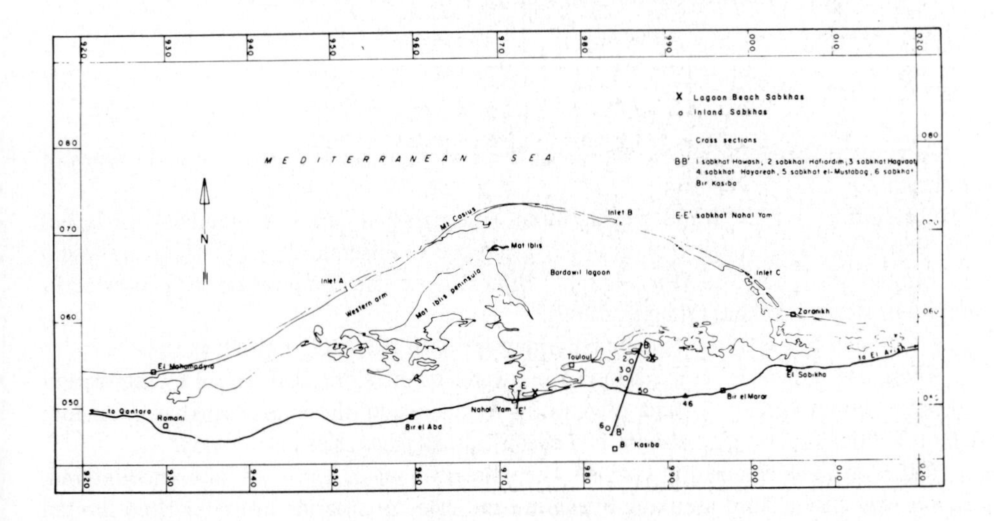

Fig. 11-4. Changes in some chemical parameters

(2) A lagoonal or early evaporitic environment where clayey sediments (with more than 20% clay fraction) with some gypsum crystals prevail. No marine diatoms or pelecypods occur in these sediments. These sediments prevail in the main lagoon. In the inner lagoons gypsum is abundant (Fig. 11-3).

According to the spatial distributions of the various sediment types at different depths, their mineralogical composition, and the types of organisms present, it is possible to reconstruct the conditions which prevailed in the lagoon area.

The lagoon sediments were deposited on an old quartz sand area separated from the Mediterranean by a sand barrier. In the earliest periods represented, relatively high energy conditions prevailed, as shown by sediments composed of quartz sand containing marine diatoms and pelecypods. At later periods, the connection with the Mediterranean was more restricted and low energy conditions prevailed, especially in the inner lagoon and main lagoon area. Clayey sediments containing gypsum and lagoonal diatoms were deposited. At present, the lagoon waters are diluted again and gypsum cannot precipitate. Marine diatoms are found again in the sediments, especially in the northern portion of the lagoon.

It is believed that most of the sediments were deposited in the nearby Tina Gulf (Fig. 11-1) by the Pelusiac branch of the Nile and brought to the lagoon by long shore west east currents in the Mediterranean.

THE BARDAWIL SABKHAS

These may be classified into two groups, depending on their present connection with the Bardawil lagoon waters:

(a) Marine Beach sabkhas, presently connected to the lagoon,

(b) Inland sabkhas, disconnected from the lagoon at present (Fig. 11-4).

The sabkha's sedimentology and mineralogy were described by Levy (1977b).

Geochemistry of Brines

Analysis of brines from the Northern Sinai sabkhas showed that two main types of brines occur:

(1) Normal marine brines, having ionic concentration ratios as predicted for brines developed merely through evaporation of sea water to different degrees. These are found in lagoon beach sabkhas (Fig. 11-4). Brines of similar composition were previously found in Mexico sabkhas (Phleger, 1969).

(2) Inland brines of the calcium chloride type with $[Ca/(SO_4 + HCO_3)]$ eq > 1.

These are formed by evaporation of sea water to different degrees and by interaction between brines (group 1) and preexisting sediments in the inland sabkhas. Calcium chloride brines were found also in the Persian Gulf sabkhas (Kinsman, 1965).

Brines whose compositions fall between the two major types are also encountered. These may result from a mixing of marine and calcium chloride brines or from limited interaction between normal marine brines and preexisting sediments. Thus, along section

B–B' (Fig. 11-4) marine brines are encountered in the north; brines of transitional composition are found in the center of the section, while in the southern part of the section typical calcium chloride brines occur (Fig. 11-4). Groundwater taken from shallow wells drilled in the sand dunes show that they are diluted calcium chloride brines (Levy, 1977a).

The mechanism leading to the formation of calcium chloride brines in the Bardawil Area is:

(1) Evaporation of normal marine water leads to precipitation of aragonite in the early stages and gypsum at later stages of evaporation ($E^* \geqslant 3.5$). This causes an increase in the molar ratio of magnesium to calcium from a value of 5 for normal sea water to about 118 for sea water evaporated to E = 17.6.

(2) Brines with high Mg/Ca ratios form dolomite on contact with previously precipitated calcium carbonate minerals according to the following reaction:

$$\underset{\text{brine}}{Mg^{++}} + \underset{\text{sediment}}{2(CaCO_3)} = \underset{\text{sediment}}{CaMg(CO_3)_2} + \underset{\text{brine}}{Ca^{++}}$$

The excessive amount of calcium entering the brine lead to the precipitation of calcium sulphate. The first stage in the formation of a calcium chloride brine would be the initiation of additional sulphate precipitation by the calcium ions entering the solution through the exchange reaction, thus the resulting brine will be deficient in sulphate, but not enriched in calcium. This is a chloride brine. Only when most of the sulphate has been precipitated will additional calcium remain in solution, resulting in the formation of calcium chloride brines. Dolomitization has previously been proposed as a mechanism for the formation of calcium chloride brines in sabkha environment by Illing et al. (1965), Butler (1969), Kinsman (1965) and Bush (1970).

The geochemical processes discussed above explain the concentrations of the major elements, Na, Mg, Ca, Cl and SO_4 in the calcium chloride brines. The deficiencies in K and Br may be explained as follows:

(a) Deficiencies in K which exceed 90% may be explained by the uptake of potassium by clay minerals or phytoplankton. The uptake of K by clay minerals has been proposed previously by Friedman and Gavish (1970) for interstitial water in transitional environments including lagoonal, deltaic, river estuarine salt marsh and cove environments. Clay minerals are not common in the upper beds of the sabkhas, yet boreholes drilled in the neighboring coastal areas have penetrated clayey sediments (Fink 1969; Zelinger, et al., 1971). Thus, although the uptake of some potassium by clay minerals cannot be excluded, it is not considered to be an important process for the depletion of potassium from the Bardawil brine.

The uptake by phytoplankton was previously proposed by Kinsman (1966) for brines from the Persian Gulf, which show a deficiency of 5–10% in their potassium content. This preferential uptake of potassium by phytoplankton was already shown by Bowen (1966). Potassium was found in concentrations up to 0.15% in Bardawil sabkha sediments

E* = degree of evaporation of the brine compared with normal sea water

containing algal mats. Thus it seems reasonable to attribute the deficiency to the action of phytoplankton.

(b) The brines are also deficient in bromine, but the bromine deficiency is much less accentuated than the potassium deficiency. The extensive occurrence of algae and algal mats in the sabkhas and the well-known fact that bromine is always found in marine plants and animals – for instance as dibromotyrosine in algae and sponges (Rankama and Sahama, 1948) – suggests that bromine is extracted from the calcium chloride brines and assimilated by the algae.

The origin of the calcium chloride brines occurring in the inland sabkhas may be from either normal sea water or groundwater. A mechanism for the formation of calcium chloride brines by evaporation of sea water and dolomitization was presented by Levy (1977a) and has previously been proposed by Kinsman (1965). The occurrence of groundwater of high salinity in coastal areas has been explained as a residual marine brine (White, 1965, Bentor, 1969). The presence of highly saline groundwater in the Gaza Strip was attributed to the same process by Fink (1970).

The possibility of a former invasion of the areas now occupied by dunes and sabkhas by sea water as shown by the occurrence of marine carbonates has been discussed (Levy, 1977b). If the origin of the groundwater is marine, the question of whether the calcium chloride brines were formed from sea water or from groundwater evaporation is not a question of origin, since both are of marine origin. Most likely the calcium chloride brines in the Bardawil region were produced by sea water flooding which left residual marine water masses that were concentrated by evaporation and chemically altered by interaction with calcium carbonate sediments to form the calcium chloride brines. In the area of marine regression covered at present by extensive sand dunes between which the sabkhas occur, groundwater accumulates and flows from south to north, thus flushing out and diluting the calcium chloride brines. The low ratio of rainfall together with the high rates of evaporation do not permit complete flushing of the saline solutions into the lagoon. These solutions are partially evaporated and concentrated in the sabkhas where they form residual calcium chloride brines. Thus, even if at present the calcium chloride brines in the inland sabkhas form by the evaporation of groundwater, the process involves recycling calcium chloride brines of marine origin diluted by rain water, which subsequently flowed to the present sabkhas, where evaporation reconcentrated them to form "recycled" calcium chloride brines.

In Bardawil the low rainfall, the high rate of evaporation, and the restricted environment (because of separation from the lagoon by sand dunes) favor the preservation in inland sabkhas of the calcium chloride brines which were formed in an earlier period.

The Distribution of Evaporite Minerals in Sabkhas

The sediments in the lagoon beach sabkhas (Fig. 11-4) consist primarily of algal mats, gypsum and detrital quartz. In sabkhas receiving open lagoon water and occasionally flooded by groundwater, e.g. sabkhat Nahal Yam, halite occurs only in summer months in small amounts, while in sabkhas receiving highly saline brines from a restricted lagoon environment, e.g. sabkhat Hawash, halite is the predominant evaporitic mineral.

The inland sabkhas are not connected to the lagoon at present; the marine sediments found in them were formed in an earlier cycle when lagoon waters flooded the southern area they occupy. At present these sabkhas are flooded after rainy periods by groundwater flowing from the southern dunes to the lagoon in the north. Sabkhat Hafiordim (Fig. 11-4) is an example of an earlier cycle lagoon beach sabkha which was later disconnected from the lagoon.

In Sabkhat el Mustabag (Fig. 11-4) the effects of flushing by groundwater are pronounced. In the summer, the sabkha is covered by a thin layer of halite. In winter, after rainy periods, the sabkha is flooded and halite is dissolved in most of the area. The rapid rates of evaporation cause halite to recrystallize a few weeks after each flooding. True dolomite was found in this sabkha. In contact with the calcium chloride brines, calcium rich dolomites were also found in other inland sabkhas.

Thus different evaporite mineral assemblages occur in different sabkhas. These assemblages are controlled by two main factors: the degree of restriction and hence the salinity of the body of lagoon water which floods the sabkhas, and the degree of flushing of the sabkha by groundwaters as shown by the degrees of dilution of the brines. Thus, if a lagoon beach sabkha is under the influence of the lagoon water alone, these waters flooding the sabkha occasionally are rapidly evaporated thus resulting in the precipitation of evaporite minerals. In such a case the evaporite minerals precipitated depend on the salinity of the lagoon waters flooding the sabkha. Thus if the flood waters are of low chlorosity (about 50 g/l Cl^-) their further evaporation in the sabkha will lead to the precipitation of gypsum, yet the chlorosity attained in this case will not permit the precipitation of halite. If, on the other hand, the lagoon waters flooding the sabkha are highly saline (about 100 g/l Cl–) as in the case when the sabkha receives its water from a restricted area of the lagoon, their further evaporation in the sabkha will lead to the precipitation of both gypsum and halite. Thus in lagoon beach sabkhas to the south of the open lagoon, gypsum is the prevalent mineral, while in sabkhas receiving their water from a restricted lagoon environment, gypsum will precipitate in the area close to the lagoon and to the south gypsum and halite precipitate together, at the southern extremity of the sabkha halite predominates.

The main effect of the flooding of the inland sabkhas by groundwater flowing from the southern sand dunes will be the dissolution of halite found in the southern portions of the sabkha but as quantities of water flooding the sabkhas are small and evaporation rates are high, groundwaters soon become saturated with respect to halite, and thus are unable to dissolve all of the halite occurring in the central sabkhas. The result of such a partial halite redissolution will be a sabkha in which gypsum predominates in the peripheries while halite together with gypsum occur in the center. Two mechanisms responsible for low concentrations of halite in lagoon sediments therefore can be identified. The absence of halite in the northern portion close to the lagoon is due to the fact that it could not be precipitated from the lagoon waters, while the absence of halite in the southern portions of the sabkha is a result of its redissolution in the flushing groundwaters.

It is suggested that the dolomite occurring in some of the inland sabkhas (Levy 1977b) was formed diagenetically through the interaction between formerly precipitated calcium carbonate and concentrated marine brines rich in magnesium (Levy, 1977a). Weyl (1960) and Deffeyes et al. (1965) also suggested that such normal marine brines might cause dolomitization of calcium carbonates. In this respect some inland sabkhas represent a supratidal evaporitic diagenesis in carbonate host sediments and are similar to the Trucial Coast sabkhas described by Kinsman (1965).

Kinsman (1965) compares arid-zone supratidal diagenesis in carbonate host sediments in the Trucial Coast to the supratidal diagenesis in non-carbonate host sediments at Laguna Oja de Libre, Baja California, Mexico. In the northern Sinai coastal plain sabkhas of both the carbonate and non-carbonate host sediment types occur in close neighborhood and are separated from each other by sand dunes.

CONCLUSION

The spatial changes in salinities of the Bardawil lagoon water as well as the changes in salinities with depth reflect the hydrographic conditions in the lagoon at the time of sampling. These conditions are dictated mainly by the degree of connection of the lagoon with the Mediterranean, the evaporitic conditons prevailing in Northern Sinai at present and the restricted connection between the lagoon and the Mediterranean lead to the high salinities (three times the normal concentration of sea water) encountered in the lagoon waters. The water of the inner lagoons shows a positive anomaly in calcium and sulfate due to gypsum dissolution from bottom sediments. Interstitial water was formed by evaporation of lagoon water to about 200 g/l Cl^-, interaction between this water and the bottom sediments, and subsequent dilution by lagoon water.

The types of sediments found in the lagoon bottom reflect the different hydrologic regimes in different periods. Thus, in the lagoon, the sand sediments occurring in the western arm suggest that they were deposited under relatively high energy conditions, while the clayey sediments in the main lagoon indicate deposition under low energy conditions. The occurrence of gypsum in some of the cores shows that hypersaline conditions were prevalent in the lagoon.

In the Sinai Mediterranean coastal plain two types of brines developed in close neighborhood. Normal marine brines result from mere sea water evaporation and occur in sabkhas in which the host sediments are non-carbonates or coarse grained carbonates not suitable for dolomitization; calcium chloride brines are found in inland sabkhas and result from interaction between the normal marine brines and host fine-grained calcium carbonates, thus leading to the formation of diagenetic dolomite. Brines of composition falling between these two major types also occur. These may result from mixing of the above two types or from limited interaction between normal marine brines and preexisting carbonate sediments. The groundwaters in the area, which have a calcium chloride composition, were formed through the dilution of preexisting calcium chloride brines by rain water.

In a supratidal evaporitic environment, as in the Bardawil coastal plain, two major factors affect the evaporite mineral association to be formed: the salinity of the lagoon water which floods the supratidal areas and the extent of flushing of these areas by groundwater. The result of superimposing these two processes within sabkha areas can lead to the formation of adjacent sabkhas containing different evaporitic mineral assemblages; on the other hand sabkhas showing the same mineral composition may be a result

Chapter 12

THE PHYSICAL STRUCTURE OF THE DEAD SEA WATER COLUMN – 1975–1977

ILANA STEINHORN AND GAD ASSAF[1]

Geoscience Group, Isotope Department, Weizmann Institute of Science, Rehovot, Israel

SYNOPSIS

The present study reports ongoing measurements of physical parameters of the Dead Sea initiated in October 1975. The results show that the permanent stratification of the Dead Sea, which was clearly observed until 1960, has undergone significant changes. Measurements of tritium content and water density indicate that the depth of the pycnocline decreased from 40–50 m in 1960 to 100–120 m in June 1976 and 130–150 m in March 1977. The stratification which until 1960 was maintained by a density gradient of 2.5% is now less than that value, indicating that the Dead Sea is approaching overturn.

INTRODUCTION

Until a few years ago the Dead Sea (Fig. 12-1) was considered to be a classic example of a permanently stratified water body. All density measurements of Dead Sea water showed differences of several percent between the deep waters and the upper layer. According to Neev and Emery (1967), the surface density in 1960 was 1.20 g/cm^{-3} (salinity of 290 g/lit) and the density of the deep waters was 1.23 g/cm^{-3} (salinity of 325 g/lit). The permanent pycnocline reached at most a depth of 40–50 m.

In the last years the situation has changed. Dead Sea level has dropped considerably (Fig. 12-2) resulting in decreasing the volume of the shallow southern basin and increasing the density of the upper 50 m or so of the lake. The salinity of surface waters during the 70's is close to the salinity of the deep waters with both containing 340 g/lit (Beyth, 1979). The stratification of the Dead Sea which in the 60's was maintained by a sharp salinity gradient is not as clear cut as before (Fig. 12-3).

Since 1975, a set of physical measurements has been performed in several localities in the Dead Sea. This preliminary report presents profiles of temperature, density and tritium content taken in the deepest region of the Dead Sea during 1975–1977.

RESULTS OF THE MEASUREMENTS

Salinity and density. The increase in salinity of the upper water mass of the Dead Sea led to a marked decrease in chemical differentiation between the two water masses. Therefore, changes in chemical composition of the water are not a reliable measure of the stratification of the lake. Since the analytical noise of total dissolved salts measurements is of the same order of magnitude as the difference between the water masses it cannot be

[1] Present address: Solmat Systems Inc., Jerusalem.

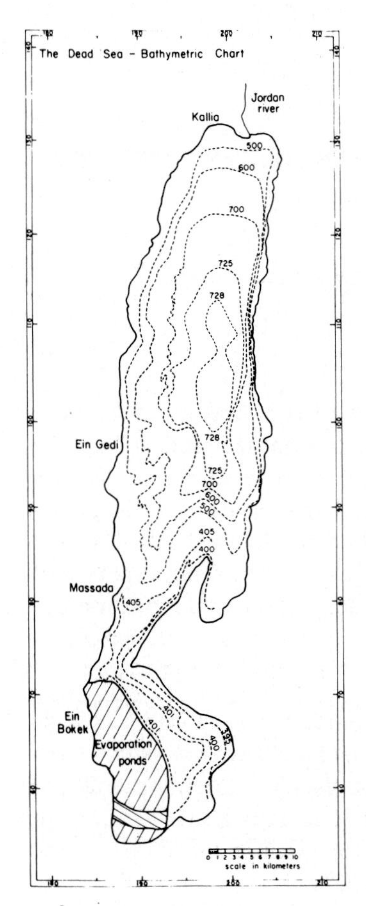

Fig. 12-1. The Dead Sea (after Neev and Hall, 1976).

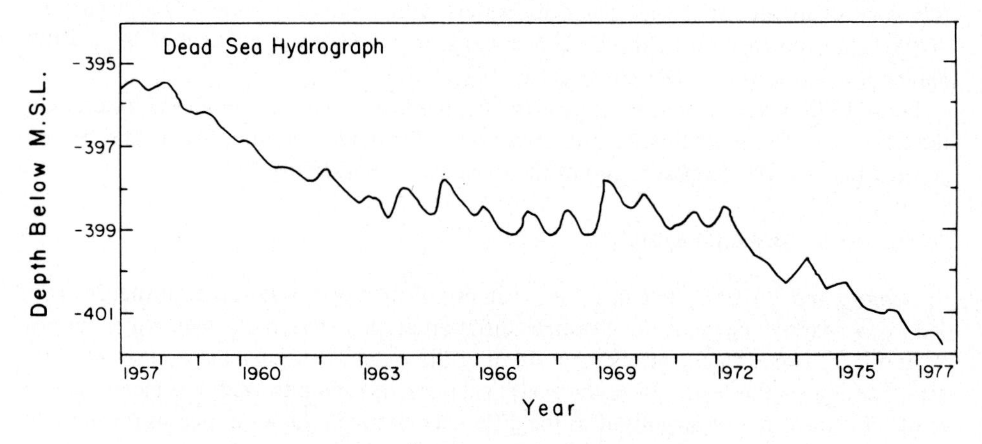

Fig. 12-2. Changes in Dead Sea level 1957–1977 (measurements by the Dead Sea Works Ltd.).

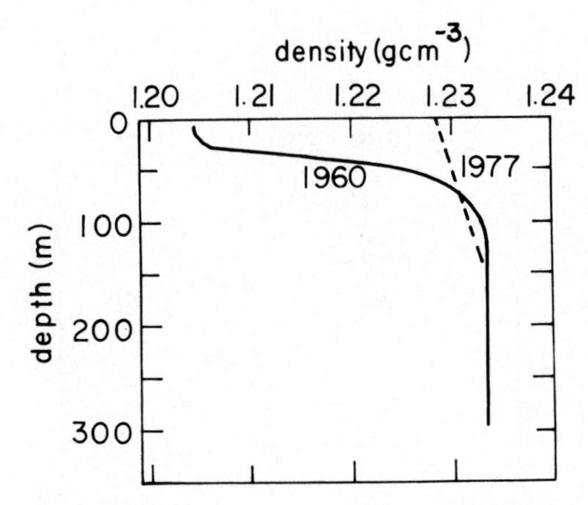

Fig. 12-3. Schematic representation of the changes in the stratification of the Dead Sea.

used to accurately identify different components of the water column. Furthermore, the difference in density between the seasons is approaching the observed difference between the upper and lower water masses. Therefore, a considerable effort was devoted to obtain highly accurate density measurements. The method which was found to be the most suitable was the classical technique of weighing a known volume of water in a pycnometer under closely controlled temperature. Starting with the March 77 measurements the overall error in density measurement is estimated at 0.0003 gm/cm^3 at most.

The salinity of Dead Sea water is operationally defined as the density at 25°C. In order to estimate the *in situ* density, the thermal expansion coefficient of Dead Sea water has to be known. Experimental work gave a value of:

$$\frac{\partial \rho}{\partial T} = -0.00043 \text{ g cm}^{-3}\,{}^{\circ}\text{C}^{-1}$$

$$\alpha_T = \frac{1}{\rho}\frac{\partial \rho}{\partial T} \approx -350 \times 10^{-6} \text{ cm}^{-6}\,{}^{\circ}\text{C}^{-1}$$

Results of salinity measurements for the water column are given in Figs. 12-4–5. In October 1975 the halocline depth was between 70 to 100 m. In January and in June 1976 the halocline reached a depth of 100–120 m. In March 1977 the halocline deepened to 130–150 m depth, but a small step in the salinity–depth profile was seen at 100 m depth, corresponding to the halocline depth nine months earlier. In July 1977 the halocline did not deepen any further. During the same period, increased evaporation accompanied by high summer temperatures, caused an overall increase in the salinity of the upper water mass. The salinity data clearly show the decrease in salinity difference between the two water masses. This has been caused by the lowering of the pycnocline and the increasing water loss from the lake.

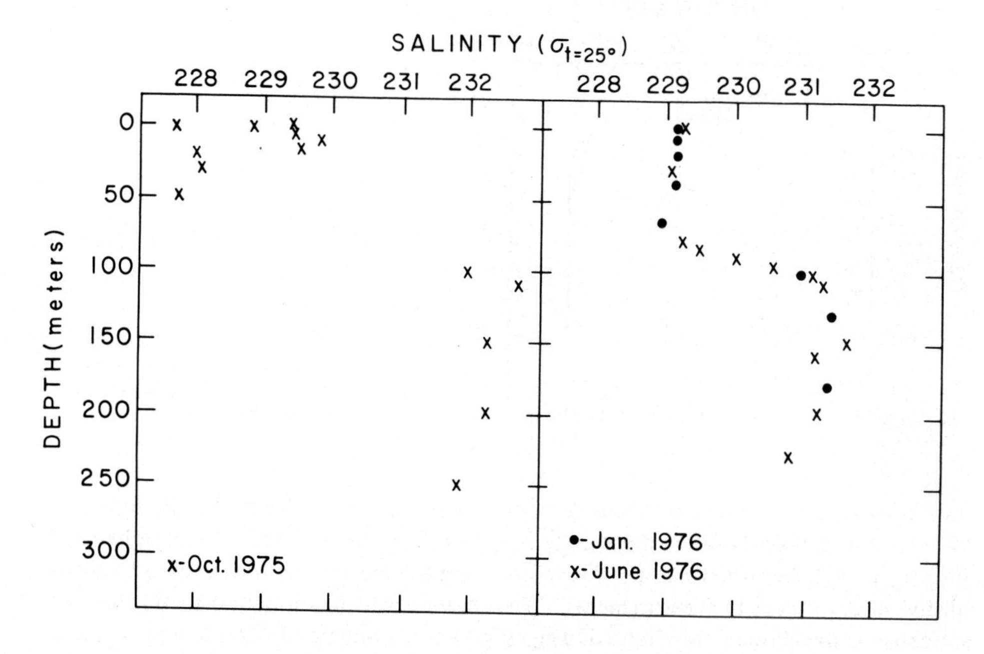

Fig. 12-4. Salinity profiles of the Dead Sea water column, 1975–1976.

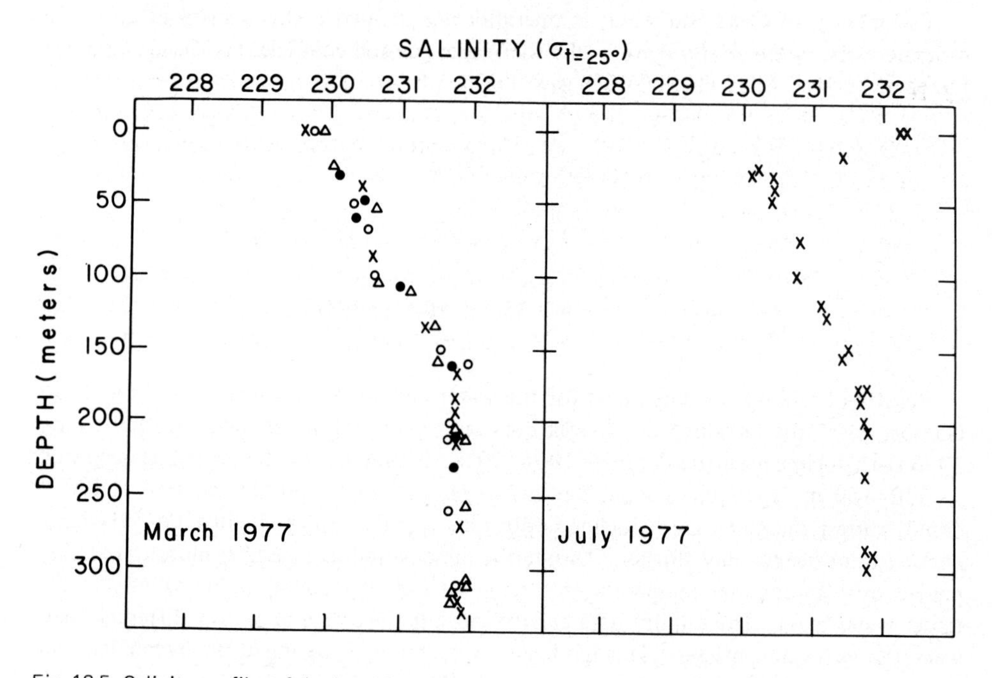

Fig. 12-5. Salinity profiles of the Dead Sea, 1977.

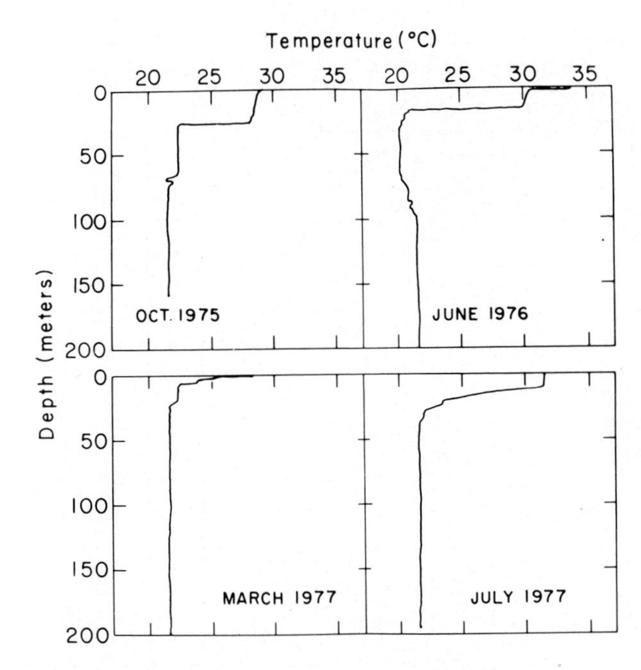

Fig. 12-6. Temperature profiles of the Dead Sea, 1975–1977.

Temperature. The temperatures of the Dead Sea water column were measured using a mechanical bathythermograph. Typical profiles, taken in the deepest part of the lake, are presented in Fig. 12-6. The surface temperature of the Dead Sea alternates between approximately 18° in winter and 37° in summer. However, these changes do not affect the water below the depth of the permanent pycnocline. The deep water has a permanent temperature of 21.3° approximately, except for a small rise in temperature due to adiabatic heating (Ben-Avraham et al., 1977).

As long as the stratification was maintained by a sharp salinity gradient, seasonal changes in temperature did not significantly affect the density differences between the upper and lower parts of the Dead Sea water column. This situation has changed in the last few years as the seasonal changes in the temperature has a considerable effect on the stability of the lake.

The in situ density. Profiles of the density, based on the salinity and temperature measurements and Eq. [1], are presented in Fig. 12-7. The density of the deep water for all profiles was estimated using the salinity data of March 1977.

The data indicate the permanent pycnocline of the Dead Sea during 1975–1977 to be at the same depth as the halocline. The permanent stratification of the water column is then held by salinity differences between the upper and the lower water masses.

In summer, a seasonal pycnocline exists in a depth of not more than 25 m due to effects of heating. At that time, a surface layer which is high in salinity is formed due to evaporation (Fig. 12-5). This layer is stabilized due to high temperatures (up to 37° at the

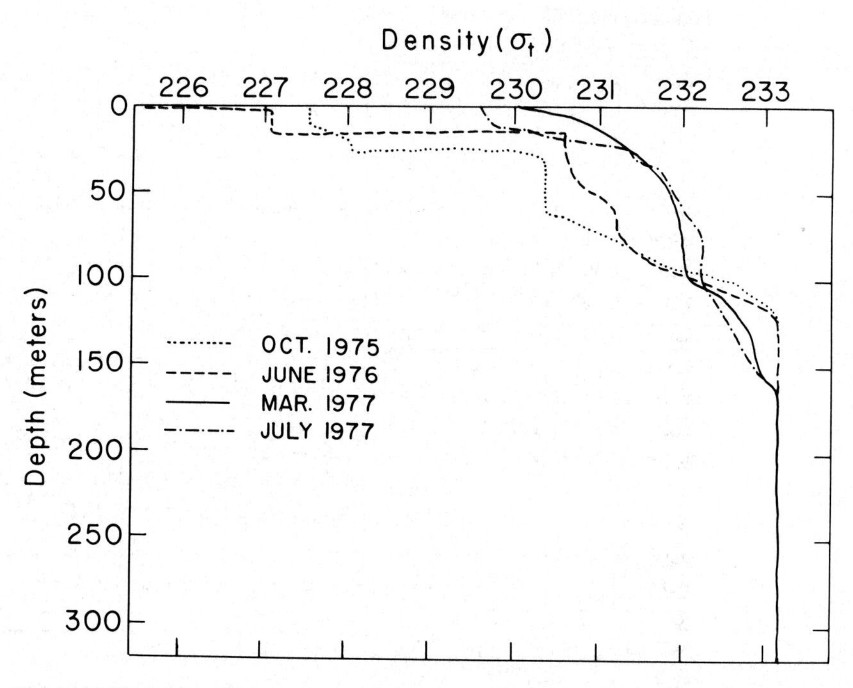

Fig. 12-7. The *in situ* density of Dead Sea brines, 1975—1977.

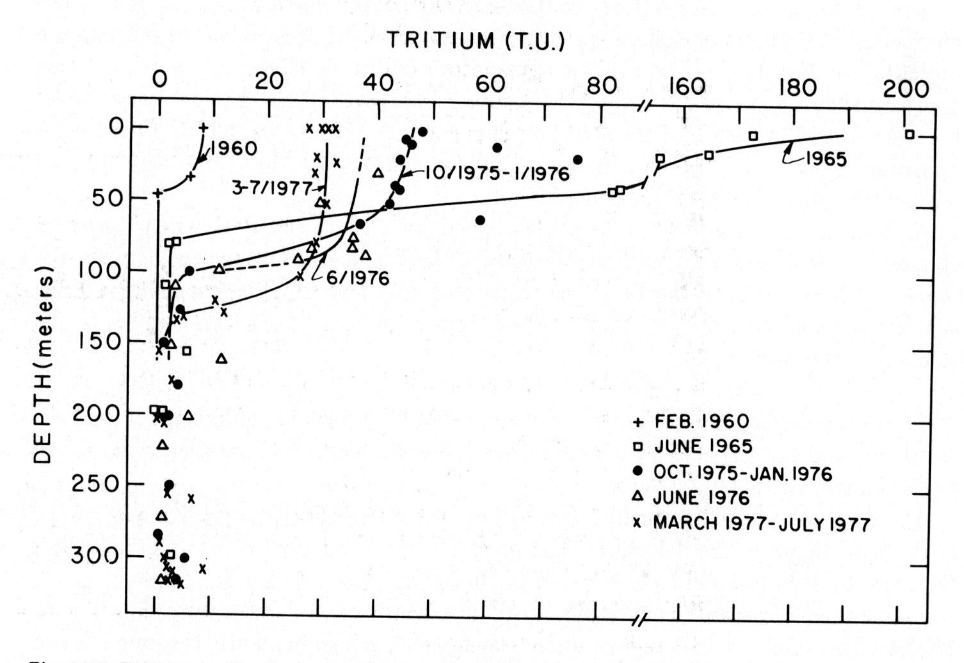

Fig. 12-8. Tritium profiles in the Dead Sea water column, 1960—1977.

surface). In winter, the reverse situation exists; as a result of flood waters, the surface layer is relatively cold and low in salinity (Fig. 12-5, 6).

One can see that the density of the Dead Sea is increasing from year to year. This is due to: a) increase in salinity caused by excess of evaporation oven inflow; b) mixing with the more saline deep waters.

Tritium Profiles. The tritium content of the Dead Sea waters can also be used as an indicator of the vertical mixing rate in the Dead Sea (Fig. 12-8).

In 1960, the tritium content of the surface waters was found to be 8.5 T.U. (Neev and Emery, 1967). During 1965, the surface waters of the Dead Sea reached the maximal value of 200 T.U. (Nissenbaum, 1969). Since then, the value has been reduced to 50 T.U. mainly as a result of the changes in the tritium levels in the atmosphere.

The deep water of the Dead Sea is low in tritium, indicating that the penetration of surface waters is rather slow. Yet, it is not clear whether the scatter of a few T.U. in the data of the bottom layers is due to contamination during sampling, or to a real effect of influx of surface waters.

However, the deepening process of the pycnocline of the Dead Sea can be seen quite clearly from the tritium data: In 1960 the upper layer of the Dead Sea reached the depth of 50 m at most, while in October 1975 the surface waters penetrated to a depth of about 70–100 m, and in March 1977 reached about 130–150 m.

The rapid reduction in the tritium content of the upper layer, from 45 T.U. in October 1975 down to about 30 T.U. in March 1977, seems to be mainly due to mixing with the deep waters of low tritium content.

DISCUSSION

The results of density and tritium measurements show that the stratification of the Dead Sea has undergone rapid changes. The permanent pycnocline, which has been clearly observed until a few years ago, is now under destruction. What is the reason for this process?

The answer is not clear as yet, but that situation is undoubtedly connected to the lowering of Dead Sea level in the last few years (Fig. 12-2). The density of the upper layer has increased from about 1.20 g/cm^3 in 1960 to 1.229 mainly as a result of excess evaporation of about 4 m. These changes in level have also affected the surface area of the southern basin, which in 1960 had the mean depth of 3 m only (see also Fig. 12-1).

Since 1966 the southern basin has been partially closed by the Dead Sea Works Ltd. and used as evaporation ponds. Dead Sea water has been pumped into the pond, and the solutions remaining after extracting the usable minerals are discharged into the northern basin as concentrated end brines.

The changes in the southern part of the Dead Sea had direct effect on the flow through the Lisan straits (Beyth and Assaf, 1977). Recently, Assaf and Nissenbaum showed, that the long-range of the pycnocline of the Dead Sea during the last 150 years was due to internal thermohaline circulation. The erosion of the pycnocline by wind mixing was counterbalanced by introduction of dense water from the southern basin to an intermediate depth within the upper water mass of the Dead Sea (Assaf and Nissenbaum, 1977). The changes in the morphology of the Dead Sea, then, seem to have a great effect on the mixing process. However, not much is known yet about the way in which the denser water from the south, including the end brines, are being mixed in the northern basin nor what the penetration depth of surface water is, nor the contribution of subaquatic springs to the composition of the deep water.

During the two years of measurements 1975–1977, and especially since the disconnecting of the Southern basin in summer 1976, the Dead Sea reached horizontal homogeneity. As of the lowering of the level of the Dead Sea continues, evaporation and cooling are now of great importance in the mixing process of the Dead Sea. In summer, the mixed layer of the Dead Sea reaches the depth of 25 m. Evaporation increases the salinity of the upper layer but it still remains highly buoyant due to high temperatures at the surface. Autumn cooling of the water results in homogeneity of the salinity of the upper layers. In winter 1977, the difference in salinity between the upper and the lower water parts of the water column was small enough for the permanent pycnocline to undergo further deepening as a result of this cooling. In spring 1977 the upper layers were stabilized enough to prevent further mixing by mechanical stirring of the wind (Fig. 12-5).

It seems now that the Dead Sea is in a critical state of its development. If the lowering of the level continues one may expect an overturn of the lake within a short period. The last measurements made in November 1977 show that the salinity differences between the water masses is not more than 0.5‰. Mass balance calculations, considering the thermal expansion of the Dead Sea water, show that to avoid further mixing in winter 1978 the amount of flood waters reaching the Sea has to exceed the value of 10–15 cm.

This situation seems to be a rare and interesting one in the history of the Dead Sea, since if overturn is going to take place shortly, it would probably be the first time in at least one hundred years.

ACKNOWLEDGMENTS

This research is a part of a cooperative study with the Geological Survey of Israel. We wish to thank Drs. M. Beyth and D. Neev, Mr. D. Argas and Mr. R. Madmon for their support during all stages of the study.

Logistic support was obtained through the courtesy of the Dead Sea works Ltd., and Dr. Z. Ben-Avraham. We thank Dr. B. Gershgorn of the Dead Sea Works who provided relevant information. Mrs. E. Negrano performed the density measurements, and Mr. I. Carmi analysed the tritium samples.

This research was supported by a grant from the United States–Israel National Science Foundation (BSF), Jerusalem.

Partial support was also received from the Ministry of Commerce and Industry.

REFERENCES

Assaf, G., and A. Nissenbaum, 1977. The evolution of the upper water mass of the Dead Sea, 1919–1976. Proc. Int. Conf. Terminal Lakes, (Ed. D. G. Greer), Utah Water Research Laboratory, Utah State University, Legan, 61–72.

Ben-Avraham, Z., R. Hanel and G. Assaf, 1977. The thermal structure of the Dead Sea. Limn. and Ocean., 22, 1076–1978.

Beyth, M., 1979. The present status of Dead Sea Brines. This volume, pp. 000.

Beyth, M. and G. Assaf, 1977. Mixing controlled by straits in the Dead Sea. Geological Survey of Israel, report M.G./8/77, 13 pp.

Nissenbaum, A., 1969. The geochemistry of the Jordan River – Dead Sea system. Unpublished Ph.D. thesis, University of California, Los Angeles, California, 289 pp.

Neev, D. and K. O. Emery, 1967. The Dead Sea. Israel Geological Survey Bull. No. 41, 147 pp.

Neev, D. and J. Hall, 1976. The Dead Sea Geophysical Survey 19 July – 1 August 1974. Final report No. 2, Geological Survey of Israel, MGD 6176.

Chapter 13

RECENT EVOLUTION AND PRESENT STAGE OF DEAD SEA BRINES

MICHAEL BEYTH
Marine Geology Division, Geological Survey of Israel, Jerusalem

INTRODUCTION

Neev and Emery (1967) describe the lake's water structure as follows: an Upper Water Mass reaching down to 40 m depth, a "fossil" Lower Water Mass from 100 m downward (to 330 m) and an Intermediate Water Mass in the 40–100 m interval. It is evident today that this stratification is gradually disappearing. Assaf and Nissenbaum (1977) described it as a permanent picnocline with extremely long range stability against wind mixing, due to the internal circulation between Southern and Northern basins. Lerman (1970) pointed out that the Lower Water Mass already had a degree of halite saturation close to

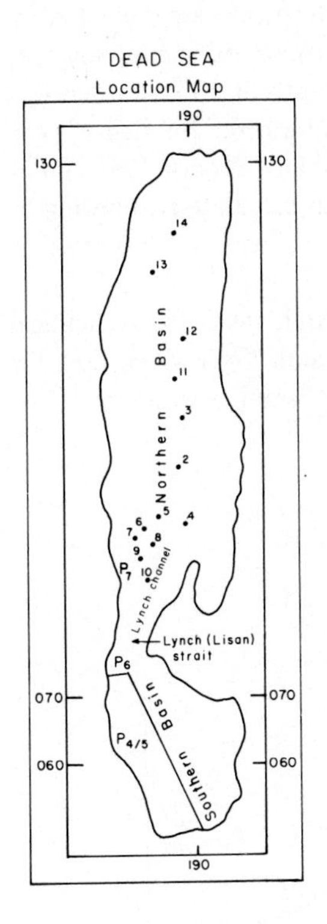

Fig. 13-1. Dead Sea – Location Map. ■Station of Cruise. 27–30 March '77 (approximate location).

1.0 at the beginning of the Sixties, indicating that it would be in equilibrium with halite crystals. He thought that this was confirmed by Neev and Emery's description of the lake's bottom sediments. Lartet (1867) commented that M. Vignes brought up numerous cubical crystals of salt and some crystals of gypsum in greyish blue clay between Ain Ghuweir and Wadi Zerka Main.

Actual precipitation of halite is known from pan 4/5 (Fig. 13-1),where it is deposited from brines of 1.240 g/l density (Adler, 1975) conforming to halite supersaturation degree (D.S.H.) of 1.15–1.21 (known as the halite point).

The drop in the overall lake level is from –397 m in the early Sixties to –402 m in September 1977 (Fig. 13-1,2) (Klein, 1961, and after Dead Sea Works Ltd.). At this level the Southern Basin is almost dry (Fig. 13-1). The areas of pan 4/5, the Southern Basin and the Northern Basin, as shown in Fig. 13-1 are 80 sq km, 110 sq km and 750 sq km respectively (measured by planimeter on 1:50,000 scale map).

As a result of the work carried out in the Southern Basin of the Dead Sea (Beyth, 1976) it is known that low Na/Cl weight ratios in brines (<0.16) indicated the start of halite precipitation (Fig. 13-3) same as low Cl/Br weight ratios (<40) (Zak, 1967).

In order to monitor the changes in lake composition, water samples were taken by Nansen bottles and analyzed for major elements along several profiles in Northern Basin, between August 1974 and March 1977. In addition, bottom sediments were collected by a modified Petterson Grab and analyzed for minerals. A bathythermograph survey was also conducted (Steinhorn, this volume) down to a maximum depth of 200 m which was the instrumental limit. Average composition and degree of saturation for halite were computed from new analyses and from data taken from Neev and Emery (1967) and Nissenbaum (1969). The degree of halite saturation (D.S.H.) was calculated according to the method of Lerman (1967, 1970) and Starinsky (1974), which gives a ±5% error as compared to direct measurements.

The field work was carried out from a Dead Sea Works research vessel. The chemical analyses were performed at the Dead Sea Works laboratories and X-ray difraction for mineral identification was carried out at the Geological Survey of Israel, Jerusalem.

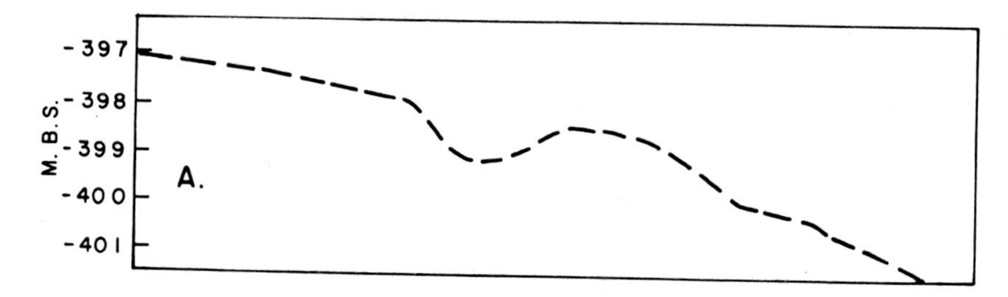

Fig. 13-2. Changes in Dead sea level. (After Dead Sea Works.)

RESULTS

The results of sampling and analyses of several periods are presented in the graphs and tables given below:

TABLE 13-1.

Average composition (g/l) of Northern Basin water masses during 1959-60 (according to Neev and Emery, 1967 and Starinsky, 1974)

	Mg^{++}	Ca^{++}	Na^{+}	K^{+}	Cl^{-}	Br^{-}	T.D.S.	Na^{+}/Cl^{-}	D.S.H.
Upper Water Mass	36.15	16.38	38.51	6.50	196.94	4.60	299.89	0.196	0.65
Intermediate W.M.	40.57	16.60	38.47	7.15	210.67	5.15	319.32	0.183	0.84
Lower W.M.	42.43	17.18	39.70	7.59	219.25	5.27	332.06	0.181	0.99

TABLE 13-2.

Average composition (g/l) of Northern Basin water masses during 1963-65 (calculated according to Nissenbaum, 1969)

	Mg^{++}	Ca^{++}	Na^{+}	K^{+}	Cl^{-}	Br^{-}	T.D.S.	Na^{+}/Cl^{-}	D.S.H.
Upper Water Mass (N=7)*	38.79	15.30	37.9	6.80	201.5	4.6	305.6	0.198	0.71
Intermediate W.M. (N=7)	42.7	16.4	39.2	7.52	216.5	5.0	328.5	0.181	0.95
Lower W.M. (N=10)	44.4	16.8	38.1	7.85	220.8	5.2	333.2	0.173	1.00

(N = Number of samples)

TABLE 13-3.

Upper Water Mass, Southern End of Northern Basin
Average composition (g/l) of Dead Sea Water groups which were determined by discriminant

	Mg^{++}	Ca^{++}	Na^{+}	K^{+}	Cl^{-}	Br^{-}	T.D.S.	Na^{+}/Cl^{-}	D.S.H.
August 1974 (N=31)	41.6	16.1	39.0	7.20	214.0	5.0	323.7	0.182	0.89
July 1975 (N=45)	43.1	17.1	40.4	7.46	222.7	5.2	336.0	0.181	1.06
June 1976 (N=31)	43.4	17.1	40.4	7.69	223.6	5.2	338.2	0.181	1.07

TABLE 13-4.

Composition (g/l) of Northern Basin Water masses during October 1975 – January 1976 (Fig. 3-I)

	Mg^{++}	Ca^{++}	Na^{+}	K^{+}	Cl^{-}	Br^{-}	T.D.S.	Na^{+}/Cl^{-}
Upper Water Mass (N=9)	43.0	16.7	41.0	7.5	∿224.	5.15	335.0	0.183
Intermediate W.M. (N=5)	43.0	16.7	40.5	7.5		5.15	337.0	
Lower W.M. (N=8)	44.0	17.0	41.5	7.82	∿227.	5.3	341.0	0.183

TABLE 13-5.

Composition (g/l) of Northern Basin brines (sampled June 25, 1976) along a profile at coord. 92119/196408

Depth (m)	Mg^{++}	Ca^{++}	Na^{+}	K^{+}	Cl^{-}	Br^{-}	T.D.S.	Na^{+}/Cl^{-}	D.S.H.
30	42.1	16.8	40.9	7.59	219.9	5.2	332.2	0.186	1.02
50	42.4	16.5	41.4	7.59	221.0	5.2	334.8	0.187	1.05
80	42.4	17.0	41.5	7.59	222.1	5.3	336.6	0.187	1.06
102	44.2	16.3	39.1	7.59	222.4	4.8	335.2	0.176	1.03
160	43.9	16.5	39.0	7.59	221.7	4.8	334.2	0.176	1.02
220	43.8	16.8	39.1	7.65	222.1	5.0	335.1	0.176	1.02
270	43.2	17.3	41.6	7.79	225.3	5.3	341.2	0.185	1.12
310	43.6	17.0	41.1	7.79	225.1	5.3	340.6	0.183	1.11
316	43.5	17.0	41.2	7.85	225.0	5.3	340.6	0.183	1.11

TABLE 13-6.

Average composition (g/l) of Northern Basin brines based on 98 samples collected from 14 profiles between March 27 and March 30, 1977

T.D.S.	Mg^{++}	Ca^{++}	Na^{+}	K^{+}	Cl^{-}	Br^{-}	SO_4^{-}	D.S.H.
339.6	44.0	17.2	40.1	7.65	224.9	5.3	0.45	1.09

The recent evolution of the Dead Sea brines from 1960 to 1976 is shown in Figs.13-3,4,5 and Fig.13-3, where a slight change is observed in Lower Water Mass and a significant one in the Upper Water Mass. The changes in D.S.H. and Na/Cl ratios indicate a clear separation between these two water masses in the early Sixties which disappear gradually from 1975 onwards.

A very detailed sampling (98 samples from 14 stations) was carried out between March 27 and March 30, 1977, when the Dead Sea water level was 401.55 below sea level (Figs. 13-, 2, 3, 7, 8). The results of the analyses (Beyth, 1977b) indicate that apart from surface waters and some technically doubtful samples, the brine's composition varied within fairly restricted limits with a density range of 1.232–1.235 (measured by hydrometer under laboratory conditions with water temperature of 23–20 degrees centigrade).

The thermal profiles (which will be presented separately) show a mixed layer down to approximately 25 m, below which the temperatures stabilize around 21.5°C in all profiles.

Typical gypsum rosettes were found in bottom sediments at stations 5, 6, 8 and 14. Well-crystallized halite was found in bottom sediment at stations 6, 8, 9, 13 and 14 (Plate 13-1). High D.S.H. values occur at stations 7 and 8, and somewhat lower values at 5 and 11. All these stations are located either along the western fault (near the salt diapirs; Neev and Hall, 1975) or along the Lynch Channel where concentrated brines from Southern Basin flow northward.

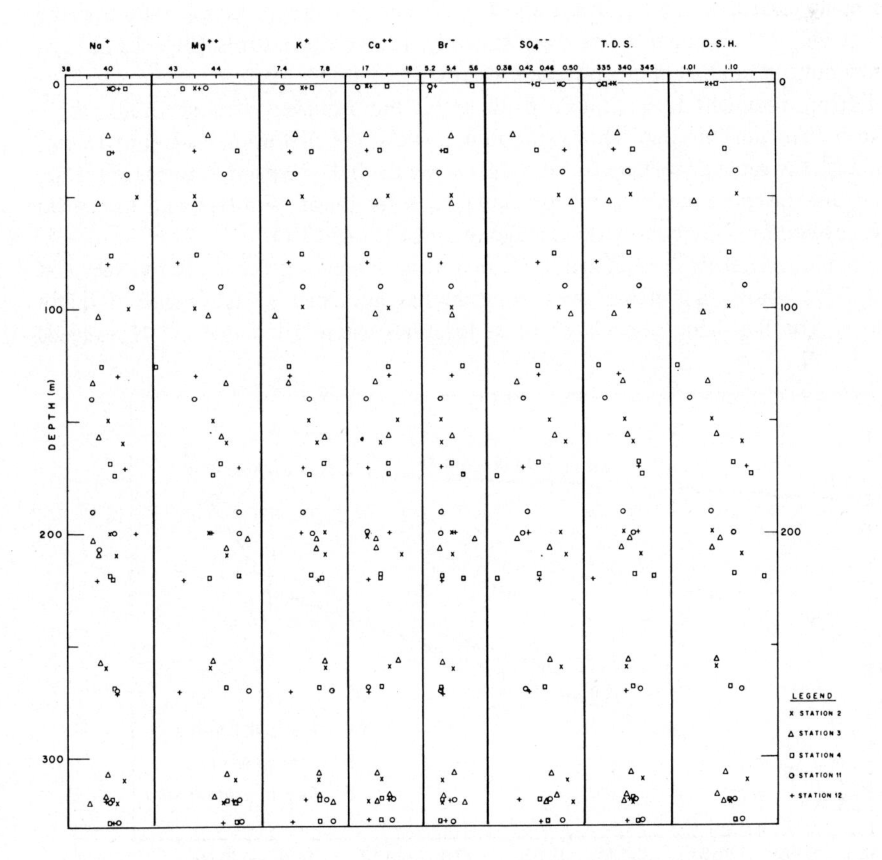

Fig. 13-3. Vertical distribution of major ions, T.D.S. (g/l) and D.S.H. degree of halite saturation in Dead Sea waters, North Basin. Samples taken during March 1977 at Stations 2, 3, 4, 11, 12.

DISCUSSION

The disappearance of the water stratification described by Neev and Emery (1967), with chemical and density boundaries at 40 m, and at 100 m, suggest that an overturn and mixing have recently taken place. This is confirmed by the homogeneous thermal profile. Further evidence for overturn is found in the presence of gypsum in bottom sediments from the North Basin. Neev and Emery (1967) and Nissenbaum (1969) described how the deposition of gypsum is prevented in the Lower Water Mass of the North Basin, as a result of the reduction of sulfate to sulfide in this oxygen-deficient environment. The presence of fresh gypsum in bottom samples may indicate aeration of the lower waters by the overturn. The direct flow of water from Southern Basin into the Lower Water Mass (Beyth, 1977a) may also cause this aeration. This is different from the situation encountered in 1963 (Nissenbaum, 1969), when waters flowing through the Lynch (Lisan) Straits were of much lower salinity (314 g/l). Nissenbaum therefore suggested at that time that brines from the Southern Basin were flowing northward, into the Upper and Intermediate Water Masses of the Northern Basin. On the other hand, it must be made clear that there is still a smell of H_2S in few deeper samples and that the amount of SO_4^{--} decreases from approximately 130 m downwards (Fig. 13-2, 5, 6) which may indicate that some stratification still exists.

The bottom sediment from Station 8 illustrates the sequence of deposition: there appear to be two possible generations of halite crystals; the first are large (up to 15 cm) and rounded, the second smaller (up to 3 cm) but euhedral. Gypsum rosettes cover the younger halite (even though gypsum precipitates at far lower salinites), and the entire sample is covered by a finely crystalline aragonite mud (Plate 13-1).

The presence of freshly crystallized halite in bottom samples, never found by Neev and Emery (1967), show that halite oversaturation has advanced to the point of halite deposition. The big halite crystals occur at localities where calculated D.S.H. exceeds

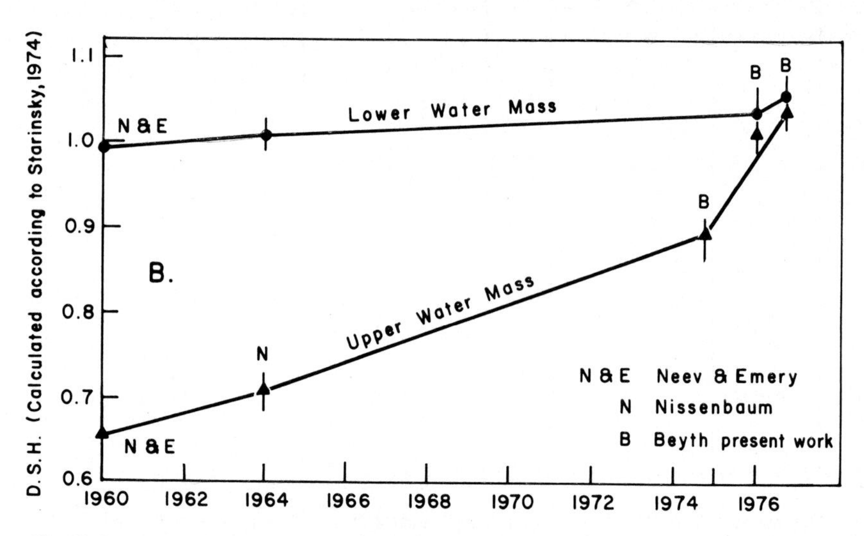

Fig. 13-4. Changes in D.S.H. in Northern Basin, Dead Sea.

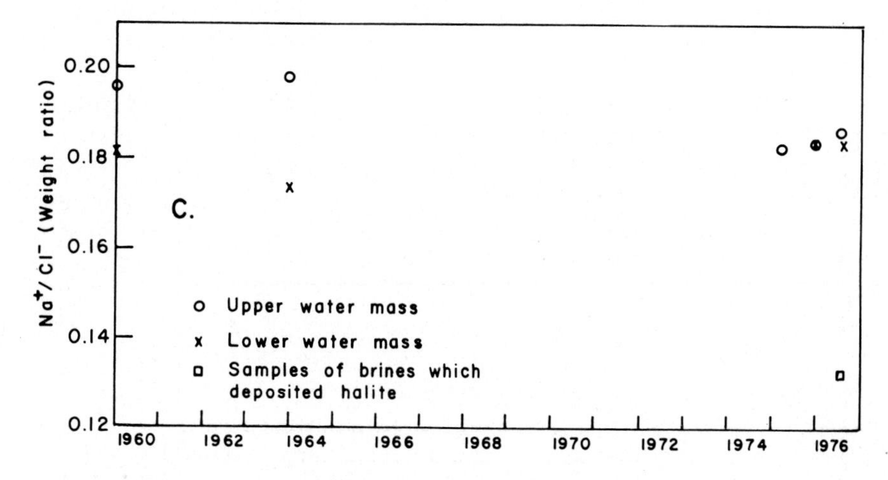

Fig. 13-5. Na^+/Cl^- ratio as an indicator of halite deposition 1960/76.

1.10 and at station 8 it reaches 1.20. On the other hand Na/Cl and Br/Cl values indicate that no massive precipitation of halite has yet occurred north of E Coord. 079000 (Fig. 13-5 and Beyth, 1977b).

SUMMARY

Recent observation of the Dead Sea brines failed to discern the stratification described by Neev and Emery (1967). This was presumed to be due to climatic effects and human activity, which resulted in a drop in water level. Calculations showed that the Dead Sea is oversaturated with respect to halite and indeed halite crystals and gypsum rosettes were found in bottom sediments. This oversaturation significantly affects the formation of "salt reefs" at the locations where End Brines from the evaporation pans mix with the Dead Sea brines (Epstein et al., 1975; Druckman and Beyth, 1977).

ACKNOWLEDGMENTS

B. Zelvianski, J. Epstein and D. Toker of the Dead Sea Works for performing the chemical analyses, permitting the publication of the data and organizing the cruises. J. Sasson the skipper of the vessel. D. Argas and R. Madmon provided technical assistance; A. Peer drafted the figures and I. Perath for editing the manuscript.

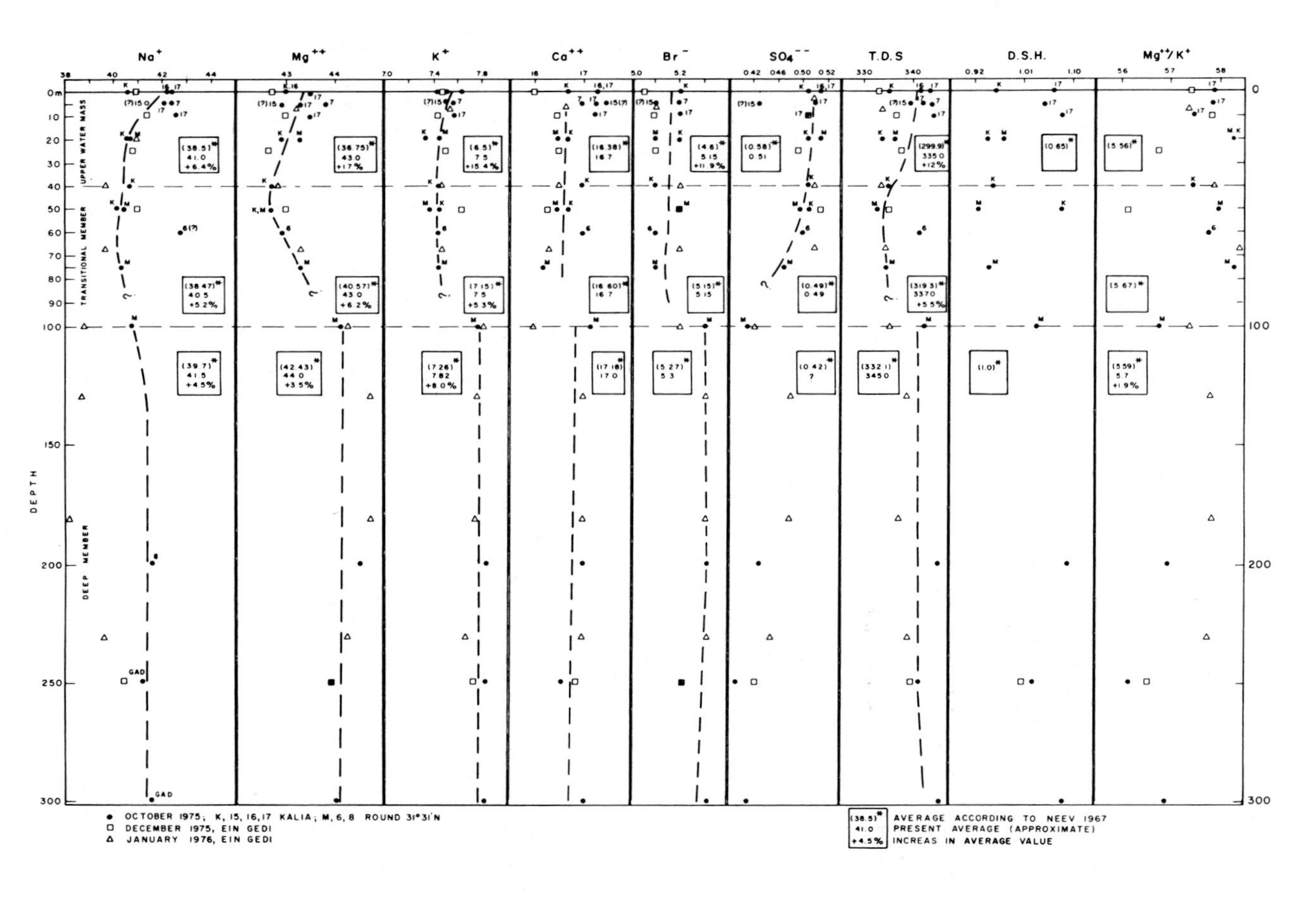

Fig. 13-6. Vertical distribution of major ions, T.D.S. (g/l) and degree of halite saturation in Dead Sea water. Samples taken during 1975 and 1976 cruises.

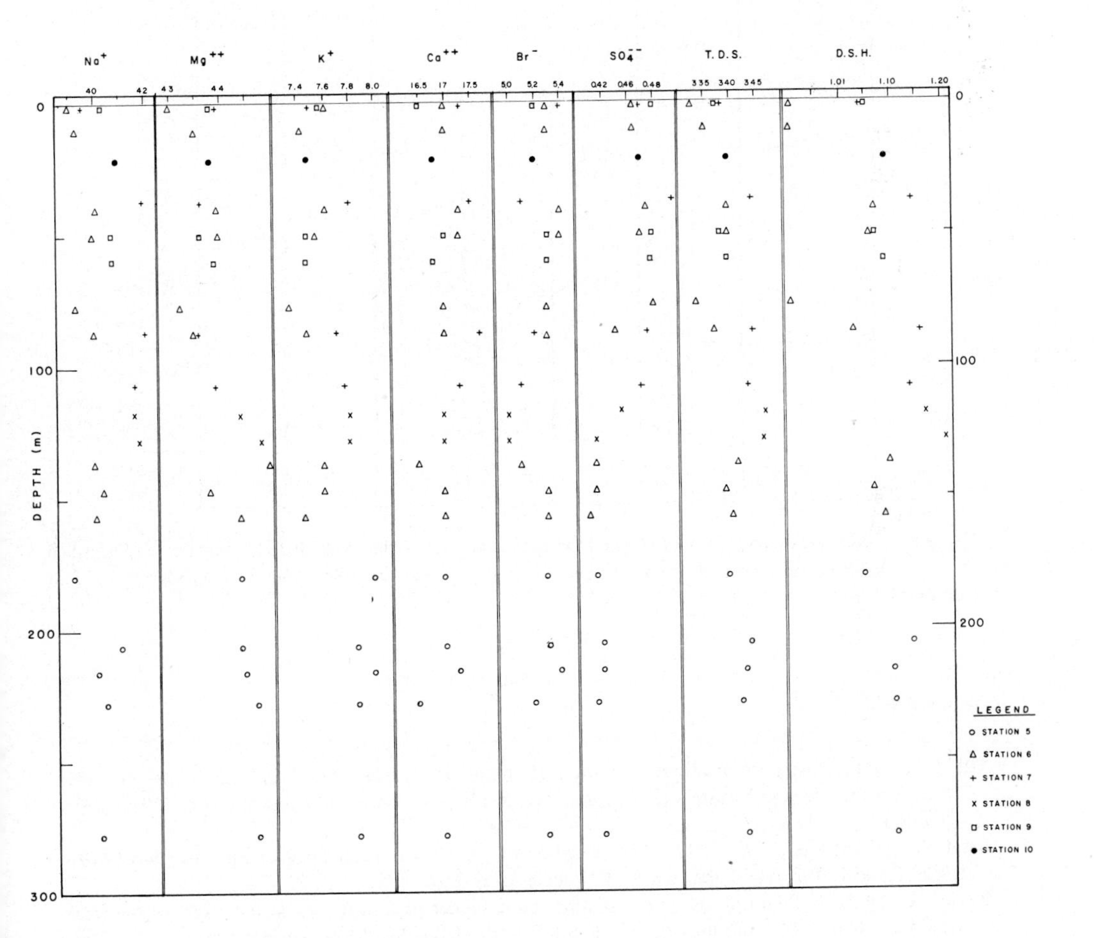

Fig. 13-7. Vertical distribution of major ions, T.D.S. (g/l) and D.S.H. of halite saturation in Dead Sea water. Southern part of North Basin and Lynch Channel. Samples taken during March 1977 at Stations 5, 6, 7, 8, 9, 10.

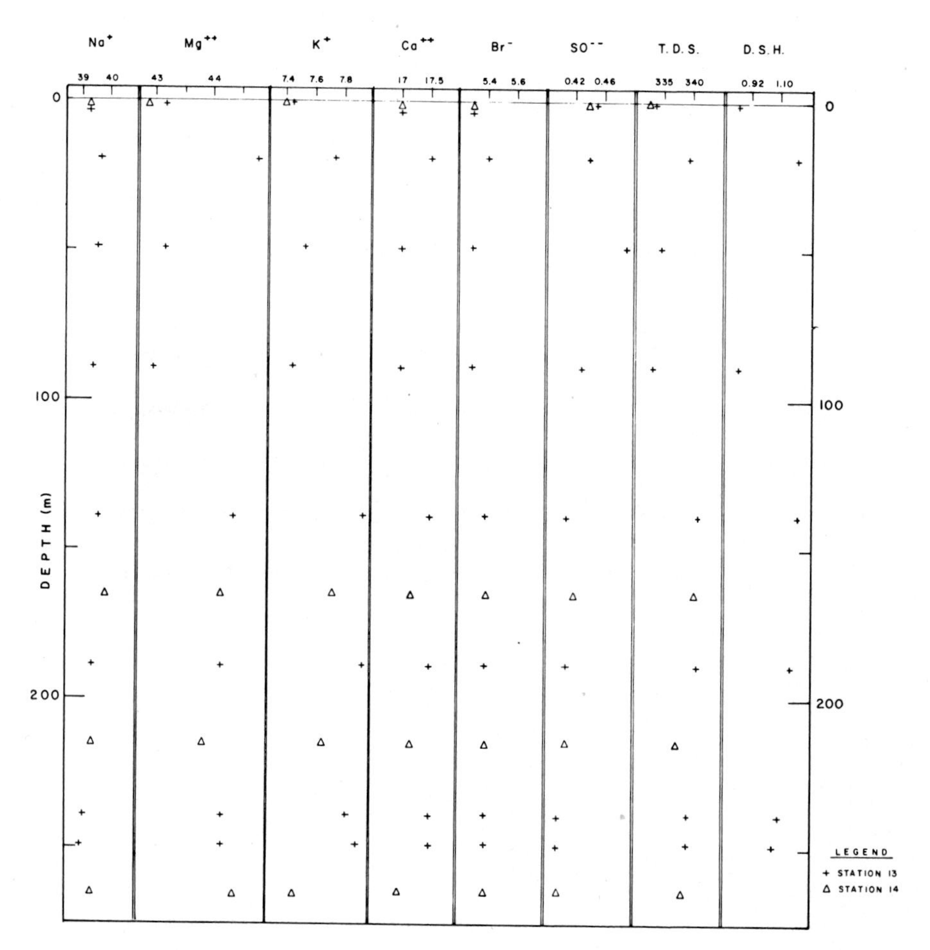

Fig. 13-8. Vertical distribution of major ions (g/l) and D.S.H. degree of halite saturation in Dead Sea waters. Northwestern part of North Basin. (Ein Feshkha) Samples taken during March 1977 at Stations 13, 14.

REFERENCES

Adler, S., 1975. Carnallite production in the solar ponds Dead Sea Works. Proc. Symposium: The Dead Sea Present and Future. (ed. R. Ben-Avi) Ministry of Commerce and Industry, 393 pp. (in Hebrew).

Assaf, G. and Nissenbau, A., 1977. The evolution of the Upper Water Mass of the Dead Sea 1819–1976; Desertic Terminal Lakes (ed. D. C. Greer), Utah Water Res. Lab., 61–72.

Beyth, M., 1976. A chemical survey to examine the influence of End Brines on the brines of the Dead Sea in the area of P6 Pumping Station: G.S.I. Report MG/1/76. 11 pp. (in Hebrew).

Beyth, M., 1977a. Recent evolution of Dead Sea Brines: G.S.I. Report MG/3/77, 11 pp.

Beyth, M., 1977b. Present stage of Dead Sea Brines: G.S.I. Report MG/11/77, 7 pp.

Druckman, Y., and Beyth, M., 1977. "Salt Reefs" A product of Brine Mixing. Lynch Straits, Dead Sea (a preliminary study), G.S.I. Report, MG/7/77, 15 pp.)

Epstein, J. A., Zelvianski, B. and Ron, G., 1975. Manganese in sodium chloride precipitating from mixing Dead Sea brines: Israel Jour. of Earth-Sci., 24: 112–113.

Lartet, L., 1867. Essay on the formation of the basin of the Dead Sea etc. in: Ritter, Carl, The Comparative geography of Palestine and the Sinaitic Penninsula: Edinburgh, T & T Clark, 1867, 4 vols., 1–1709 pp. (in Vol. III, p. 351 ff, Lartet, 1865a).

Lerman, A., 1967. Model of chemical evolution of a Chloride Lake. The Dead Sea: Geochim. Cosmochim. Acta, 31: 2309–2330.

Lerman, A., 1970. Chemical-Equilibria and Evolution of Chloride Brines: Mineralogical Society of America Special publication No. 3: 231–306.

Neev, D. and Emery, K., 1967. The Dead Sea: Depositional processes and environments of evaporites: G.S.I. Bull. No. 41, 147 pp.

Neev, D. and Hall, J. K., 1975. The Dead Sea geophysical Survey, July 19th–August 1, 1974. Final report No. 2, G.S.I. Report MG/2/75, 21 pp.

Nissenbaum, A., 1969. Studies in the geochemistry of the Jordan River–Dead Sea System: unpublished Ph.D. thesis, University of California, Los Angeles, 286 pp.

Schnerb, Y. and Zelvianski, B., 1970. Manual of analytical methods, Dead Sea Works (in Hebrew). 131 pp. Dead Sea Works, Beer Sheva.

Starinsky, A., 1974. Relationships between Ca-chloride Brines and sedimentary rocks in Israel: Unpublished Ph.D. thesis, the Hebrew University, Jerusalem (in Hebrew), 126 pp.

Zak, I., 1967. The geology of Mount Sedom: unpublished Ph.D. thesis, the Hebrew University, Jerusalem (in Hebrew with English abstract). 208 pp.

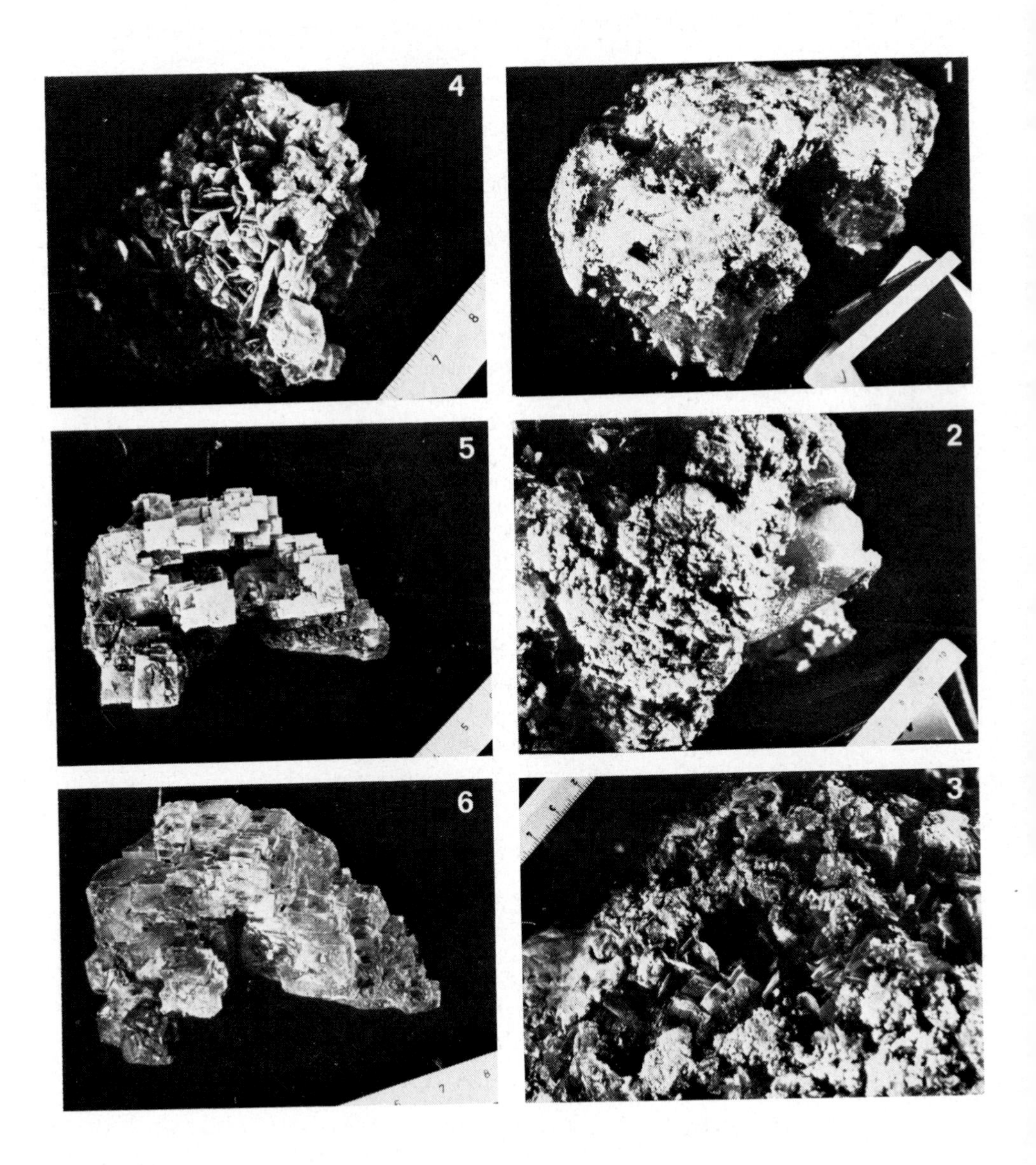

PLATE 13-1
1, 2 & 3. Bottom sediment from Station 8 (128 m) illustrating the sequence of deposition with two possible generations of halite crystals; first large and rounded (Fig. 2 central right) and second smaller euhedral with gypsum rosettes (Fig. 3 center) with a cover of finely crystalline aragonite mud. 4. Gypsum rosettes from Station 14 (249 m). 5 & 6. Halite from Station 13 (266 m).

Chapter 14

PROVENANCE, DISTRIBUTION AND ALTERATION OF VOLCANIC SEDIMENTS IN A SALINE ALKALINE LAKE

RUTH G. DEIKE AND BLAIR F. JONES

U.S. Geological Survey, Reston, Virginia, 22092 USA

INTRODUCTION

The closed-basin, shallow, saline, alkaline Lake Abert in the volcanic terrane of south-central Oregon provides an ideal setting to examine mineral-water interaction. Studies of the geochemical evolution of the lake water and interstitial brines of lake and playa sediments suggests that reactions with clay minerals are removing major cations from solution, and are thus exerting a significant influence on the solute budget of the lake system (Jones and VanDenburgh, 1966; Jones, et al., 1969; Phillips and VanDenburgh, 1971; VanDenburgh, 1975). The work presented here forms a background for comparison of the composition and structure of clay minerals in the lake sediments with those supplied by a wide variety of Tertiary volcanics and their weathering products. In this study we have looked for indicators of source for the starting materials in mineral-water reaction, i.e. the provenance of the clay-sized lake sediment.

Qualitative comparison of the lake sediment mineral phases with those of the exposed volcanics and their weathering products alone does not indicate what components of the rocks have contributed most to sedimentation or to the geochemical evolution of the lake water. By quantifying all the components of one size fraction in the lake sediment we hoped to quantify provenance, and in addition to look at which minerals were altering and how quickly, and to determine the distribution and morphology of possible authigenic minerals. Available samples originally obtained in study of the sediment-water interface and later, the pore fluid composition (Jones, et al., 1969; Lerman and Jones, 1973; VanDenburgh, 1975) gave us a reasonably good areal and shallow depth distribution, and ^{14}C age determinations provided estimates of rates of sedimentation.

For the provenance study we chose the 75–150 μm fraction, which is convenient to quantify by grain count with petrographic microscope. In addition, because the size of roughly 2/3 of the lake sediment is $<37 \mu m$, the 75–150μm fraction includes some of the coarsest material. This fraction contains mineral and rock fragments which presumably are not yet chemically altered or broken into finer material and thus can most easily be related to source lithologies.

FACTORS AFFECTING SEDIMENT INPUT TO LAKE ABERT

Physical Setting

Lake Abert and nearby Sumer Lake, located in the northwestern-most part of the Great Basin, are the last remnants of pluvial Lake Chewaucan, one of many large closed basin lakes which occupied the down faulted valleys of the Basin and Range province during the Pleistocene (Phillips and VenDenburgh, 1971). Several stages of lake rise and dessication left shoreline as much as 80 m above present water surface of Lake Abert and distributed sediment over a broad flat valley floor.

Lake Abert itself now occupies the eastern arm of pluvial Lake Chewaucan. The present water body is elongate, roughly triangular and covers an area of approximately 150 km^2. It overlies a sediment wedge on the downthrown surface of the eastward tilted Coglan Buttes fault block. The east side of the lake is bounded by Abert Rim, the steep escarpment of the adjacent normal fault block. The Chewaucan River, which supplies nearly 90 percent of the total discharge to Lake Abert (VanDenburgh, 1975, p. 7), enters the lake from the south and west just below its confluence with Crooked Creek, the only other perennial stream in the Lake Abert drainage basin.

An outline of Lake Abert, Summer Lake, the peripheral playa flats and the complete pluvial Lake Chewaucan drainage basin is given in Fig. 14-1.

Climate

The climate in the Lake Abert drainage basin is semi-arid. VanDenburgh (1975) finds that precipitation occurs primarily as snowfall during the winter months and as thundershowers during the summer. The bulk of the discharge in the Chewaucan River system is normally associated with seasonal snowmelt, which, unless unusually rapid, does not give rise to high sediment yield. Conversely, the high sediment concentrations often resulting from intensive local thunderstorms are most likely to occur in drainage off Abert Rim.

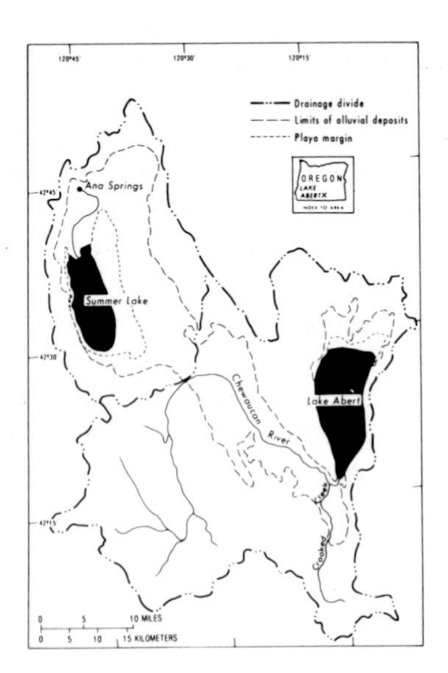

Fig. 14-1. Location and extent of the drainage basin of pluvial Lake Chewaucan in south-central Oregon and its modern successors, Abert and Summer Lakes. The ancient lake shorelines are roughly equivalent to the boundary of alluvial deposits; present playa margins are shown also.

The mechanical production of sediment is augmented by extreme diurnal temperature and humidity fluctuations. Phillips and VanDenburgh (1971, p. 14) find that during the summer months humidity ranges from 10 to 30 percent at mid-day and approaches 100% at night. The daily temperature range often exceeds 30°C.

Without specific study it would be impossible to estimate the amount of sediment input to the Lake Abert basin by wind. VanDenburgh (1975) suggests that a net *loss* of solutes might result from salt blown out of the combined Summer and Abert basins.

Sediment Transport and Deposition

Runoff moving toward Lake Abert from all areas except the Abert Rim, particularly during low lake stages, must traverse very low gradient courses near the lake. This can be expected to result in the deposition of coarser sediment sizes near local sources. At any rate, little coarse sediment reaches Lake Abert; the maximum sediment size found in lake cores and playa pits was a few millimeters. On a weight basis, an estimated two-thirds of the lake and playa sediment was finer than sand size.

Flood flow of over 30 m^3/sec. (1100 cfs) on the Chewaucan River below Crooked Creek (collected by G. E. Tyler, April, 1969) contained less than 100 ppm suspended sediment, largely plant debris. Low sediment concentrations in Chewaucan River inflow must be attributable in no small part to the low gradients of the old pluvial lake bottom, especially the Chewaucan marsh, acting as a sump for sediment carried out of the mountain front.

The salinity of Lake Abert has apparently influenced sedimentation also. Within recent decades the total solute concentration has seldom been less than 20,000 mg/l and has reached much higher levels (VanDenburgh, 1975, p. 20). Sodium chloride and carbonate species make up 90% of the solutes; the bulk of the alkaline earth and silica, as well as much of the potassium and sulfate brought in solution by inflowing waters, have been lost to the sediment through chemical reaction accompanying evaporative concentration (Jones, et al., 1969; VanDenburgh, 1975). Major solute balances have been affected by diffusion into and out of the sediment in response to lake level fluctuations (Jones and VanDenburgh, 1966; Lerman and Jones, 1973). The solute flux has also brought about flocculation of fine sediment contributed to the lake, so that suspended particulates in Lake Abert are low in concentration, and are largely algal debris. This is in marked contrast to the milky clay suspension usually found at nearby Summer Lake, which has about one-third the solute concentration of Lake Abert and contains a higher percentage of carbonate species.

The sedimentation rate in the Lake Abert basin was estimated from data on the ^{14}C content of inorganic carbonate at depth in peripheral playa sediments leached of pore fluid before ^{14}C determination. Rates are derived from relative ages of carbonate asumed to have formed within the lake approximately at the time of deposition of the enclosing sediment. This assumption is supported by calcite coatings on rocks at least 100 m above the present lake, and calcite activity products which remain close to or exceeding saturation even in dilute inflow (a condition which probably existed in the past). Samples from depths of 61–67 and 122–152 cm in the northeast playa as well as 15–30 and 61–67 cm in the northwest playa suggest sedimentation rates of .027 and .010 cm/yr at the two localities.

SOURCES OF SEDIMENT: LITHOLOGIC DISTRIBUTION

The Lake Abert drainage lies entirely within the Great Basin physiographic province, but it is within 100 km of the High Cascades to the west, and the High Lava Plains to the north. The basin is entirely underlain by Cenozoic volcanics or volcanoclastic rocks with affinities to both other provinces. The geology has been mapped in reconnaissance by Walker (1963) (Fig. 14-2). More detailed studies of correlative rock units from outside the basin have been given by Walker (1960, 1963, 1967), Stewart, et al. (1966), and Walker and Swanson (1967, 1968a, 1968b). Basalt flows predominate areally, but andesite, andesitic or dacitic tuffs and breccias, ash-flow tuffs and rhyolitic plugs also are common. With regard to sediment production, these rocks can be grouped into two principal types: 1) holocrystalline flows of locally highly feldspathic basalt to andesite (Fig. 14-3a) that are broken down primarily by fracture into blocky talus, boulders and other coarse sediment; and 2) andesitic to rhyodacitic pyroclastic rocks characterized by much cryptocrystalline or glassy matrix (Figs. 14-3b, 4 and 5), which give rise to mostly fine sediment through disaggregation.

The exposure of the principal volcanic rock types divides the Lake Abert drainage basin into major areas of sediment provenance. The most obvious is the Albert Rim to the east of the lake. Here two units of largely Miocene rocks have been described by Walker (1963). The lower unit consists of basalt and basaltic andesite flows, which may

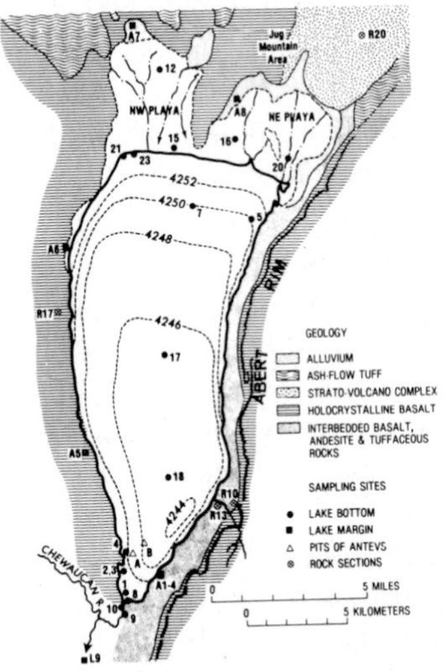

Fig. 14-2. Principal volcanic lithologies exposed in the Lake Abert drainage basin as simplified and generalized from Walker (1963). Note unit inclusion of interbedded tuffaceous material with andesites and basalts exposed in lower Abert Rim. Core, pit and lake margin sampling localities are shown along with the locations of thin section photomicrographs (Figs. 14-3–5). The two localities sampled by Antevs at near-dryness of Lake Abert (Shrock and Hunzicker, 1935) are also shown.

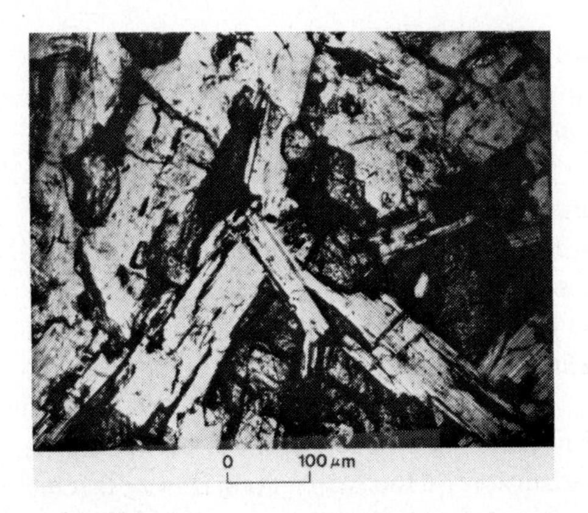

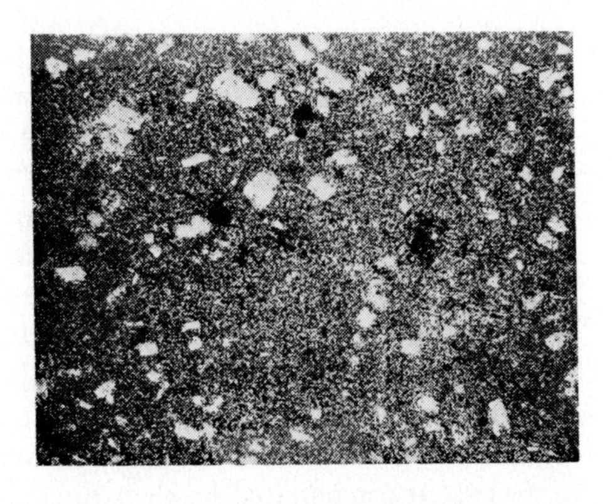

Fig. 14-3a. Photomicrograph showing subophitic texture typical of feldspathic basalts exposed in the upper Abert Rim and over most of the western Lake Abert drainage basin (locality R-17). Source of monomineralic grains of plagioclase, pyroxene, and opaques. A few clasts of fine-grained pyroxene and opaques show up as lithic matrix fragments (LMF).

14-3b. Photomicrograph of pilotaxitic andesite found in Jug Mountain vent rock and exposed in middle to lower Abert Rim (locality R-13). A few percent LMF are derived from microlitic matrix.

be ophitic, porphyritic, or pilotaxitic in texture and sometimes are capped by zones of dense red and black scoria. Interbedded with the lavas are devitrified dacitic tuffs, and tuffaceous sediments underlying talus on the gentler slopes of the rim escarpments. In the vicinity of Poison Creek (Fig. 14-2) this material crops out near the top of the lower unit, and contains abundant pumice altered to smectite (Fig. 14-4). Near the north end of Lake Abert, tuff is exposed near the base of the rim escarpment and also immediately east of the northeasternmost playa.

Fig. 14-4. Photomicrograph of glass and/or cryptocrystalline ground mass of pumiceous tuff exposed in lower unit of Abert Rim (locality R-10). The Si:Al composition and smectite content of this groundmass is like that of cryptocrystalline matrix fragments (CCMF).

The upper unit of the Abert Rim, which forms the conspicuous cliffs at the top of the scarp and underlies most of the drainage area west of the lake, is composed of medium-grained holocrystalline basalt (Fig. 14-3a) containing labradorite, subophitic pyroxene, magnetite, and relatively coarse olivine. These flows are generally uniform, with variation chiefly in amount of pyroxene, alteration of the olivine, and the size of feldspar laths.

The same type of basalts which cap the Abert Rim underlie the drainage area to the west of the lake. The principal differences are in the complete alteration of the olivines in the west shore rocks and the presence there of a few interstitial voids filled with bright yellow non-fibrous nontronite (?) apparently similar to that described by Summers (1976) in the Columbia River basalts.

Thus sediment derived from the west and northwest of Lake Abert can be considered to be largely comminuted basalt, with the exception of the major drainageway leading into the playa flats from the northwest. In this direction lies extensive outcrop of Pliocene rhyolite ash-flow sheet deposits, variably welded. This constitutes the most likely source for clear or incipiently altered detrital glass, despite the limited area of Lake Abert drainage involved.

Seemingly, some of the most erodible material in the Lake Abert drainage would be in the small strato-volcanic complex near Jug Mountain northeast of the lake. This feature is a complex mass, largely of andesite compositoin, but consisting of agglomerate, tuffs, and breccias of widely varying surficial character intruded by dense, dark vent dikes. Generally, mafic minerals are scarce and much of the groundmass is cryptocrystalline or glass (Fig. 14-5), often variably iron-stained. The incoherent nature of so much of this material suggests strongly a good source for poorly characterized, fine-grained sediment readily transported by overland flow in the highly ephemeral drainage northeast of Lake Abert.

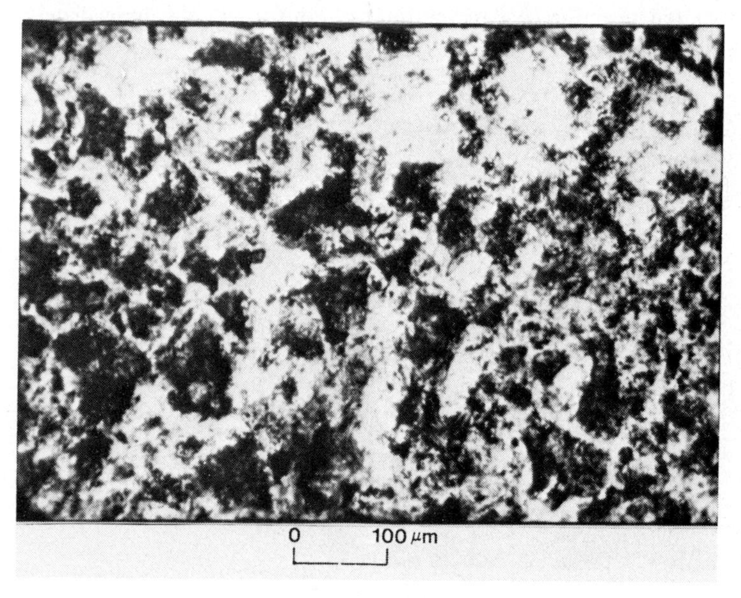

Fig. 14-5. Photomicrograph of groundmass from pyroclastics (locality R-20) associated with the Jug Mountain strato-volcanic complex. The size and morphology of cryptocrystalline masses are very similar to that of cryptocrystalline matrix fragments (CCMF) such as those found in sediments in the northeastern part of Lake Abert (see Fig. 4-6).

As mentioned earlier, the Chewaucan River appears to contribute very little sediment to Lake Abert, despite its domination of the total fluid input. Actually, the lithologies available to the lower river basin are very similar to what has been described in the southern peripheral lake area; interbedded basalts, andesites, and tuffs from the lower Abert Rim on the east, with holocrystalline basalts to the west and the upper part of the Abert Rim section. A rare, but significant sediment event might be expected in the Crooked Creek drainage, but again the contributive lithologies can be viewed as extensions of those closer to the lake.

MINERAL DISTRIBUTION IN LAKE SEDIMENT

Methods

Hand-driven cores and dug pits up to two meters deep from the Chewaucan River "estuary," mid-lake and playa (see Fig. 14-2), obtained for other purposes (see Introduction) were also used for the sediment study. Inasmuch as examination of only these sites left the central east and west near-shore areas relatively unsampled, analyses of peripheral and ancient lake deposits, particularly along the west shore, have been used to extend the distributions and interpretations from mid-lake core material.

Samples from twenty-three cores and dug pits were divided into 99 sections 15 cm long. The samples were sized at 37 μm (400 mesh) by wet sieving which also removed soluble salts. The <37 μm fraction was further separated by soil column settling and centrifugation into 2–37 μm, 0.1–2 μm and <0.1 μm fractions. Weight percent of the different size fractions was determined for selected samples by pipette analysis. In other samples, each centrifuge size fraction was dried at 60°C for several days and then weighed.

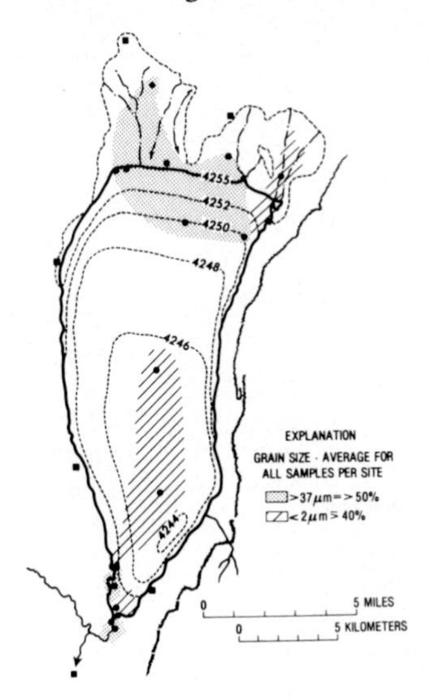

Fig. 14-5a. Distribution of coarse and fine sediment in Lake Abert showing coarser material near input source. Based on data listed in Table 14-1.

The 75–150μm fraction chosen for petrographic analysis was separated from the bulk >37μm fraction by dry sieve and mounted in index oils under coverglass. Different operators counting duplicate mounts of 10 random samples agreed within ±7% maximum difference for the averages of any one component. The average errors were ±2% for minerals present from 0–10%; ±7% for those from 11–20%; and ±6% for those >20%. In a second round of at least 100 total grain counts lithic matrix fragments (LMF) were examined to delineate major types. A refractive index oil of 1.62 was used in order to readily separate pyroxene from plagioclase. Special search was made for quartz in the fine sand fraction, but few grains were found. X-ray evidence for quartz in finer sizes was equivocal. Percentages of individual components were contoured over the lake from an average for all samples in each core.

X-ray diffraction samples of sand fractions were ground, pressed mechanically into aluminum sample holders and then roughened with a glass slide to minimize preferred orientation effects on intensity. The finer fractions were mounted as a water smear on glass slides. X-ray diffraction patterns were run on a Norelco diffractometer with Cu Kα radiation at 45 kv and 25 ma discriminated with a graphite crystal monochrometer and recorded at 1°/minute. From hand-ground random powder mounts of the 75–150 μm

TABLE 14-1.

Particle size distribution for Lake Abert sediments in weight percent. Sample locations shown on Fig. 14-2.

Sample	depth (cm)	>37 μm	2–37 μm	0.1–2 μm	<0.1 μm[1]
1e	37–49	7	41	47	6
4d	26–36	13	38	51	n.d.
5c	21–36	13	27	60	n.d.
9c	18–36	56	25	20	n.d.
12b	9–31	54	25	20	1
14b	6–31	83	12	5	<1
15b	3–46	65	18	15	<1
16a	0–3	44	18	37	n.d.
16c	18–36	37	18	36	n.d.
17i	26–52	4	51	43	2
21ca	0–15	87	7	6	n.d.
21cb	15–31	70	17	12	1
21cc	31–45	54	23	21	2
21g	76–91	44	31	21	n.d.
21ja	52–67	54	18	27	n.d.
21jb	67–98	5	38	53	4
21jc	98–113	5	50	38	7
23ca	0–15	79	15	6	n.d.
23cc	31–46	59	34	7	n.d.
23cg	91–107	43	17	40	n.d.
23de	204–219	6	46	40	7

[1] n.d. – not determined

fraction, clay mineral basal reflections in the 4°–8° 2Θ (11–22Å) region were measured by planimeter. These patterns were run utilizing a theta compensating slit to reduce background in the low 2Θ angle region.

Grain Size Distribution

Size distribution in weight percent for size fractions of sand (>37 μm), silt (2–37 μm), clay (0.1–2 μm), and, for some samples, fine clay (<0.1 μm) is shown in Table 14-1 and Fig. 14-6 for 21 samples from 11 localities. Because a greater number of sized sample cuts per site were taken from playa locations an overall average is somewhat biased. Nevertheless such an average should approximate the lake sediment closely enough to be useful for discussion. Our particle size data at correlative depths are in good general agreement with the more limited size data obtained from one meter pits dug by E. Antevs in the dry lake bed in 1931 (Shrock and Hunzicker, 1935).

The overall average lake sediment appears to be nearly equally divided between >37 μm, 2–37 μm, and <2 μm fractions. Therefore, the 75–150 μm fraction taken for petrographic analysis is toward the coarse end of the size spectrum. Dispersion away from source areas appears to be small, and the areal distribution of various components of the 75–150 μm fraction should be indicative of provenance.

The size data for the selected samples is shown geographically in Fig. 14-6. Sand-sized material is dominant in the northwest part of the basin, whereas clay composes a significant part of the lake sediment in the northeast. This distribution suggests a relationship between clay content and the rock types exposed along Abert Rim and in the Jug Mountain area.

Note that NW playa sediments (Table 14-1, sites 21 and 23) appear to be generally coarser than those in the Chewaucan Estuary (sites 9 and 10). This verifies the indications that even at high discharge the Chewaucan River brings in little coarse sediment. Where found in the lake then, the coarse material is probably derived from local runoff.

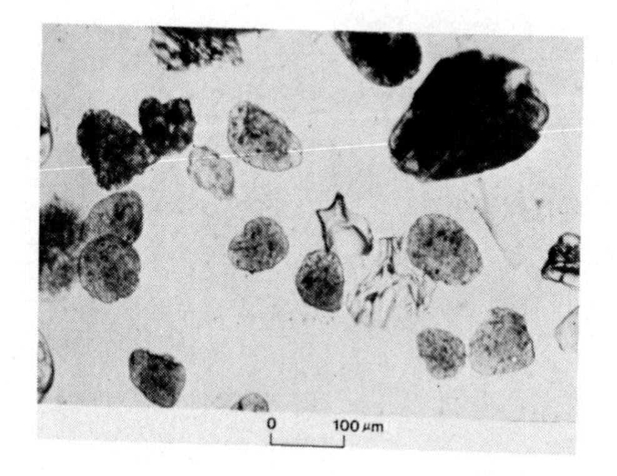

Fig. 14-6. 75–150 μm fragments from site 5c (21 to 36 cm) showing cryptocrystalline matrix fragments (CCMF). Many are rounded, but whether from abrasion or aggregation is not certain (see SEM enlargement, Fig. -9). Note also several types of glass shards and a large scoriaceous matrix fragment.

Mineralogic Components in the Fine Sand Fraction

Distribution of major minerals, plagioclase, calcite, pyroxene and clays based on petrographic grain counts of the 75–150 μm fraction was verified by x-ray diffraction analysis of the >37 μm size fraction; minerals comprising less than 10 percent or so of the x-ray sample were seldom specifically identifiable. Actually, x-ray data from the 2–37 μm fraction also correlate with the petrographic findings, although diffraction maxima from the major components mask most other reflections.

Mineral components identified and counted in the 75–150 μm fraction are listed in Table 14-2 according to their overall average abundance. Minerals that occur in only a few samples and/or in small quantities include garnet (andradite), hypersthene, hornblende, aragonite, apatite, zircon, and olivine.

The bulk of the 75–150 μm fraction is composed of lithic matrix fragments (LMF), calcite, and plagioclase, but samples from different lake environments show wide variation in the amounts of individual components. The distribution of the two principal detrital components, LMF and plagioclase can be related to the two predominant rock types exposed in the drainage basin, whereas the distribution of calcite is related to its precipitation in the lake.

Lithic Matrix Fragments

On the average more than one-third of the 75–150 μm fraction of Lake Abert sediment is lithic matrix fragments (LMF), of which at least 75% resembles the cryptocrystalline or glassy matrix of tuffaceous and other pyroclastic rock types (Figs. 14-3b, 4 and 5). The rest is composed chiefly of nearly opaque groundmass from scoriaceous material, associated with a few crystalline aggregates from other fine-grained basaltic or andesitic rocks.

TABLE 14-2.

Mineral Components in 75–150 μm fraction, average for all samples (95).

Component		Average %
Lithic Matrix Fragments (LMF)		34
Clear Cryptocrystalline Matrix Fragments (CCMF)	75	
Other, mainly scoria	25	
Calcite		30
Hydrocalcite	4	
Plagioclase		18
Clear Glass Shards		8
Diatoms		3
Pyroxene		2
Opaques		2
Brown Glass Shards		<1
Mica Flakes		<1
Amphibole		<1

LMF material is more common adjacent to the northeast portion of the Abert Rim, and in the lower lake. Here, samples of the 75–150 μm fraction average >50% LMF. The most likely source of this material is from the tuffaceous and pyroclastic rock units interbedded with basalts on the lower Abert Rim.

The most abundant LMF have a clear glassy groundmass and we refer to this subtype as clear cryptocrystalline matrix fragments (CCMF). These are typically subrounded to elliptical masses of yellowish-brown glass containing small crystalites of feldspar, perhaps pyroxene, and disseminated equigranular opaques (see photomicrograph, Figs. 14-6 and 7). Some of the CCMF contain clumps of apparently altered cryptocrystalline groundmass. Under crossed nicols the groundmass commonly appears microgranular; some grains have a characteristic irregular clay-like birefringence, some are isotropic, and some have very low first order birefringence. Index of refraction of various CCMF in six samples indicate values between 1.516 and 1.520, which suggests a silica content of about 67% based on measurement of natural volcanic glass shards (George, 1924).

A scanning electron photomicrograph (Fig. 14-8) of a clear cryptocrystalline matrix fragment (site 17i) shows the irregular, almost layered nature of the material. Note also the rounded surface of a typical whole CCMF grain shown in Fig. 14-9, most likely resulting from abrasion during transport, but possibly due to aggregation. Qualitative elemental composition from energy dispersive analysis of the areas included in Figs. 14-8 and 9 are shown with each scanning electron micrograph, and in Table 14-3 similar information for several other CCMF fragments is listed. In general, silica is high and alumina low with Mg, K, Ca, and Fe present in decreasing order of abundance. These CCMF particles are found in nearly every sample, and range from 0–100%, averaging 53% of the total LMF group, and 20% of the bulk 75–150 μm fraction.

In three cores from the lower lake CCMF fragments incorporating diatom frustules are abundant 30–60 cm below the sediment-water interface. In some samples these fragments make up half the LMF group. The diatom-bearing fragments were not found above the

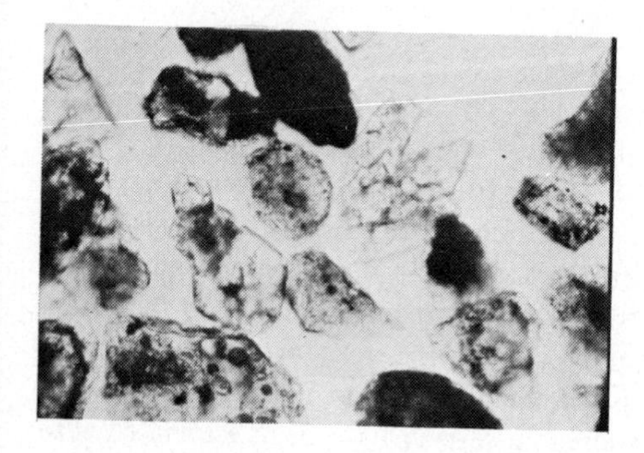

Fig. 14-7. A typical assemblage of 75–150 μm components from playa site 15 (3 to 46 cm depth) which includes several opaque-bearing fragments, scalenohedral calcite, and glass shard with inclusions. Subrounded and angular CCMF fragments are near the center of the photograph.

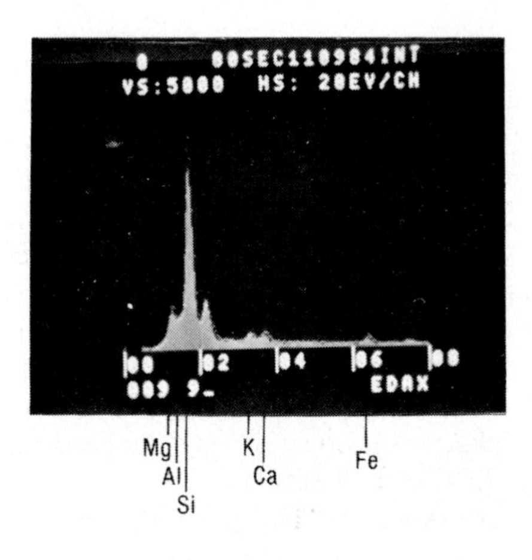

Fig. 14-8. SEM enlargement of the surface of a rounded cryptocrystalline matrix fragment from locality 17, 26 to 52 cm depth.

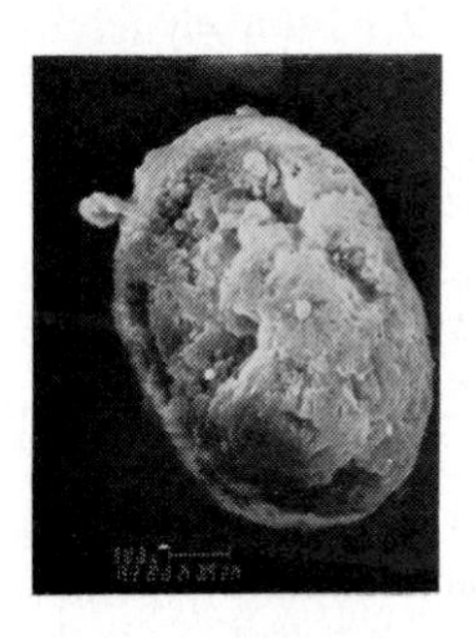

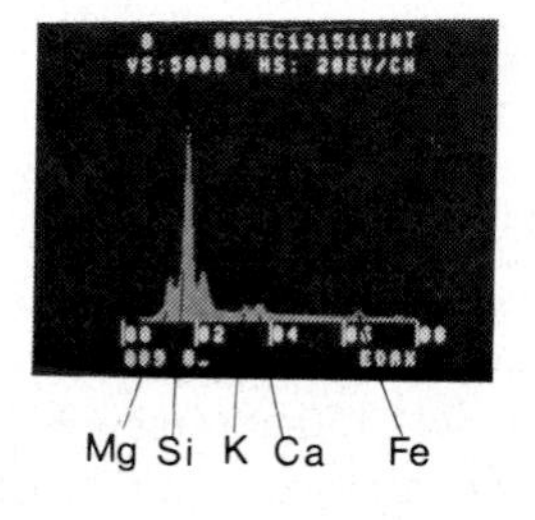

10μ

CCMF—N.E. RIM PROVENANCE

Fig. 14-9. SEM photograph showing rounded morphology of CCMF fragments. Chemical composition scan is representative of the whole fragment.

30 cm level nor in any mid-lake or playa core samples. The diatoms are all very similar to a species which has been associated with older lacustrine sediments.

As shown in the photomicrograph, Fig. 14-10, the frustules are an integral part of a matrix of flocculated material with a silica rich composition (from energy dispersive analysis, see Fig. 14-11). A typical fragment enlarged in the SEM photograph, Fig. 14-11, shows a complete diatom shell and at least one shell fragment cemented to the glassy matrix. Such fragments seem very vulnerable to erosion during transport. These particles were most likely produced by incorporation of the diatom tests in floccules of colloidal silica and ultrafine clay. Intervals of frustule abundance may have been associated with dilution at high lake stage or flood waters from the Chewaucan. CCMF particles incorporating abundant diatom shells are common in sediment associated with dilute spring inflow to a highly saline playa pond northwest of Lake Abert today.

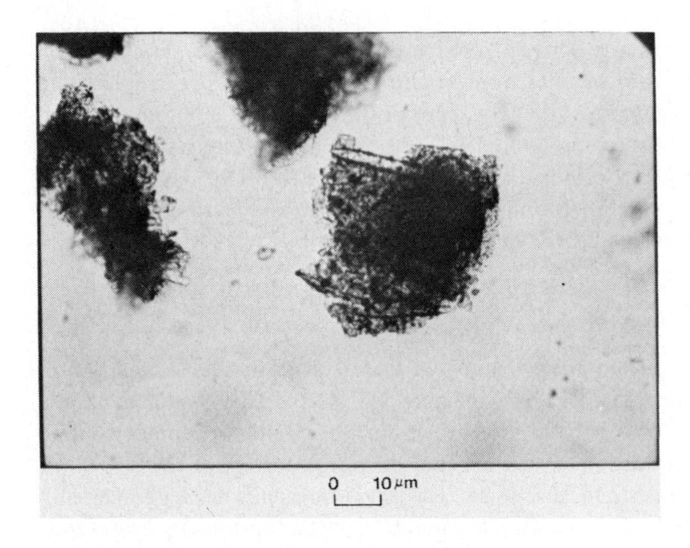

Fig. 14-10. CCMF particle with included diatom frustules.

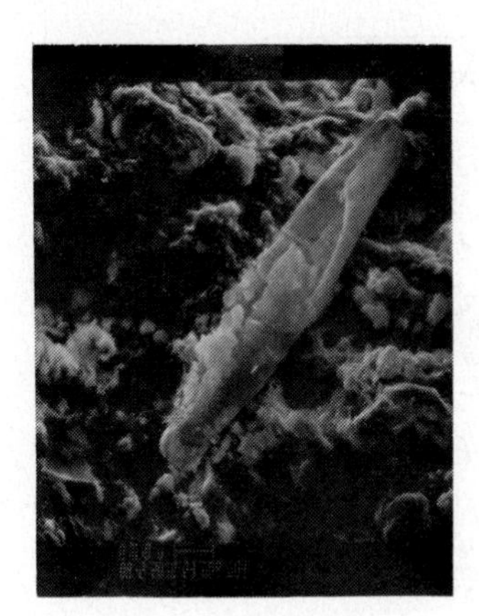

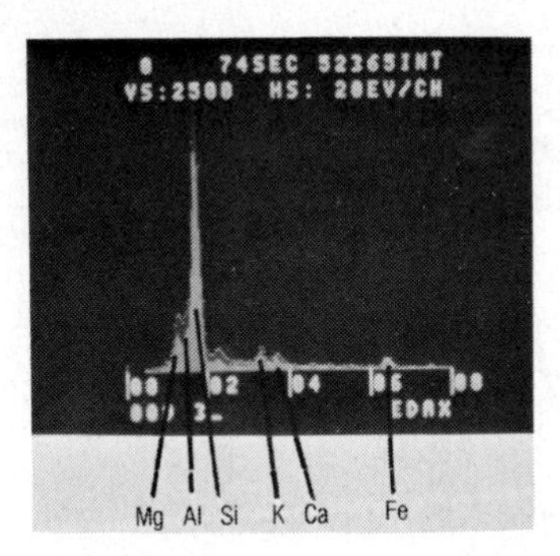

Fig. 14-11. SEM enlargement of diatom frustule attached to CCMF fragment shown in Fig. 14-10. Chemical composition is for floccule without diatom; compare with composition of CCMF, Fig. 14-9.

Some of the CCMF particles are more angular and clearly have not been subject to much transport. These angular fragments make up as much as 100% of the total LMF in samples from well below the sediment surface at localities 5 and 7 suggesting a sustained local source to the northeast. The angular varieties have the same optical characteristics as the rounded CCMF particles, and have thus been included in the same distribution. Both highly angular and diatom-bearing CCMF are similar in appearance and optics, including the indefinite birefringent properties; this suggests both primary (outcrop) and secondary (lake) aggregation of particles too fine for microscopic resolution.

We feel that CCMF are obviously derived from local pyroclastic sources. For example, they are particularly abundant in samples from localities 5, 20, and deeper samples from site 16, all in the northeastern part of the lake. Much of the sediment input here is from the stratovolcanic complex near Jug Mountain. In Fig. 14-5, the glassy matrix of agglomeratic pyroclastic rock is shown in a thin section (sample site R-20, Fig. 14-2) from this area. This material is plentiful in the stratovolcanic complex, and like the tuff of Abert Rim, it physically resembles the CCMF. (Fig. 14-12)

Scoria Matrix Fragments

Lithic matrix fragments derived from mafic scoria are shown in the distribution in Fig. 14-13. Scoria makes up most of the remaining 25% of the LMF material. These generally angular fragments are granular, red-brown to black, and sometimes contain feldspar microlites. The distribution, which is based on the percentage of the total LMF group, appears primarily related to local sources. The scoria fragments decrease markedly toward the center of the lake, due possibly to alteration (reduction) in the lake water. Chewaucan "estuary" bottom sediments contain more scoriaceous material than in the lake center, as do lake sediment samples from the southwest lake shore.

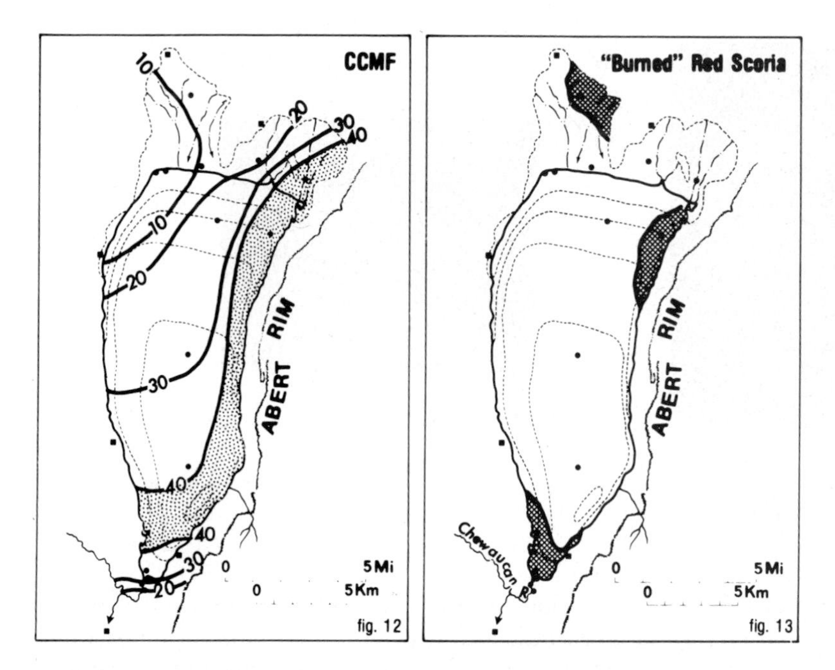

Fig. 14-12. Distribution of CCMF as percent of 75–150 μm fraction based on average for all samples at each location. Stippled area denotes CCMF content greater than 40%. Greater abundance of CCMF close to Abert Rim and the stratovolcanic complex to the NE is apparent.

Fig. 14-13. Distribution of scoria fragments as percent of 75–150 μm fraction based on average for all samples at each locality. Stippled area represents scoria abundance >5%. Occurrence is apparently related to very local sources.

Detrital Monomineralic Grains

The most common detrital mineral in the Lake Abert sediment is plagioclase, which accounts for nearly one-fifth of the 75–150 μm fraction. It occurs most often as single crystal fragments. The crystals are generally angular, but somewhat more equidimensional than laths observed in thin sections of the basalts. In the distribution map, Fig. 14-14, the largest percentages of plagioclase in the 75–150 μm fraction are found along the northwestern and western shores. Though far less in percentage, pyroxene and iron oxides were distributed similarly to plagioclase. The most likely source of all these components is holocrystalline basalts underlying most of the western Lake Abert drainage basin.

Extinction angles measured for albite twinning in plagioclase fragments from lake sediment in the northwest part of the basin largely correspond to the composition of An_{60} reported by Stewart, et al., (1966) for microlites in feldspathic basalts of the Warner Valley to the east (with equivalents in the Abert Rim and north of Lake Abert), and approximately An_{70} measured in thin sections of the basalts from the top of Abert Rim and outcrops west of the lake. A few measurements indicate that the average An content of plagioclase decreases in samples from the northeastern and southern ends of the lake. This can be expected because of the more siliceous material contributed from volcanics of the southern Abert Rim and the stratovolcanic complex near Jug Mountain.

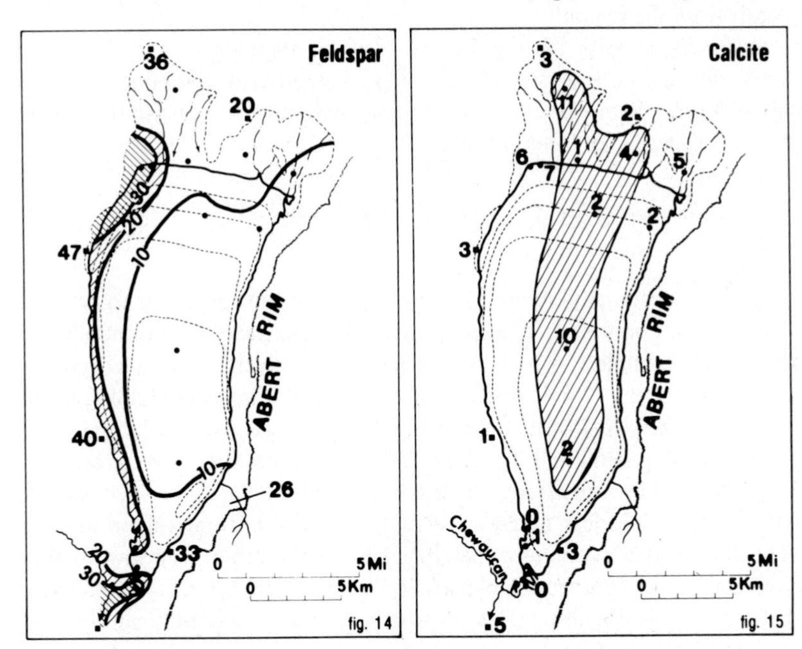

Fig. 14-14. Distribution of feldspar as percent of 75–150 μm fraction based on average for all samples at each locality. Patterns denote areas of greater feldspar content. Greatest abundance is apparently related to a western source in holocrystalline basalts.

Fig. 14-15. Distribution of calcite as percent of 75–150 μm fraction based on average for all samples at each locality. Numbers indicate the percentage of "hydrocalcite." Shaded area represents calcite >40%.

Calcite

On the average for all samples a little more than one quarter of the 75–150 μm fraction is calcite, only a small percentage of which is detrital (Truesdell, et al., 1968). Phillips and VanDenburgh, (1971, p. B49) attribute the exceedingly low amounts of calcium in solution in the lake waters to continuing precipitation of calcite. The calcite occurs most abundantly in mid-lake samples (Fig.14-15) as cryptocrystalline masses, crystal aggregates, and single scalenohedra of pure $CaCO_3$. A rhombic habit is rarely observed. Crystal aggregates are often brown and cloudy with fine included material, whereas single crystals are usually gray and clear. Calcite is prevalent in the central playa as well as at mid-lake, but it progressively decreases in sediments deeper than about one meter. Fibrous bundles of aragonite were present at a depth of 3–40 cm at site 15, but were not found elsewhere.

Despite supersaturation of the solutions, all or most of the calcite in the south end of the lake could be detrital. It occurs in very small quantities or even not at all in the finest fractions of the 24 samples from the seven southernmost localities in Lake Abert. This includes the Chewaucan River "estuary" and the lake sediments within about five kilometers of the Chewaucan River inflow. Where present in the 75–150 μm fraction the amount corresponds to about the 2% calcite estimated by Stewart, et al., (1966) for groundmass in basalt from the region.

By contrast, near mid-lake (site 18, Fig. 14-15) calcite averages 45% of the fine sand fraction, and x-ray diffraction analysis indicates it is a significant component of size fractions as small as 4 μm. Some kinetic process apparently prevents calcite from accumulating at or near the Chewaucan River inflow.

Hydrocalcite

Thin, curved fragments generally about 100 μm by 20–50 μm suspected to be pure $CaCO_3$ (from scanning electron microscope-energy dispersive analysis) are common in the fine sand fractions of 27 samples. This material may originally have been a hydrous calcium carbonate incrustation. X-ray diffraction analysis of samples containing up to 27% (by grain count) of these fragments does not correspond with data for hydrocalcite reported in the literature (e.g. Hull and Turnbull, 1973, and Marschner, 1969), but it is possible that hydrocalcite has dehydrated in the time the samples have been stored. Typical fragments show a colorless to gray layering of scalenhohedral calcite crystals (Fig. 14-16). The birefringence of individual fragments varies from very low to that typical of calcite, and some fragments are partially biaxial, although the interference figure may be difficult to distinguish. The fragments often have a pebbly surface and many have tiny round holes and protuberances set in a way suggesting formation on a substrate.

Variation in abundance with depth in the sediment suggests that specific climatic and/or hydrochemical conditions are associated with this form of calcite. In mid-lake at site 17, hydrocalcite occurs at the sediment-water interface and at a depth of about 60 cm, but not in between. On the northwest playa at sites 21 and 23, samples with quantities

$>$17% occur at depths of 15 cm, 160 cm and 200 cm. At central playa sites 16 and 20 abundances $>$17% are found only at depths of 120–150 cm.

Unusually textured $CaCO_3$ fragments are apparently independent of total calcite abundance except that they are also absent near the mouth of the Chewaucan River and in the "estuary." They are also independent of pore fluid concentrations, as quantities $>$15% (by grain count) are coexistent with pore fluid ranging from 25–150 g/kg. Areal distribution is shown with that of calcite in Fig. 14-15.

Glass Shards

The general abundance of glass shards increases to the northwest, in the direction of the most extensive outcrop of the Pliocene ash-flow tuffs mapped by Walker (1963). They account for an average of 8% of the fine sand fraction. There are several shard types present, and at some localities thin layers are composed almost entirely of shards. Preliminary observations suggest that whereas some shards are detritus derived from Pliocene tuffs, the shard layers may have resulted from ash-fall. Shard composition and petrography closely resemble Mazama tephra (Kittleman, 1973), and shards from several Lake Abert sites may be related to a similar ash fall source, but ^{14}C ages of co-existing authigenic calcite require an event younger than the Mazama eruption.

Zeolites

We believe most, if not all, zeolites found in Lake Abert are detrital. Phillipsite and clinoptilolite are the only zeolites identified in significant quantity by either x-ray diffraction or petrographic analysis. Occasional fibrous zeolite varieties were noted but not positively identified by petrographic analysis of fine sand fractions. Clinoptilolite occurs widely, but in very small quantity throughout the lake, as inferred from a small but persistent x-ray reflection at 9.9° 2Θ.

Partially zeolitized glass shards are common throughout lake bottom sediments, but they may be readily found together with clear shards showing no zeolitization.

Phillipsite occurs abundantly but locally in coarse-grained northwest playa samples that are also rich in unaltered glass. SEM analysis by Arthur J. Gude, III (USGS, Denver, CO, 1975) of a playa sample from a depth of 180 cm revealed apparent combined growth of phillipsite and clinoptilolite at the expense of glass shards; the identifications were confirmed by x-ray diffraction. This suggests glass alteration to phillipsite by weathering on outcrop before transport and deposition.

Many studies (e.g. Hay, 1966; Surdam and Sheppard, 1976) have indicated saline, lacustrine lake environments receiving glassy volcanics are ideal for the formation of zeolites. However, the irregular distribution of zeolites at Lake Abert suggest detrital input from other points of origin.

Based on the intensity of the principal x-ray reflections, phillipsite apparently occurs in some mid-lake samples only to a depth of 18 cm. Sediments above 18 cm have accumulated during playa conditions at lower lake stages. This suggests that phillipsite might be unstable at the higher lake levels probably associated with sediments below this depth.

Diatoms

The 75–150 μm fraction of Lake Abert sediment averages about 3% single diatom frustules counted as separate "grains." The greatest abundance of frustules is in the Chewaucan River "estuary" bottom and in lake sites near the Chewaucan River mouth. These samples contain a mixture of species common to more dilute water (Thomas J. Harvey, pers. comm., 1976).

Abundant diatom frustules in both "estuary" bottom sediments and lower lake samples at site 8 most likely represent higher lake stage conditions providing more dilute waters, and preservation of the shells. Diatom tests are plentiful in southern lake sites 8, 1 and 4, at two levels roughly 30 cm and 60 cm. At the 30 cm level, planktonic species were tentatively identified which Hecky and Kilham (1973) note occur over a wide range of alkalinites but thrive in relatively dilute water. The abundance of these frustules thus would indicate a deeper lake and therefore more dilute water conditions than at present. Diatom-bearing CCMF (described previously) are also particularly abundant at the 30 cm level in samples from near the river mouth, and as noted, the diatoms may have been incorporated with the lacustrine sediments through flocculation in a more saline environment.

The occurrence at southern lake sites of a few sediment intervals containing abundant diatoms and diatom-bearing CCMF interbedded with sections that are nearly diatom free suggests a variation in chemical conditions accompanying difference in lake stage. As indicated by Jones, et al., (1969) silica concentrations in pore fluids near the sediment-water interface under present conditions are consistent with the dissolution of diatom tests, together with diffusion effects. However, the tests may well be preserved during conditions of higher lake stage, when more dilute lake waters can be expected to be associated with lower pH, and thus with lower silica solubility. Hecky and Kilham (1973, p. 69) suggest the relation is not simple and organic compounds may stabilize the amorphous silica of diatom frustules. Thus diatom preservation, or lack thereof, may be an indication of lake chemical history, but it is very complex. Increase of pore fluid silica concentration would be subject to diffusion gradients, as is the case for the major solutes (Lerman and Jones, 1973), and would be significantly dependent on the dissolution rate of the frustules. Diffusion inhibited by the low permeability of very fine sediments would act to slow frustule dissolution rates.

RELATIONSHIP OF CCMF TO CLAY DISTRIBUTION

The optical counts indicate that lithic matrix fragments, particularly the clear cryptocrystalline type (CCMF), and specifically the diatom-bearing particles, are in greatest abundance in sediments associated with Chewaucan River inflow. Non-diatomaceous CCMF are abundant in sediments from both the Abert Rim and the Jug Mountain area northeast of Lake Abert. The CCMF are therefore apparently derived from the andesitic pyroclastics of the Chewaucan River drainage, in the lower to middle Abert Rim scarp (lower unit of Walker, 1963; see Fig. 2), and the stratovolcanic complex near Jug Mountain. Further, glassy groundmass which could provide CCMF type material is minimal in

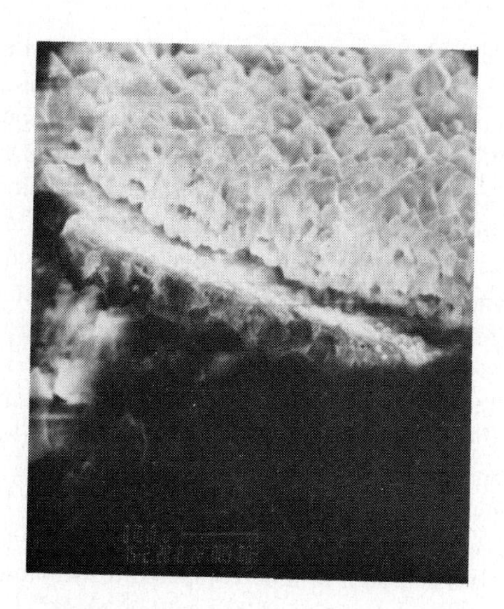

Fig. 14-16. Scanning electron micrographs of typical "hydrocalcite" fragment from Lake Abert sediment and close-up of layered crystals of scalenohedral calcite.

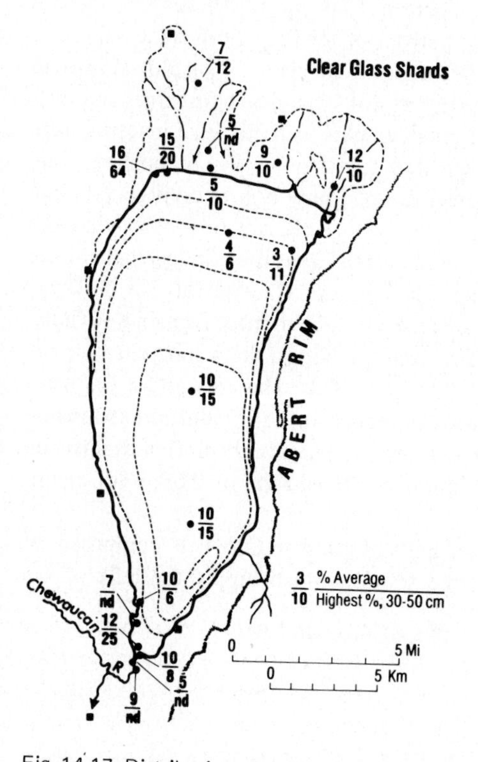

Fig. 14-17. Distribution of clear glass shards as percent of 75–150 μm fraction of Lake Abert sediment. Besides average for each locality, the highest value for the interval of 30–50 cm depth is also given.

those feldspathic and holocrystalline basalts exposed on top of Abert Rim and west of the lake. Of those volcanics most capable of supplying CCMF, the most proximal to the entire length of Lake Abert is the tuffaceous unit discontinuously exposed in the Abert Rim escarpment. This unit contains reverse graded pumice completely altered to smectite. Centrifuged suspensions of crushed tuff blocks yield much more clay-sized material than similar suspensions of any other rock type in the area.

There is evidence that generally LMF-rich lake sediment samples contain more clay-size and smectitic material. In Fig. 14-18 three bulk sample parameters are plotted on a trilinear diagram against the amount of plagioclase, clear glass, and lithic matrix fragments found in 75–150 μm fractions from each sample. Bulk parameters are as follows: sand-dominant samples are those which contain (by weight) more >37 μm fraction than <37 μm; clay-size dominant, are those in which the <2 μm fraction is >40% by weight; and smectite-rich are those samples which yield x-ray diffraction patterns with substantial reflections in the 14 Å region (other 14 Å clay species were not identified in these lake sediments). Considering only the size data, sand dominant samples are rich in plagioclase and poor in LMF compared to samples dominated by the clay-size fraction. Samples which were not sized, but contain abundant expandable clay minerals also have high LMF values. The amount of glass appears unrelated to the other categories. This suggests a relationship between lithic matrix fragments and the origin of expandable clay minerals.

To further explore this relationship the relative percent LMF and glass in the 75–150 μm fraction was plotted directly against the weight percent of the 0.1–2 μm fraction, which contains principally mixed layer clay and microcrystalline glass. The plot shown in Fig. 14-19a clearly indicates an association of LMF but not clear glass with larger amounts of 0.1–2 μm material. This association could result if LMF and clay-size material were both derived from a common source and deposited together in lake sediments. The tuffaceous units which are the most likely sources are not only rich in cryptocrystalline material but contain abundant smectite. Thus there is a significant possibility that the CCMF are also partially composed of smectite, and further, that they disaggregate in the lake environment, thus contributing to the lake clay fraction. To test this, 75–150 μm fractions containing varying amounts of clear cryptocrystalline matrix fragments (CCMF) were lightly crushed and analyzed by x-ray diffraction, utilizing a theta-compensating slit to reduce background in the low angle region (see Methods). The area under the basal reflections in the 4° to 8° 2Θ range was assumed to represent the amount of expandable clay present. For each sample, the peak area is shown in Fig. 14-19b plotted against the percentage of CCMF. The points have correlation coefficient (r) of 0.69, with a confidence level better than 1%.

The statistical correlations support textural indications that CCMF is composed of partially devitrified glassy matrix from the same sources which supply the bulk of the clay to Lake Abert sediments.

MAJOR CATION CHEMISTRY OF THE FINE SAND

The recognition of lithic matrix fragments as major components of the Lake Abert sediments complicates the consideration of effects of geologic materials on lake water

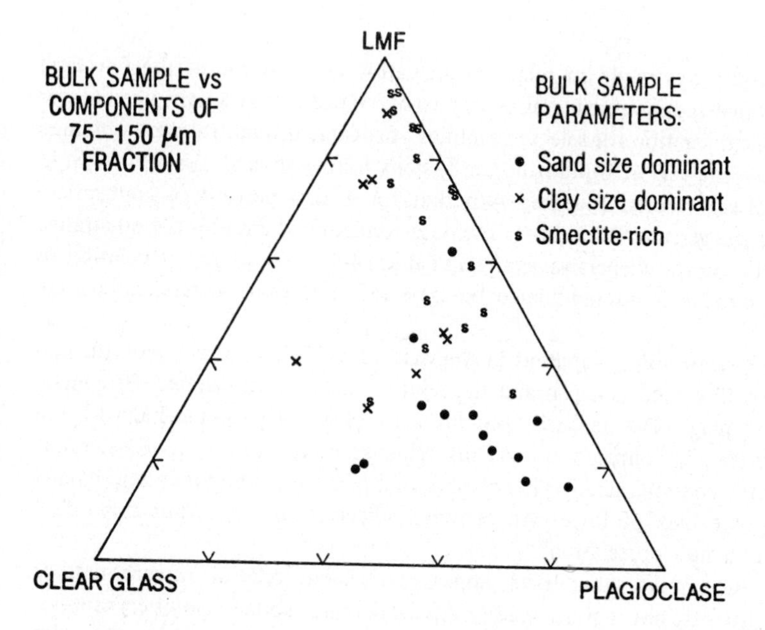

Fig. 14-18. Triangular plot showing relationships between the bulk sample parameters, size and smectite content, and relative amounts of the three components, plagioclase, glass and LMF (lithic matrix fragments) in the fine sand (75–150 μm) fraction of the same sample. Note that coarser samples are highest in plagioclase whereas the finer and smectite-rich samples contain more LMF.

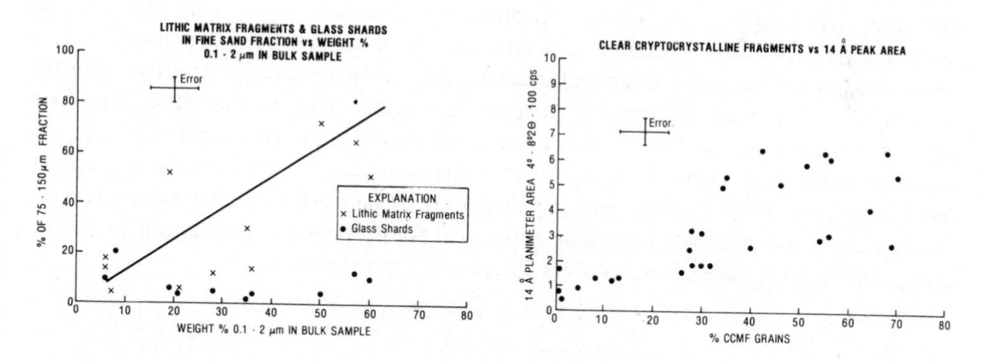

Fig. 14-19a. Plot showing relationship of LMF (lithic matrix fragments) and glass shards in the 75–150 μm fraction to the amount of 0.1–2 μm material in the same bulk sample. Note that samples containing higher percentages of LMF are also richer in the fine fraction. Glass shard amounts, however, remain constant as fine material increases.

Fig. 14-19b. Plot showing relationship between amounts of CCMF in 75–150 μm fraction and peak area for the 4–8° 2Θ (11–22Å) region (at 100 counts per second). Note the general increase in CCMF with basal reflection intensity.

chemistry. An attempt was made to relate major cation composition of the fine sand fraction to the principal sediment mineral components. Major cation weight percentages in the total 75–150 μm fraction for selected samples of sufficient quantity were obtained by atomic absorption analysis of lithium borate fusions following acid dissolution (S. L. Rettig, analyst). These results, together with that for weight percent of appropriate mineral components, are given in Table 14-4. The data confirm and amplify the qualitative results obtained with energy dispersive analysis (Table 14-3). No *unique* association of major cation with mineral component is to be expected, but some generalizations are possible.

The most obvious relationship apparent in the data of Table 14-4 is between the calcium content of the fine sand fraction and the relative quantity of calcite. Of course, Ca is also present in plagioclase feldspar, but the amount will vary somewhat with the plagioclases from different source areas. Thus feldspar contributions to Ca content should be larger from coarse-grained (phenocrysts of higher An content) mafic basalts than fine-grained (microlites of lower An content) siliceous andesites, but this effect cannot be split out in simple correlation.

Magnesium content, on the other hand, appears to bear no relation to carbonate or mafic mineral constituents, but is most readily associated with abundance of cryptocrystalline matrix fragments. Concentration of both Mg and CCMF suggests an irregular increase with depth.

Potassium concentrations, and sodium even more so, have a restricted range of values and show no apparent relation to specific sediment mineral components at all.

VanDenburgh (1975) has worked out the solute balance for Lake abert based on historical record for the lake and Chewaucan River, considering solute contributions from all other sources, such as intermittent streams, groundwater from seeps and springs, precipitation and aerosols, plus stored solutes in lake bottom and peripheral playa. He concluded (1975, p. 18) that of the "13,000 tons added to the lake in an average year, less than half (6,000 tons) are constituents that dominate in the lake (sodium plus equivalent chloride and carbonate-bicarbonate). Most of the remainder includes such components as silica, calcium and magnesium that are permanently removed from the solute cycle by organic processes and inorganic reactions." The fact that these losses have been continuous over at least recent times is indicated in his statement (1975, p. 19) "relative percentages of major constituents in the lake have not exhibited an appreciable net change since 1912." From petrographic study of lake sediment fine sand fractions, calcium and some silica can be readily accounted for in the occurrence of calcite, plagioclase, and diatoms, but the fate of other solute losses is not so obvious.

DISCUSSION

Examination of the fine sand fraction of Lake Abert sediment and comparison with source materials has suggested a number of processes occurring in the lake or associated groundwater environment, and provided background for the consideration of others.

Average grain-size distribution for representative Lake Abert sediment samples clearly indicates that the finest size fractions accumulate in the center of the basin, as expected,

TABLE 14-3

Chemical and Mineral Analyses for the 75–150 μm fraction of selected sediment samples.[1]

sample[2]	Depth (cm) (in sediment)	% Ca[3]	% Mg	% Na	% K	% Calcite[4]	% Feldspar[5]	% Pyroxene	% CCMF[6]	% LMF[7]
10a	0–12	4.4	1.4	2.0	2.0	4.0	33.3	7.3	23	40
8a	0–9	3.7	1.3	2.6	2.0	0	15.3	0.6	54	61
8c	23–41	2.1	3.7	1.0	1.9	3.3	13.0	0.6	68	75
8e	56–70	2.4	2.8	1.4	1.9	2	10.6	0	66	75
5a	0–12	6.7	1.9	2.5	1.4	5.0	38.6	15.3	n.d.	24
17a	58–70	29.4	0.5	1.0	0.5	67.3	8.6	1.3	17	18
18a	0–15	n.d.	0.8	1.0	0.7	58.0	6.2	0.6	29	30
7a	0–8	12.4	3.3	1.1	1.4	44.6	1.3	0.6	44	46
5c	21–37	8.1	3.8	1.5	1.9	33.3	6.0	2.0	46	51
19a	113–122	18.7	1.5	1.6	0.9	49.1	18.8	11.4	3	6
20a	2–9	22.3	0.9	1.7	1.0	36.0	15.3	1.3	15	22
20c	18–26	33.6	0.6	1.2	0.7	40.6	11.3	2.0	19	26
20g	82–91	13.8	2.2	2.0	1.3	18.0	16.0	2.6	47	49
20h	113–122	4.6	4.8	1.1	1.6	13.3	5.3	1.3	64	67
23cb	15–31	13.5	1.0	2.4	1.4	22.3	34.3	8.6	n.d.	14
23cf	76–91	9.0	2.0	2.1	1.7	25.3	34.0	4.6	3	15

1 Samples listed from south to north.
2 Sample numbers are site locations shown in Fig. 14-2.
3 Analyses by S. L. Rettig (USGS, Reston, VA) by lithium borate fusion
4 Mineral Percentage relative to those components counted (see Methods)
5 Feldspar includes mostly An_{50}–An_{80}
6 CCMF are Clear Cryptocrystalline Matrix Fragments
7 LMF are Lithic Matrix Fragments (of several types, see text)

TABLE 14-4.

SEM–EDAX compositions for 75–150 μm particles from Lake Abert sediment samples.[1]

Sample no. and depth in sediment[2]	Fragment	Relative constituent abundance
4d	CCMF[3]	Si ≫ K, Ca, Mg, Fe
26–37 cm[2]	CCMF	Si > Mg > Al > K > Ca > Fe
	1 μm^2 on CCMF surface	Si ≫ Al > K > Ca > Mg
18d	CCMF	Si > Mg > K > Fe > Ca > Al
40–43 cm	glass	Si > Al > K > Ca > Fe
17i	CCMF	Si > Mg > Al > Ca, K > Fe
26–52 cm		
702	CCMF	Si > Al > Ca > K, Mg > Fe
0–8 cm	glass	Si > Al > Ca > Fe
5C	CCMF	Si > Mg > Ca > K > Fe
21–37 cm	glass	Si > Al > Ca, K, Mg > Fe
16b	CCMF	Si > Al > Mg > K > Fe
61–79 cm		

1 SEM operation and analyses by Richard R. Larson, USGS, Reston, VA.
2 Locations given in Fig. 14-1.
3 CCMF = cryptocrystalline matrix fragments

but also that little coarse material is carried into the lake by the Chewaucan River inflow. Despite shallow water depths, the notably fine character of the lake sediment, which apparently increases downward in at least the top meter, argues for preferential disaggregation of the more siliceous fine-grained and cryptocrystalline rock types in the basin.

More than one-third of the fine sand (75–150 μm) fraction of Lake Abert sediment is composed of lithic matrix fragments, that is, grains composed of submicroscopic material which is chemically and petrographically similar to hyalocrystalline pyroclastic groundmass. The distribution of these fragments suggests that the major source is andesitic pyroclastics exposed to the east and south of the lake. The rocks apparently contributing the most to the fine-grained lake sediment are tuffs composed of cryptocrystalline and glassy material which contain significant amounts of expandable or mixed layer clay (Walker, 1963, 1968b; and reported herein). Near these lithologies the lake sediments show a substantial increase in both cryptocrystalline fragments and clay-size material. Major cation analysis of the fine sand fraction indicates that magnesium correlates with abundance of lithic matrix fragments and increases with depth in the sediment.

By contrast, in other areas of the lake, sediments associated with sources in holocrystalline feldspathic basalts show an increase in feldspar detritus, but considerably less of both cryptocrystalline material and clay-sized fraction.

Cryptocrystalline particles of another sort are found in the 75–150 μm fraction of Lake Abert sediments in proximity to dilute river flow. These particles, characterized by rudimentary basal reflections on x-ray diffraction analysis, have incorporated diatom frustules, and therefore are apparently produced by flocculation of clay-sized material.

Corrosion or alteration of feldspar and volcanic glass in the 75–150 μm size range was expected but not found. SEM examination of these components in the oldest ($\sim$5000 years) samples from more than 2 meters depth in the sediment showed sharp edges and fresh surfaces. A slight decrease of feldspar percentage with depth at some sites is as likely attributable to changes in sediment regime related to lake stage as to the pore fluid dissolution suggested by Jones, et al., (1969).

The distribution of pyroxene and opaques in the lake sediment, like detrital feldspar, is related to the crystallinity of the contributive lithologies. At the same time, olivine is a major component of the basalts, but is rarely found in the 75–150 μm fraction. The paucity of these minerals relative to source rocks is indicative of the importance of primary silicate hydrolysis and reducing conditions in lake bottom materials (Jones, 1966). Lake sedimentation appears to have been accompanied by complete dissolution of olivine, substantial loss of pyroxene, and reduction of iron oxide probably to form sulfide or become incorporated in clay minerals.

Glass shards in the lake sediments are either concentrated in certain horizons, most likely related to regional ash falls, or disseminated, suggesting derivation by erosion of local ash flow tuffs. Some glass is altered to zeolite, but association of fresh glass shards with clinoptilolite and phillipsite in the deepest playa sediments sampled, and the highly irregular distribution of phillipsite, suggests that the zeolites formed prior to deposition in Lake Abert sediment.

Analyses from the last 50 years (VanDenburgh, 1975, and unpublished data) indicate Lake Abert waters have been continuously supersaturated with respect to calcite, yet sediments in all size fractions at the southern end of the lake near the principal inflow from the Chewaucan River contain very little calcite. This suggests that freshly nucleated calcite is deposited further out in the lake, or that there is temporary inhibition of calcite precipitation. Further work on this is in progress.

Diatom frustule abundance can be related to the influence of dilute water inflow from the Chewaucan River and higher lake stage. Cemented or included frustules provide a key indication for the reaggregation of ultrafine sediment, and hint at a complex history for the pore fluid chemistry. Their importance in governing the silica budget of Lake Abert has been discussed by Jones, et al., (1969, p. 261) and Phillips and VanDenburgh (1971, p. 47). They apparently act as temporary sinks of SiO_2 which undoubtedly influence the kinetics of longer term silicate diagenesis.

The indications of higher magnesium contents in cryptocrystalline lithic matrix fragments as compared to unaltered volcanic glass, and the association of these fragments with clay content of the sediment is suggestive of multiple reactions between natural waters and sub-microscopic material. The establishment of sediment provenance through distribution data on mineral components in the fine sand fractions specifies that such reactions must include not only dissolution of primary silicates, but sorption and some diagenesis of clays and oxy-hydroxides. Study of the clay fractions is proceeding to further detail the nature of these reactions.

ACKNOWLEDGMENTS

We wish to express our indebtedness to a number of our colleagues in the U. S. Geological Survey, especially to Robert Eugene Smith for his invaluable assistance with all phases of the mineralogic work, and particularly his keen observations and attention to detail in the optical study. We are also grateful to Shirley L. Rettig for chemical analyses, to Meyer Rubin for ^{14}C determinations, Richard R. Larson for his expertise on operation of the SEM, and to L. N. Plummer, Rufus Getzen, and David R. Dawdy for review of the manuscript. Finally, we thank A. S. VanDenburgh, D. R. Dawdy, G. W. Walker, and Glenn E. Tyler, State Watermaster, Lake Co., Oregon (retired), for their continued advice; E. D. McKee for encouraging our contribution to the Symposium (Internat. Geol. Cong. 1976) at which this work was originally presented; and Betty L. Hudner, Carolyn E. Moss, and Gloria E. Agnew for patient help during the manuscript preparation.

REFERENCES

George, W. O., 1924. The relation of the physical properties of natural glasses to their chemical composition: Jour. Geol., v. 32, p. 353–372.

Hay, R. L., 1966. Zeolites and zeolitic reactions in sedimentary rocks: Geol. Soc. Amer. Special Paper 85, 130 p.

Hecky, R. E. and Kilham, P., 1973. Diatoms in alkaline saline lakes: ecology and geochemical implications: Limnol. and Oceanog. v. 18, p. 53–71.

Hull, H. and Turnbull, A. G., 1973. A thermochemical study of monohydrocalcite: Geochim. Acta, v. 37, p. 685–694.

Jones, B. F., 1966. Geochemical evolution of closed basin waters in the western Great Basin. Northern Ohio Geol. Soc., 2nd Sympos. on Salt, p. 181–200.

Jones, B. F. and Van Denburgh, A. S., 1966. Geochemical influences on the chemical character of closed lakes: in Symposium of Garda, Hydrology of Lakes and Reservoirs: Internat. Assoc. Sci. Hydrology Pub. 70, p. 435–446.

Jones, B. F., Rettig, S. L. and Eugster, H. P., 1967. Silica in alkaline brines: Science, v. 158, no. 3806, p. 1310–1314.

Jones, B. F., VanDenburgh, A. S., Truesdell, A. H. and Rettig, S. L., 1969. Interstitial brines in playa sediments: Chem. Geol., v. 4, p. 253–262.

Kittleman, L. R., 1973. Mineralogy, correlation, and grain size distribution of Mazama tephra and other postglacial pyroclastic layers, Pacific Northwest: Geol. Soc. Amer. Bull., v. 84, p. 2957–2980.

Lemke, R. W., Mudge, M. R., Wilcox, R. E. and Powers, H. A., 1975. Geologic setting of the Glacier Peak and Mazama ash-bed markers in west-central Montana: U. S. Geol. Survey Bull. 1395–H, 31 p.

Lerman, A. and Jones, B. F., 1973. Salt transport between sediments and brine in saline lakes: Limnol. and Oceanog., v. 18, p. 72–85.

Marschner, H., 1969. Hydrocalcite ($CaCO_3 \cdot H_2O$) and nesquehonite ($MgCO_3 \cdot 3H_2O$) in carbonate scales: Science, v. 165, p. 1119–1121.

Phillips, K. N. and VanDenburgh, A. S., 1971. Hydrology and geochemistry of Abert, Summer and Goose Lakes and other closed basin lakes in south central Oregon: U. S. Geol. Survey Prof. Paper 502–B, 86 p.

Shrock, R. R. and Hunzicker, A. A., 1935. A study of some Great Basin lake sediments of California, Nevada, and Oregon: Jour. Sed. Petrology, v. 5, p. 9–30.

Stewart, D. B., Walker, G. W., and Wright, T. L., 1966. Physical properties of calcic labradorite from Lake Co., Oregon: Amer. Min. v. 51, p. 177–197.

Summers, K. V., 1976. The clay component of the Columbia River palagonites: Amer. Min., v. 61, p. 492–494.

Surdam, R. C. and Sheppard, R. A., 1976. Zeolites in saline, alkaline lake deposits (abs): Programs and Abstracts, Zeolite '76, Tuscon, Arizona, June 1976, p. 65–66.

Truesdell, A. H., Jones, B. F., O'Neil, J. R. and Rettig, S. L., 1968. Chemical reaction and diffusion in the estuary of saline Lake Abert, Oregon (abs): Trans. Amer. Geophys. Union, v. 49, no. 1, p. 367.

VanDenburgh, A. S., 1975. Solute balance at Abert and Summer Lakes, south-central Oregon: U. S. Geol. Survey Prof. Paper 502–C, 29 p.

Walker, G. W., 1960. Age and correlation of some unnamed volcanic rocks in south-central Oregon, in, Geological Survey Research 1960, U.S. Geol. Survey Prof. Paper 400-B, p. 298–300.

Walker, G. W., 1963. Reconnaissance geologic map of the eastern half of the Klamath Falls (AMS) Quadrangle, Lake and Klamath Counties, Oregon: U. S. Geol. Survey, Mineral Inv. Field Studies Map MF–260.

Walker, G. W., 1967. Contraction jointing and vermiculitic alteration of an andesite flow near Lakeview, Oregon, in, Geological Survey Research 1967: U. S. Geol. Survey Prof. Paper 575-D, p. 131–134.

Walker, G. W., and Repenning, C. A., 1965. Reconnaissance geologic map of the Adel quadrangle, Lake, Harney, and Malheur Counties, Oregon: U. S. Geol. Survey Misc. Invest. Map I–446.

Walker, G. W. and Swanson, D. A., 1968a, Laminar flowage in a Pliocene soda rhyolite ash-flow tuff, Lake and Harney Counties, Oregon, in Geological Survey Research, 1968: U. S. Geol. Survey Prof. Paper 600–B, p. 37–47.

Walker, G. W. and Swanson, D. A., 1968b, Summary report on the geology and mineral resources of the Poker Jim Ridge and Fort Warner areas of the Hart Mountain National Antelope Refuge, Lake County, Oregon: U. S. Geol. Survey Bull. 1260–M, 16 p.

Waring, G. A., 1908. Geology and water resources of a portion of south-central Oregon: U. S. Geol. Survey Water-Supply Paper 220, 86 p.

Chapter 15

LAKE MAGADI, KENYA, AND ITS PRECURSORS[1]

H. P. EUGSTER

Department of Earth and Planetary Sciences, Johns Hopkins University, Baltimore, Maryland 21218

INTRODUCTION

Lake Magadi is an active salt lake located in the Rift Valley in Kenya. It contains a thick deposit of trona, $NaHCO_3.Na_2CO_3.2H_2O$, which has been exploited commercially for many years. Through a number of studies, Lake Magadi has become something of a prototype for saline alkaline lakes.

The Eastern, or Gregory Rift Valley in Ethiopia, Kenya and Tanzania has given rise to some 20 closed basin lakes (see Fig.15-1), some of which are quite fresh, like Naivasha or Turkana, while others, like Magadi or Natron, are very saline. Bedrocks are predominantly volcanic, and bicarbonate and carbonate are the principal anions in the waters, with the concentrated brines being very alkaline. Magadi is located just south of the equator at an elevation of 600 m, a low point of the Rift Valley. It is the only lake in the Rift in which bedded evaporites have accumulated.

The chemistry of the spring and lake waters was first investigated systematically by Stevens (1932) and later by Eugster (1970) and Jones et al. (1977). Baker (1958, 1963) described the geology of the Magadi area and also summarized the unpublished data of Stevens (1932). Silicate gels forming near the lake shore were described by Eugster and Jones (1968) and two new minerals, magadiite and kenyaite, were found in the sediments of the Pleistocene precursor of the lake (Eugster, 1967). Magadiite was linked with the bedded cherts found in the older sediments (Eugster, 1969), while the authigenic minerals, particularly the zeolites, were investigated by Surdam and Eugster (1976).

As Fig.15-2 shows, there are no perennial rivers entering Lake Magadi and inflow is by perennial springs and ephemeral runoff. In fact, there is a stretch of 100 km between Lakes Magadi and Naivasha where the Rift Valley is devoid of perennial surface flow. For sedimentological considerations, it is important to realize that the lake is located in a long, narrow fault trough with the steep sides formed by trachyte lava flows. This is in contrast to another well-known alkaline lake: Eocene Lake Gosiute of the Green River Formation of Wyoming.

During the last million years or so, the Magadi trough has been occupied by several lakes in succession, Lake Oloronga being the oldest, while High Magadi Lake was the immediate precursor of the present lake. The stratigraphy of the tectonic and volcanic events in this part of the Rift has been worked out by Baker et al. (1971), Baker et al. (1972), Baker and Mitchell (1976) and Baker et al. (1977), and it is now possible to correlate the sedimentation events with this time frame.

[1] Dedicated to the memory of J. A. Stevens (1905–1979) of Kiambu, Kenya – scientist and friend.

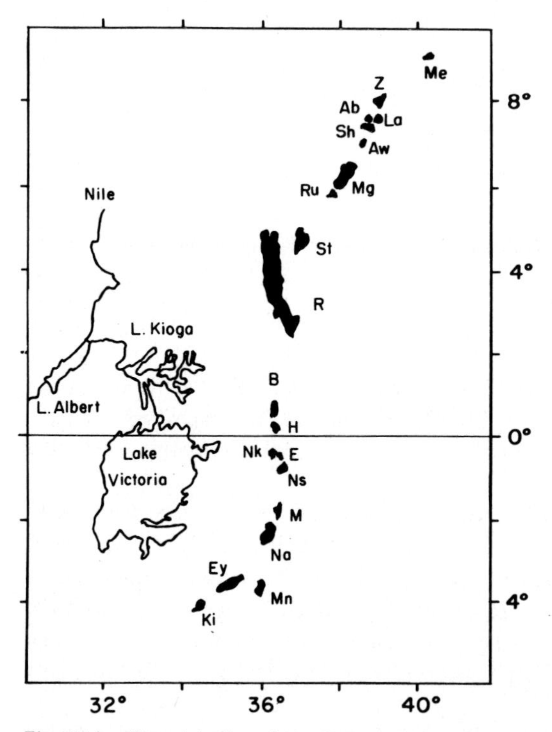

Fig. 15-1. Closed basins of the Eastern Rift Valley, Ethiopia, Kenya and Tanzania. From north to south: Me: Metahara, Z: Zwei, Ab: Ablata, La: Langano, Sh: Shala, Aw: Awassa, Mg: Margherita, Ru: Ruspoli, St: Stephanie, R: Turkana, Ba: Baringo, H: Bogoria, Nk: Nakuru, E: Elmenteita, Ns: Naivasha, M: Magadi, Na: Natron, Mn: Manyara, Ey: Eyasi, Ki: Kitangiri. Lake Victoria is given for reference.

It is the purpose of this report to present a coherent account of the Pleistocene to Recent history of Lake Magadi, its precursors and their deposits. It will include not only summaries of the published accounts, but also much of the unpublished material collected in 1966, 1968, 1970 and 1973. In this way, the reader will have a better chance to assess what Lake Magadi has to teach us about hydrochemistry, mineral reactions and sedimentation in an arid environment.

After setting the stratigraphic framework, we will proceed from the youngest to the oldest, because we know most about the active lake in terms of hydrochemistry and evaporite deposition. Discussion of the High Magadi Beds will afford us an opportunity to describe the cherts and their sodium silicate precursors. The Oloronga Beds fill in the important early history of the basin and make correlations with other contemporaneous deposits possible, such as the Peninj Group in the Natron Basin to the south (Isaac, 1965), the Legemunge Beds (Baker and Mitchell, 1976), the Naivasha Basin (Richardson and Richardson, 1973), and the Elmenteita-Nakuru Basin (Butzer et al., 1972) to the north. Authigenic mineral reactions are discussed next, because they affect lake beds of all ages. Finally, environments of sedimentation will be considered and the Magadi Basin can then be compared and contrasted with other closed basins.

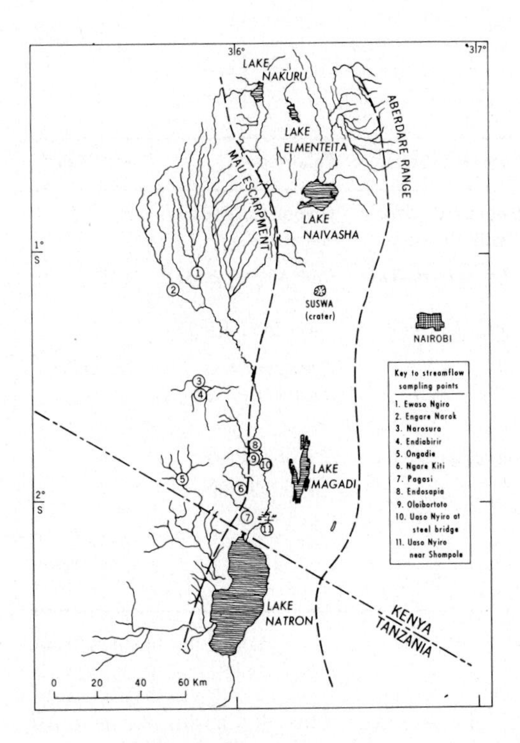

Fig. 15-2. Hydrology of the area between Lake Nakuru and Lake Natron (from Jones et al., 1977). All perennial streams are shown, 8, 9 and 10 being the important rim streams for Magadi. The dashed lines indicate the rift boundary.

THE STRATIGRAPHIC FRAMEWORK

The most comprehensive account of the geology of the Magadi area was published by Baker (1958). His report not only covers the Magadi trough, but a 30 minute quadrangle which includes most of the Rift Valley in that latitude. It is concerned with the metamorphic rocks of the Basement System, with the volcanic rocks of the Rift Valley proper, as well as with the lacustrine and fluviatile sediments.

Baker (1958) gives a short account of the previous geological and geochemical studies at Magadi. They include an early report by Parkinson (1914) and two unpublished reports, one by J. A. Stevens (1932) and the other by B. N. Temperley (1951). Baker (1958) distinguishes four major stratigraphic units among the lake deposits of the Magadi area: Oloronga Beds, Chert Series, High Magadi Beds and Evaporite Series, with major faulting and erosion separating Oloronga Beds from the Chert Series and minor faulting between Chert Series and High Magadi Beds (Baker, 1958, p. 9 and p. 26). In a subsequent report concerned with the area south of Magadi, Baker (1963) reversed the positions of the Oloronga Beds and the Chert Series and placed both episodes of faulting, grid faulting and minor faulting, after Oloronga time. In both versions, kunkar limestone (caliche in our terminology) directly underlies the High Magadi Beds. Eugster (1969) later showed that many of the cherts of the Magadi trough in fact belong to the High Magadi Beds and are

TABLE 15-1.

Magadi Stratigraphy.

Baker (1958)	Baker (1963)	Eugster (1969)	This Report	Age, Years
Evaporite Series (no deformation)	*Evaporite Series* (no deformation)	*Evaporite Series* (slight titting)	*Evaporite Series* (slight tilting)	0
High Magadi Beds (Kunkar limstone) (minor faulting)	*High Magadi Beds* (Kunkar limestone) (grid faulting)	*High Magadi Beds* (grid faulting)	*High Magadi Beds* (minor faulting)	9,100[1]
Chert Series (grid faulting) (Kunkar limestone)	*Oloronga Beds* (erosion)		*Trachyte flows* (Kunkar limestone)	780,000[2]
Oloronga Beds	*Chert Series*	*Oloronga Beds*	*Oloronga Beds* (grid faulting)	
Plateau Trachytes				up to 1,300,000[3]

[1] Butzer et al. (1972); [2] Fairhead et al. (1972); [3] Baker and Mitchell (1976).

time-equivalent with the magadiite horizons of those beds, thus eliminating the need for a separate Chert Series. Many of the dips of beds used by Baker (1958, 1963) to indicate faulting were reinterpreted as primary depositional dips. Table 15-1 summarizes the published versions of Magadi stratigraphy.

Subsequent field work has revealed the following new aspects:

1) The Oloronga Beds, named by Baker (1958) and located by him primarily on the scarp to the SW of Lake Magadi, can be traced directly to the floor of the valley just south of the lake. They contain several chert horizons, some of which were mapped by Baker (1958) as Chert Series and by Eugster (1969) as High Magadi Cherts.

2) Oloronga Beds also occur extensively to the north and north-east of Lake Magadi as well as in the area of the NW lagoon.

3) Transgressions of High Magadi Beds over Oloronga Beds are exposed extensively throughout the Magadi basin. The surest way of recognizing the transgressive surface is by the presence of a thick caliche accumulation (kunkar limestone) which caps the Oloronga Beds. No caliche exists in the High Magadi interval.

4) The youngest trachyte flows, those descending to the Magadi floor N of the NE lagoon, post-date Oloronga deposition, making the Oloronga Beds at least 800,000 years old. As the High Magadi beds have been dated at 9,100 years (Isaac et al., 1972, Butzer et al., 1972), this implies a very long depositional hiatus which is represented by the thick caliche horizon.

5) Dips of beds observed in the Oloronga Beds can all be interpreted as primary depositional slopes and there is no unequivocal evidence for major post-Oloronga faulting. This places the major phase of the extensive grid faulting observed in the trachyte flows in pre-Oloronga time.

Lake Oloronga must have been the first lake of the Magadi basin, and it did not form until the major plateau trachyte extrusions had been completed and the flows had been extensively block-faulted. Its deposits can probably be correlated with those of the Peninj Group west of Lake Natron just south of Lake Magadi (Isaac, 1965). Lake Oloronga was the largest and perhaps freshest of the three lakes, though at times its waters must also have been quite alkaline, judging from the Magadi-type cherts present. It is reasonable to assume that it continued to exist after the 780,000 year old lava flow for an extended period of time, and that some of the Oloronga outcrops are contemporaneous with the Legemunge Beds to the north (Baker, 1958), some of which have been dated at 400,000 years (Baker, et al., 1976). In fact, the lake could have existed until as recently as 20,000 years ago, the last high-water level in E. Africa before High Magadi time ($\sim$ 10,000 years, see Butzer et. al., 1972). As mentioned, however, the thick caliche cap on top of all Oloronga sections points to a long, dry hiatus.

High Magadi Lake filled the basin again to a level of about 13 m above the present lake level and its shore lines are still clearly visible, particularly on the east shore. It was smaller and probably shallower than Oloronga Lake and it also fluctuated from relatively fresh to saline, as indicated by the several magadiite and chert horizons. Its sediments, although similar to those of Lake Oloronga, can readily be distinguished. The date of 9,100 years (Baker et al, 1972) comes from the *Tilapia* horizon in the lower part of the section (Baker, 1958; Eugster, 1969).

High Magadi Lake did not exist for a long time. Its place was taken by a lake similar to the existing lake, from which the Evaporite Series was deposited. This series is contiguous with the trona deposition taking place at the present time.

The Evaporite Series occurs only in the presently active lake, while the Oloronga and High Magadi sediments can be studied both in outcrop and drill cores. Correlation between outcrops and cores is difficult, because of the very different appearance of the sediments when they are soaked in brine. Using mineralogic arguments, Surdam and Eugster (1976) have made some tentative assignments, which are given in Table 15-2, with the drill core locations shown in Fig. 15-3. The thickness shown for F6 must be fairly close to maximum thickness. Fig. 15-4 shows a schematic cross-section and illustrates the relative positions of the stratigraphic units. It also illustrates the fact that the original lake bottom was not flat, but interrupted by a number of narrow, long, N-S trending lava horsts, over which the sediments were draped, some with primary depositional slopes of

TABLE 15-2.

Stratigraphic units in drill cores, depth in m (From Surdam and Eugster, 1976).

	F_6	J_{10}	C_{16}	H_4
Evaporite Series	0 – 18	0 – 10	0 – 8	0 – 28
High Magadi Beds	18 – 82	10 – 32+	8 – 11	28 – 43+
Oloronga Beds	82 – 133+		11 – 17+	

over 20°. It is these dips which led Temperley (1951) and Baker (1958) to infer post-depositional deformation ("Chert Series").

As we shall see, the sediments are mainly of volcanic derivation, with illite, the only clay mineral, occuring in traces only (Surdam and Eugster, 1976). Most of the sediments are reworked and water-laid volcanic tuffs, with the rest consisting of cherts and evaporites (trona). Also, we must remember that many of the High Magadi sediments are probably largely reworked Oloronga material.

With the stratigraphic framework in mind, we can now turn to a discussion of the presently active Lake Magadi.

HYDROCHEMISTRY AND BRINE EVOLUTION

Chemical composition for representative types of water inflows are given in Table 15-3.

A series of perennial alkaline springs issue at the perimeter of the Magadi basin. Their temperatures vary from ambient to 86°C and their concentrations from 6 to 35 g dissolved solids per kg. Temperature and concentration do not vary sympathetically, however the Na:Cl ratios are surprisingly constant (Eugster, 1970). For the analyses reported by Jones et al. (1977), the molar ratio is 3.21. A similar ratio, but with greater scatter, also applies to the fresh-water streams entering the Rift Valley from the western rim (see Fig. 15-2). Except for the Ewaso Ngiro, which feeds Lake Natron, all of these streams disappear in the alluvial fans before they reach the floor of the valley. Because of this similarity in Na/Cl these streams have been postulated to be the major recharge for the deep ground water reservoir which feeds the perennial springs. Fig.15-5 shows the Na:Cl ratios of the major inflow sources and lake brines. There is a compositional gap between river streams and springs, which is occupied by only three points, labelled as ground waters. They were sampled from shallow pits dug into lake beds N, E and S of the lake and they presumably represent regional shallow ground water fed by local runoff.

The springs drain into a number of shallow, perennial water bodies peripheral to the lake, which Baker (1958) called lagoons (see Fig. 15-3). Brine evolution from springs to lagoon brines to lake brines takes place principally by surface evaporation and can be followed directly. As fig. 15-5 shows, the Na:Cl ratio remains constant until the most concentrated brines, which exhibit an enrichment in Cl. This is readily accounted for by precipitation of trona, $NaHCO_3.Na_2CO_3.2H_2O$, depleting the residual brine in Na.

The hydrologic model proposed by Eugster (1970) and Jones et al. (1977) rests on the observation that Na or Cl are not removed preferentially during the evaporative concentration processes until the very end. Hence they can be used as tracers to define how much of the other constituents are lost. Important losses occur for the major constituents Ca and Mg, HCO_3 and CO_3, SiO_2 and K and these losses can be used to elucidate the concentration mechanisms by which the brines evolve.

Between rim streams and shallow ground water, Ca and Mg have not increased substantially, while Na and Cl have been concentrated some 30 times. This means that over 90% of the alkaline earths initially present in the waters have been removed. Further removal occurs between shallow ground water and springs. The latter, because of their high pH (9–10), are virtually devoid of alkaline earths. It is reasonable to assume that the loss is

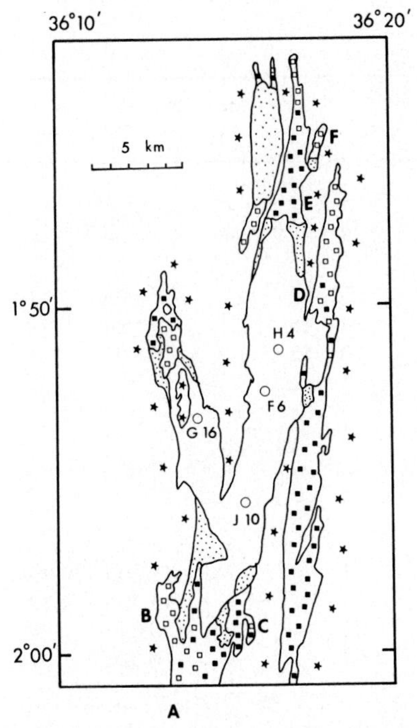

Fig. 15-3. Lake Magadi sketch map. Stars: trachyte flows, open squares: Oloronga Beds, solid squares: HIgh Magadi Beds, dots: lagoons (open water), open: trona, circles: locations of drill holes. Bold letters: sections of Fig.15-20, A–F.

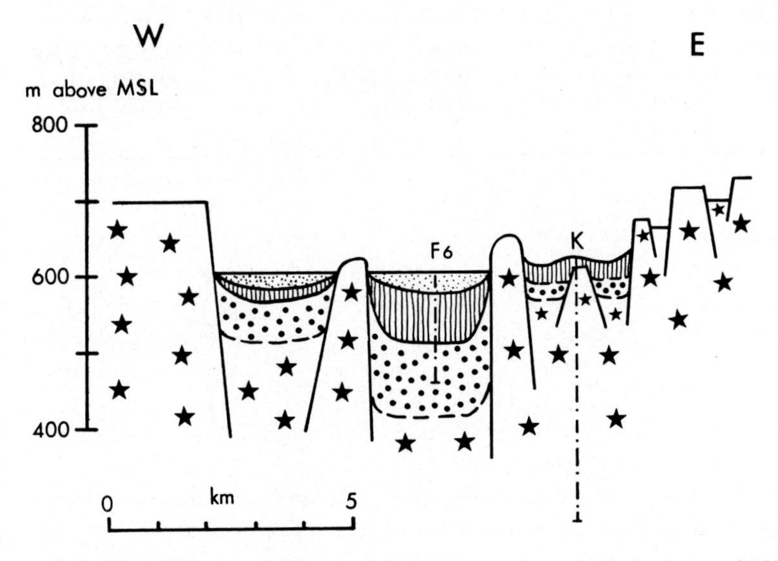

Fig. 15-4. Cross-section through Lake Magadi with a vertical exaggeration of 1500. Stars: trachyte flows, large dots: Oloronga Beds, vertical lines: High Magadi Beds, small dots: Evaporite series. F6 and K are drill holes located on Fig. 15-3.

TABLE 15-3.

Representative water compositions (from Jones et al., 1977)

Date	Location And/Or Sample Number	Temp. (°C)	Field pH	Lab pH	SiO_2	Ca	Mg	Na	K
	Dilute Inflow								
5/73	Oloibortoto	19	7.20	7.35	19	6.5	3.6	5.3	2.3
	Uaso Nyiro:								
5/73	At steel bridge	23	7.42	7.7	29	12	2.7	39	15
5/73	2 Km N of Shompole	23	7.8	7.7	38	15	2.1	60	20
5/73	Near Shompole	24	7.62	7.6	38	12	2.3	63	17
	Ground Waters								
1/68	Precip. seepage M288	–	–	8.3	–	9.7	2.3	40	10
1/68	Ngejepa wells, M217	–	–	9.3	60	1.9	.9	1,000	33
	Hot Springs								
1973	M1018	83	9.44	9.05	85	–	–	10,500	198
7/70	M547	41.8	9.60	9.45	46	–	–	12,700	127
5/73	M1008	37	9.66	9.30	55	–	–	3,870	49
5/73	M1011	36	9.44	9.25	63	–	–	6,320	98
	Surface Brines								
7/70	M536	40	10.9	10.45	1,040	–	–	124,000	1,790
5/73	M1014	32	10.4	9.65	496	–	–	75,000	1,390
	Borehole Brines								
1/71	J10 (surface)	–	–	10.30	1,540	–	–	122,000	2,900
1/71	J10-31.7 m	–	–	9.85	288			69,400	1,170
1/71	K-297 m	–	–	10.00	–	–	–	40,500	–

HCO_3	CO_3	SO_4	Cl	F	BR	B	PO_4	Dissolved Solids	Density
46	–	3.5	3.7	.1	.013	.0	.0	67	–
148	–	8.6	8.2	2.2	.038	.01	.01	190	–
209	–	12	13	3.2	.067	.0	.02	267	–
194	–	14	15	3.6	.083	.03	.00	260	–
128	–	7.9	11	1.7	–	.04	.02	175	–
1,000	–	25	375	11	.6	.68	.85	2,000	–
10,400	4,450	168	4,890	146	20	8.3	5.5	25,600	1.024
8,820	7,030	206	6,540	132	20	5.1	17	32,700	1.025
4,080	1,590	122	1,550	50	6.8	3.0	2.8	9,600	1.008
7,230	2,120	196	3,000	69	17	4.3	3.0	15,500	1.013
–	93,000	1,890	74,100	1,560	260	89	–	313,000	1.310
7,650	54,300	1,240	49,400	830	179	64	59	187,000	1.174
–	62,800	2,880	114,000	1,920	427	–	130	310,000	1.278
6,160	49,300	1,060	42,100	807	148	52	63	167,000	1.153
11,000	29,200	–	20,600	452	–	–	–	97,000	

due to precipitation of alkaline earth carbonates by surface or capillary evaporation. Surdam and Eugster (1976) noted the presence of abundant aggregates and veinlets of anhedral calcite in the tuffaceous lake beds.

The behavior of HCO_3 and CO_3 during brine evolution is shown in Fig. 15-6, using Cl as a tracer. No loss of carbonates during concentration is represented by the dashed line (45°). Ground waters, springs and lake brines fall well below that line, indicating a substantial, but gradual loss of $HCO_3 + CO_3$ with respect to Cl during evaporative concentration. Four different mechanisms have been invoked to account for this loss, all of which are probably active through some of the concentration range: degassing of CO_2, alkaline earth precipitation, sodium carbonate precipitation in efflorescent crusts and trona precipitation. The dilute inflow is supersaturated iwth respect to atmospheric CO_2 by about one order of magnitude (Jones et al, 1977), while the lake brines are distinctly undersaturated, because trona precipitation removes HCO_3^- and CO_3^{--} in equimolar proportions and hence drastically depletes the residual brines in HCO_3. Surface (efflorescent) crusts formed by evaporation of runoff or by capillary evaporation of shallow ground water are important in removing not only Ca and Mg, but also HCO_3 and thus CO_2. Such crusts are the result of complete desiccation and hence contain soluble as well as insoluble constituents. Subsequent contact with dilute rain water or runoff dissolves only the most soluble constituents, such as NaCl from halite and Na_2CO_3 from thermonatrite ($Na_2CO_3.H_2O$) while calcite, magnesian calcite, dolomite, silica and silicates remain largely untouched and hence their constituents are removed from the waters. We believe that such efflorescent crusts, which are present as thin white coatings on lake beds and alluvial flats of the Magadi area, are the main agents for solute acquisition during

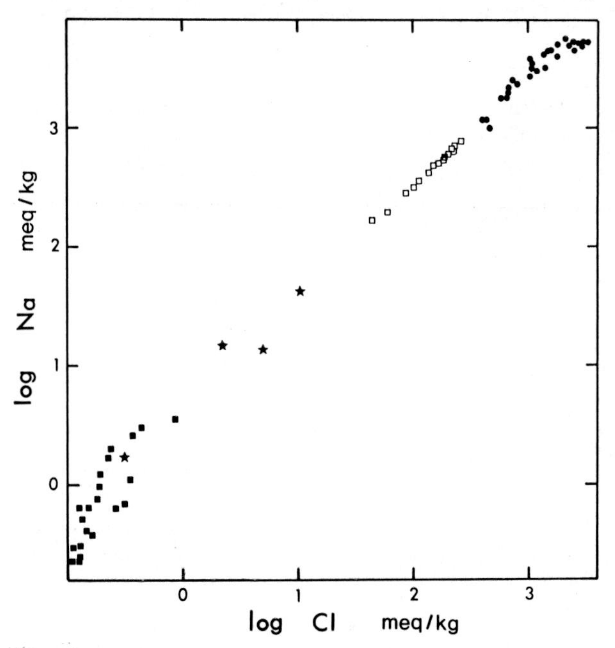

Fig. 15-5. Na vs Cl. Solid squares: dilute inflow, stars: dilute ground water, open squares: hot springs, dots: lagoon and lake brines, as well as drill hole brines. Jones et al. (1977).

evaporative concentration and brine evolution. Through complete precipitation and differential dissolution they allow the most soluble constituents, such as Na^+, Cl^- and CO_3^{--}, to become preferentially enriched in the waters. They form wherever the ground water table is close enough to the surface for capillary evaporation to be effective. Direct capillary evaporation of ground water (the evaporative pumping of Hsu and Siegenthaler, 1969) is also important in increasing the solute load but it should affect all solutes equally. Finally, the effectiveness of trona precipitation for carbonate removal form the most concentrated brines is shown dramatically in Fig. 15-6, where at the highest concentrations ($HCO_3 + CO_3$) actually decreases with increasing Cl.

Potassium is lost even more extensively than ($HCO_3 + CO_3$), as Fig. 15-7 shows. The loss occurs between rim streams and springs, with the dilute ground water again occupying an intermediate position. Once circulation is entirely on the surface, from springs to the most concentrated lake brines, no further K loss is noticeable. K loss during subsurface circulation has been accounted for by Eugster and Jones (1977) by exchange on reactive surfaces. Since clays are absent from the basin (Surdam and Eugster, 1976), these surfaces must be from volcanic glasses or the silicate gels described by Eugster and Jones (1968) and found in many of the core samples by Surdam and Eugster (1976).

K loss means relative Na enrichment. This is shown clearly in Fig. 15-8. In the springs, the atomic Na/K ratio is over 100, a 25-fold increase over that of the rim streams. This is the highest enrichment factor we have encountered in any of the closed basins studied (Eugster and Jones, 1977), testimony to the effectiveness of the trachyte glasses and gels for K removal. From springs to lake brines there is no further K loss; quite the contrary; there is a preferential K enrichment in the most concentrated brines due to trona crystallization. This late-stage enrichment of K in residual brines also takes place in marine evaporites and for similar reasons.

The gradual removal during the subsurface branch and the late-stage enrichment produces a sigmoid curve. An even more pronounced sigmoid curve is exhibited by silica (see Fig. 15-9). The SiO_2 content remains essentially constant at 50 ppm between inflow and springs, while chloride increases some 500 times. Then, as pH rises above 9, SiO_2 remains in the brine and is concentrated at the same rate as Cl is. This is possible through formation of other species in addition to H_4SiO_4 such as $H_3SiO_4^-$ and $H_2SiO_4^{--}$ at high pH.

The SiO_2 plateau between inflow and springs is probably governed by the solubility of the trachyte glasses and silicate gels and the removal mechanism is presumably connected with capillary evaporation and the surface crusts mentioned above. Since 99.7% of the silica initially present in the inflow is lost, a substantial amount of opaline cement must be precipitated. However, it would be difficult to identify it among the abundant volcanic glasses and gels. Removal of silica by diatoms is not important, since even in the High Magadi beds diatoms are not particularly abundant (Eugster and Chou, 1973).

Among the other constituents, sulfate and fluoride are most interesting. As pointed out by Eugster (1970) and Jones et al. (1977), Magadi brines are greatly depleted in SO_4 with respect to inflow. This is shown in Fig. 15-10. The most obvious mechanism for this removal seems to be bacterial sulfate reduction, which is apparently effective over the whole concentration range. Interestingly enough, some of the deeper bore hole brines are most impoverished in sulfate.

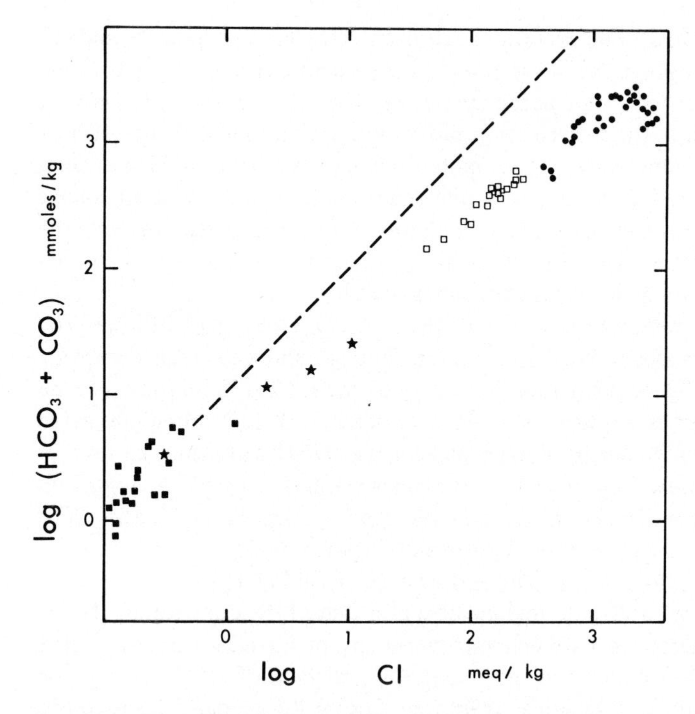

Fig.15-6. $HCO_3 + CO_3$ vs. Cl. Symbols as in Fig. 15-5.

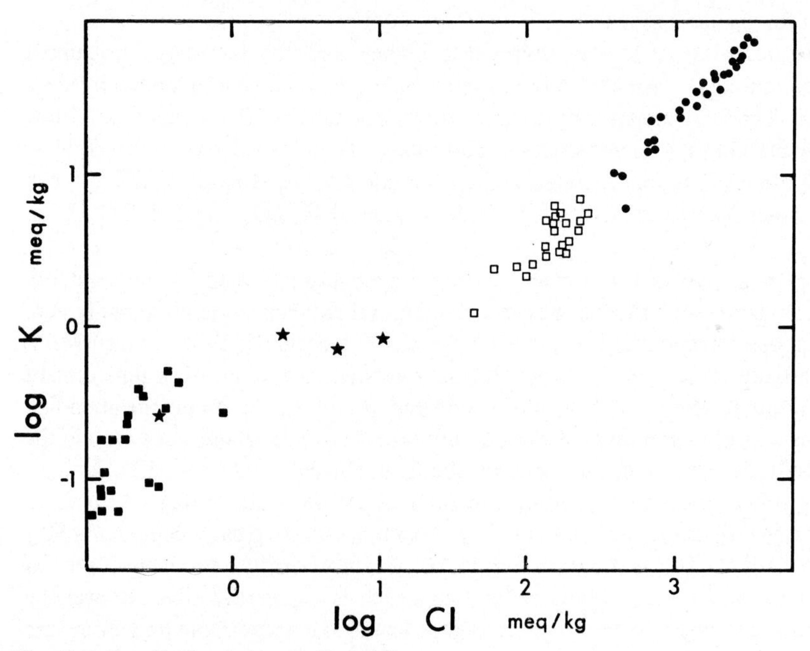

Fig. 15-7. K vs Cl. Symbols as in Fig. 15-5.

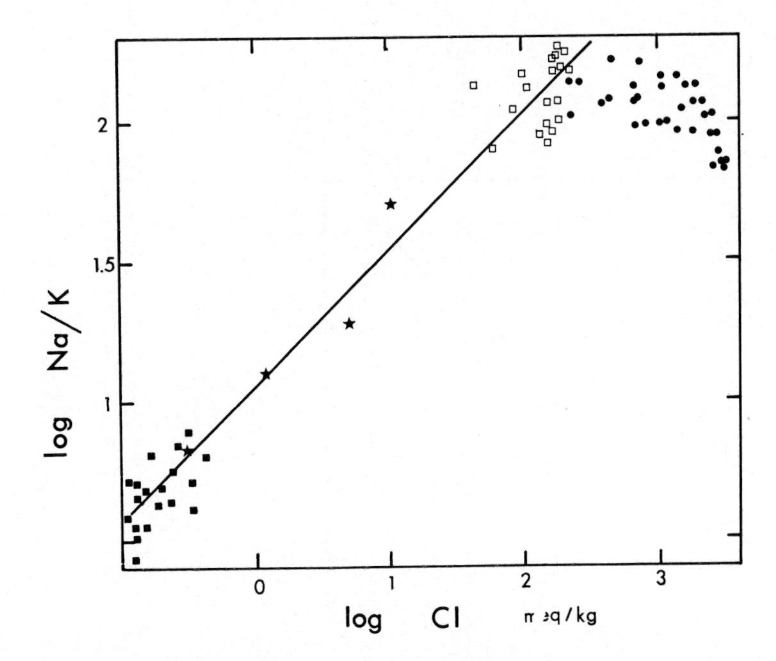

Fig. 15-8. Na/K (atomic ratio) vs Cl. Symbols as in Fig. 15-5.

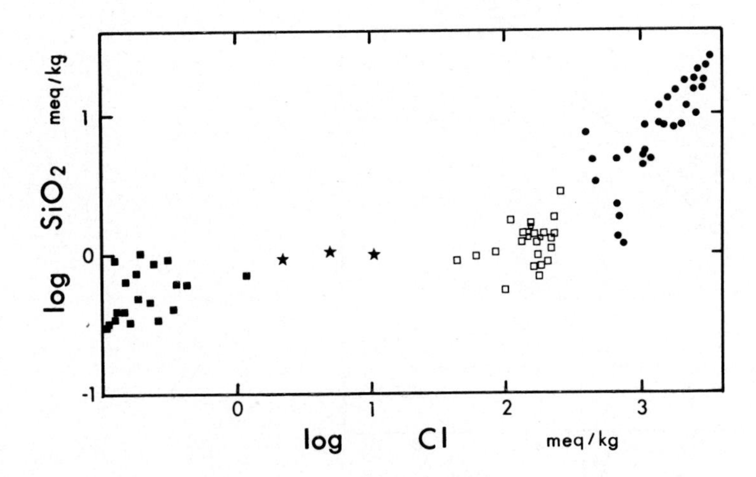

Fig. 15-9. SiO_2 vs Cl. Symbols as in Fig. 15-5.

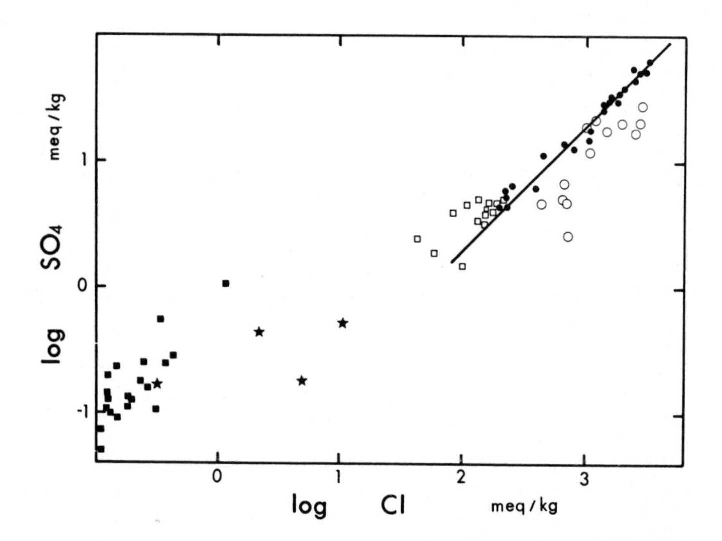

Fig. 15-10. SO_4 vs. Cl. Symbols as in Fig. -5, except that drill hole brines are shown as open circles.

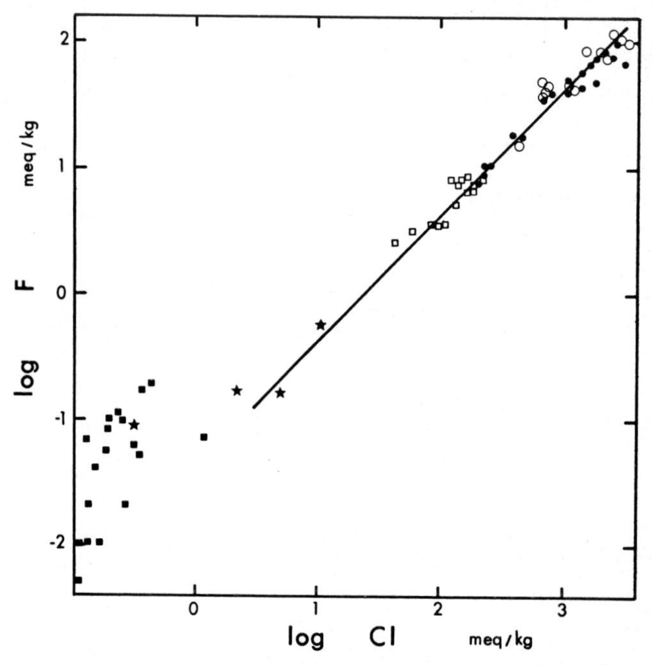

Fig. 15-11. F vs. Cl. Symbols as in Fig. 15-10.

Fluoride is enriched in the Magadi brines to an extraordinary degree (Fig.15-11). Villiaumite, NaF, occurs in considerable amounts in the trona deposits (Baker, 1958) and is the principal impurity. Fluoride is high not because of unusual inflow sources, but because Ca is generally not available to remove F as fluorite, because it has been removed earlier as carbonate. Where calcite is present, such as in the Oloronga beds in the drill core, fluorite forms by interaction with the interstitial brines in considerable quantities (Sheppard et al., 1970; Surdam and Eugster, 1976).

Bromine follows chloride closely during evaporative concentration. There is no fractionation, because halite saturation has not been reached.

A schematic circulation system (see Fig. 15-12) has been postulated for the Magadi waters based on compositional data (Eugster, 1970; Jones et al., 1977). It involves inflow from dilute rim streams descending into the Rift Valley and ephemeral runoff within the valley floor, feeding local, dilute, shallow ground water bodies. It also stipulates the presence of a hot, saline ground water body at some depth within the fractures of the trachyte lavas. This body is inferred from the costancy of the spring temperatures and compositions over 40 years. The hottest spring N of Little Magadi is taken as the most representative sample of this ground water body. It has the same Na/Cl ratio as the dilute inflow, which represents its eventual recharge.

To arrive at the various spring temperatures, compositions and concentrations, three inputs are necessary, which are mixed in different proportions for each spring: shallow ground water (cold, dilute), deep ground water (hot, saline) and recirculated surface brine (cold, very saline, very high pH). In this way, it is possible to account for the properties of each spring (Eugster, 1970).

Fig. 15-12 shows some of the major flow paths as well as representative concentrations for the various water bodies.

From springs to lagoon brines to lake brines to residual brines, processes and products are reasonably clear, because they occur at or near the surface. Surface evaporation and

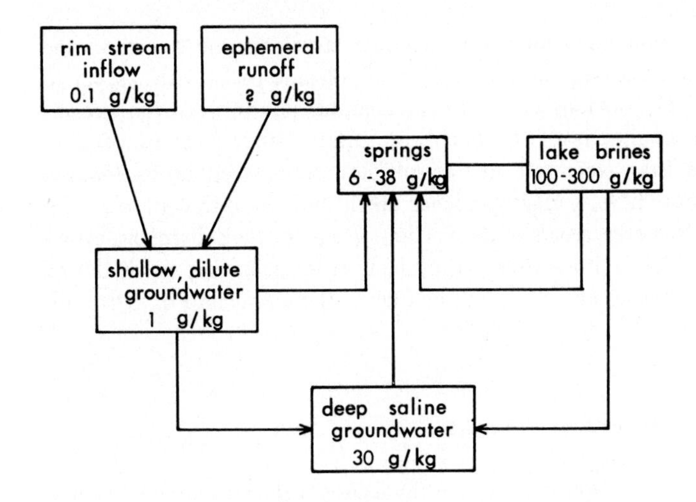

Fig. 15-12. Schematic circulation system.

fractional crystallization of trona are the major processes. On the other hand, the paths from inflow to ground waters to springs are more obscure and must be inferred. Major mechanisms for increasing the solute load are surface and capillary evaporation and fractional dissolution of efflorescent crusts. The latter depends on rain water and dilute runoff, but it is capable, through percolation to add substantial solute loads even to deeper ground water bodies. We believe it is the major mechanism for solute acquisition at Magadi which is located in an area of very little surface waters and surface flow. On the other hand, if runoff and local shallow ground water bodies were not available, efflorescent crusts could not form either.

Precipitation of efflorescent crusts by complete desiccation of surface or ground water and subsequent fractional dissolution has a decisive influence on brine evolution. Na and Cl are contained only in very soluble minerals, such as halite, thermonatrite ($Na_2CO_3 \cdot H_2O$) or trona and hence no fractionation occurs between Na and Cl. Alkaline earths, on the other hand, are removed as carbonates and silica probably as an opaline cement. K, on the other hand, is screened out by a different process: adsorption on active surfaces.

The final question relating to hydrochemistry relates to the composition of the inflow. As Hardie and Eugster (1970) have shown very clearly, the eventual brine composition is inherited from the inflow and can be predicted with reasonable confidence. The most recent brine evolution scheme of Eugster and Hardie (in press) uses alkaline earth carbonates as the major precipitates for Ca and Mg rather than sepiolite, although the latter is still possible at low carbonate and high silica levels. Lake Magadi type brines (pH up to 11) obviously form from very soft inflow waters, in which bicarbonate predominates over Ca + Mg. Thus carbonate precipitation leads to Ca and Mg depletion and HCO_3 enrichment. This is what Garrels and Mackenzie (1967) found for the Sierra Nevada waters. The rim streams which enter the Rift Valley are in contact with Precambrian metamorphic basement and with Pleistocene basalts (Baker, 1958), ideal lithologies for producing such waters. Presumably because of the presence of sulfides, sulfate is quite high in some inflow, but it is later removed to a large extent by biogenic reduction. In this manner, a classic alkaline brine results.

There is some question about how much the trachyte lavas contribute to the brine evolution. Jones (1966) has stressed the importance of volcanic rocks with respect to weathering rates, but the only obvious difference in composition between rim streams and ephemeral runoff in the valley itself is probably in sulfate, because the runoff is in contact only with trachyte lavas, which are free of sulfides. As pointed out by Jones et al. (1977), these waters also have a slightly higher alkali and fluoride content. The presence of volcanic glass, however, seems to be very important for the subsequent evolution. No basin we know of has a more effective mechanism for screening out K during subsurface circulation than Magadi and this must be due to the abundance of active surfaces in the volcanic sediment.

TRONA PRECIPITATION AND THE EVAPORITE SERIES

The alkaline brines which form by evaporation of the lagoon waters eventually become saturated with respect to trona, $NaHCO_3 \cdot Na_2CO_3 \cdot 2H_2O$, which has been accumulating at Magadi in considerable amounts since about 9,000 years ago. Baker (1958) reported the presence of trona in four drill cores. In one case, abundant trona was found to a depth of 40 m. More recent drilling confirmed this observation (Surdam and Eugster, 1976, Table 6). The interval during which extensive trona crystallization has been taking place has been assigned to the Evaporative Series which also includes the currently forming deposits. As the core data show, trona deposition has not been continuous during this time, but was interrupted several times, particularly during the early period, by deposition of volcanogenic sediments. What drill data are available indicate that two depositional centers formed early, one north and the other south of the causeway. They presumably were separated by a delta extending from east across the lake. Sediment sources could have been the alluvial valley SE of the main lake area as well as the stream descending from the eastern scarp (Olkeju Ekisichiyo). At the present time, trona deposition is taking place at its maximum rate and trona forms throughout the lake, except for the marginal lagoons. Volcanogenic sediment input is restricted and only wind-born material reaches the center of the lake.

Trona deposition is a complex process and not yet fully understood. In part, this is due to the fact that trona solubility is strongly dependent upon the CO_2 content of the solution, or P_{CO_2}. The relevant equilibria have been discussed by Bradley and Eugster (1969) and Eugster (1966, 1971) and the two most important diagrams are shown in Figs. 15-13 and 14. Fig. 15-13 accounts for the fact that trona is the dominant sodium carbonate phase at the surface and the only one capable of accumulating in thick bedded deposits. The stability field of trona straddles the P_{CO_2} of the atmosphere ($\sim 10^{-3.5}$ atm.) in the pure carbonate system at temperatures above about 20°C. Nahcolite, $NaHCO_3$, another common sodium carbonate, can form stably only when P_{CO_2} is raised to a level at least 10 times that of the atmosphere. This is most easily accomplished by biogenically produced CO_2 within anaerobic sediments. For that reason, nahcolite in ancient deposits commonly occurs in concretions, many of which probably grew during diagenesis through bacterial CO_2 addition to the interstitial brines. Nahcolite is also present in the Magadi drill cores (Baker, 1958, p. 48).

Natron, $Na_2CO_3.10H_2O$, forms at temperatures below those of the trona field. In a study of the drying-up of Owens Lake, Cal., Smith and Friedman (1975) found that natron formed as the first precipitate, but it was later converted to trona through the addition of CO_2. I have observed natron to crystallize from the brine at Magadi during cool nights. Crystals several mm in diameter grew on the underside of thin trona skins which covered the lake surface. These crystals redissolved during the morning as the brine temperature increased into the trona stability field.

According to Fig. 15-13, thermonatrite, $Na_2CO_3.H_2O$, can only form from solutions of either very high temperatures or very low P_{CO_2}. At Magadi, thermonatrite has been found to occur in local pools on the lake surface, (J. A. Stevens, pers. comm.), where it grows from brines which have been previously depleted in HCO_3^- by trona precipitation. The Magadi Soda Company exploits this process for the NaCl production. Carbonate-rich lake brines are pumped into shallow evaporating ponds where they are exposed to rapid evaporation. Thermonatrite accumulates as a thick bottom deposit. This increases the NaCl concentrations sufficiently for halite to precipitate. Halite then forms a layer on top of the thermonatrite and can be gathered simply by scraping it up.

Fig. 15-13 only applies to salts in equilibrium with a saturated solution. Thermonatrite is also an important constituent of the efflorescent crusts, where it forms by complete desiccation of capillary solutions under conditions not governed by Fig. 15-13.

All extensive bedded deposits of sodium carbonates are trona. Good examples are those of Pleistocene Searles Lake (Smith, in press) and the Eocene Green River Formation of Wyoming (Bradley and Eugster, 1969). This means that whatever the initial precipitation processes and products are, equilibration with the atmosphere will eventually prevail for surface deposits. The detailed precipitation mechanisms, however, are quite interesting and can be illuminated with the aid of Fig. 15-14. The pronounced effect of P_{CO_2} or the HCO_3 - CO_3 ratio on the solubility of the sodium carbonates is clearly evident from the bicarbonate quotient contours, the bicarbonate quotient (b.c.) being simply

$$\frac{gHCO_3}{gHCO_3 + gCO_3}$$

in the solution. Nahcolite itself is not highly soluble in bicarbonate solutions (b.c. = 1), but its solubility increases some four-fold, as the CO_3^{--} content of the solution, and with it the pH, rises. Simultaneously, of course P_{CO_2} decreases. As the bicarbonate quotient, and with it the P_{CO_2}, drop sufficiently, nahcolite is replaced by trona. The solubility of trona also increases with decreasing HCO_3^- content until thermonatrite, the most soluble of the sodium carbonates, is reached. By that time, the brines are essentially free of HCO_3 (b.c. or 0.02 or 0.03; $P_{CO_2} \sim 10^{-4.5}$). Jones et al. (1977) have calculated P_{CO_2} levels of some residual Magadi brines and have found values as low as $10^{-4.64}$. This corresponds closely to the trona-thermonatrite boundary and the maximum solubility of trona. From such brines, trona can precipitate only by the addition of CO_2 or HCO_3^-. Addition from the atmosphere is slow, but there is still considerable biologic activity in

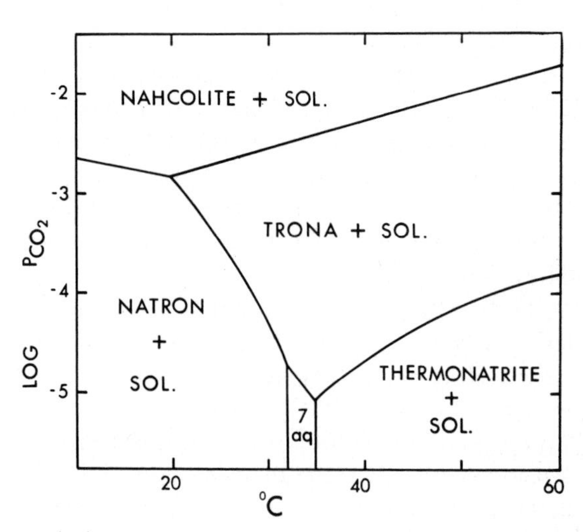

Fig. 15-13. Stability relations of the sodium carbonate-bicarbonate minerals. Sol.: solution, 7 aq: $Na_2CO_3 \cdot 7H_2O$ + Solution. Data from Eugster (1966) and Hatch (1972).

these concentrated brines as indicated by their red colors and this may be sufficient to continue trona precipitation.

Figs. 15-13 and 14 have been drawn for the chloride-free system. Since the Magadi brines carry a considerable amount of Cl, corrections are necessary. Some of these corrections have been discussed by Bradley and Eugster (1969). We are now in a position to consider trona precipitation as it takes place at Lake Magadi at the present time.

During the dry season, trona is exposed on the surface throughout the lake basin. Baker (1958) has given a good description of this surface. It is hard and composed of an open mesh-work of interlocking trona crystals. Initially, the brine level is very close to the surface, but it may drop somewhat during especially dry periods. The porosity of the deposit is hard to estimate, but it is probably higher than that of the upper salt at Searles Lake, which is given as 44% (Haines, 1959). While the surface is exposed, it may receive a considerable load of wind-blown material. Expansion polygons, several m in diameter, cover the whole surface, but the ridges are not as extreme as those of some playas (see for instance Hunt et al., 1966; Stoertz and Ericksen, 1974; Eugster and Hardie, in press). This is probably because the brine level drops rapidly and thus does not provide the extensive lateral growth necessary for thrusting of the polygons.

The rainy season, which usually occurs in the period February–April, usually provides enough runoff to flood the lake, sometimes to a depth of over 1 m. This is an important event for trona precipitation. The runoff, even though it may have acquired a considerable solute load through dissolution of efflorescent crusts, is initially fairly dilute. But in contact with the trona surface, it becomes a concentrated brine rich in CO_3^{--} and HCO_3^{-},

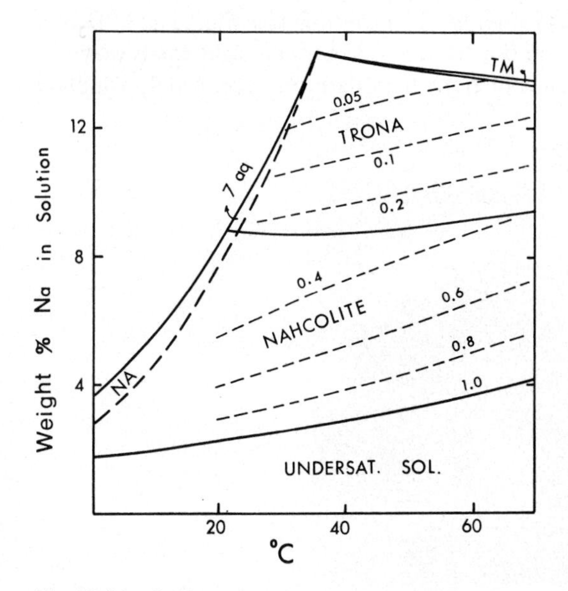

Fig. 15-14. Sodium content of solutions in equilibrium with carbonate-bicarbonate minerals (from Bradley and Eugster, 1969). Dashed contours are for the bicarbonate quotient (see text).

because trona provides one mole each. Hence its bicarbonate quotient will be near 0.5. The solution is located in the nahcolite field (see Fig. 15-14) and is considerably supersaturated with respect to atmospheric CO_2. Flooding homogenizes the lake surface and also incorporates the lagoon brines, which are the loci for brine preconcentrations.

Throughout the dry season, from May to October, and December–January, evaporation is intense, and the level of the surface brine begins to drop. There are no systematic data available, but it is likely that with the removal of H_2O, CO_2 is also lost to the atmosphere. Undersaturated surface brines (see Jones et al., 1977), seem to eventually equilibriate with the atmosphere (bicarbonate quotient near 0.15). Saturation with respect to trona sets in when the brine is no more than a few centimeters deep. As evaporation continues, trona grows in two modes: as long pointed blades attached to the substrata growing radially upward and as a thin film floating on the surface. The surface film contains a minor amount of trona, but it is important because it inhibits evaporation considerably and raises the brine temperature through the greenhouse effect. Below such skins we have measured temperatures of over 60°C. Wind often disrupts the skin and heaps the fragments in small ridges along the edges of local brine pools. As the brine level drops, the blades from below pierce through the floating skin and that point often becomes a locus for secondary nucleation. Eventually, the brine level falls below the base of the newly formed trona crystals and the cycle is completed. At this point, brines are virtually devoid of HCO_3 through fractional crystallization of trona and thermonatrite would be the next phase to form by simple evaporation. However, CO_2 addition appears to be fast enough to prevent this from happening. Trona continues to precipitate from the interstitial brines, reducing the porosity, at the rate CO_2 is supplied, mainly from biogenic sources. It is important to realize that at this stage evaporation need not take place for trona precipitation.

The trona deposit as a whole is banded on the 2–5 cm scale (see Fig. 15-15). Tightly-spaced rosettes of clear trona blades pointing upward make up the light bands which are separated from each other by thin bands of trona with dark material, mainly windborn

Fig. 15-15. Bedded trona from the surface of Lake Magadi. Hammer for scale.

dust. Though each light-dark couplet presumably forms in one season, it would be unsafe to compute the age of the Evaporite Series by counting couplets. The maximum age, assuming 2 cm couplets, would be 40 m: 2 cm = 2000 years, almost certainly too short a time. We know now that during each flooding much of the surface trona redissolves and eventually is precipitated as a new couplet, only a small amount of which is actually new trona. The rest is simply recycled material. During this continuous recycling, whole couplets can easily be wiped out, especially if they were deposited on slightly higher ground.

Extent of recycling and trona accumulation rates were discussed by Eugster (1970) and the main conclusions seem to still stand:

> "We are now in a position to estimate the time elapsed since the onset of the evaporite formation. In the four drill holes described by Baker (1958), the trona accumulation is anywhere from 7–40 m thick. Using an average of 20 m, a lake area of 75 km^2 and a density of 2, we get a total of 2 x 10^9 tons of Na_2CO_3 for the whole basin. Considering all the marginal areas, interstitial brines etc., this is probably a lower limit estimate. Stevens (in Baker, 1958) estimated total daily input for each lagoon by measuring lagoon area and evaporation rate and by assuming a balance between input and evaporation during the dry season. He arrived at 4300 tons Na_2CO_3 per day for the whole basin, or about 2.5 x 10^6 tons/year. As we have seen . . . , the bulk of this input is recirculated. For instance, the southwestern lagoon, which accounts for 37 percent of the total input, contains the most powerful springs (B 15 and B 16), which recirculate 73 percent of their solutes and only contribute the remaining 27 percent as new to the basin. If we choose an average ratio for new to recirculated material of 1:3 for the whole basin, then the yearly input of solids new to the basin is 0.5 x 10^6 tons/year. In this case, assuming a constant supply rate, the deposit took (3 x 10^9)/(0.5 x 10^6) = 6000 years to accumulate. However, it is not likely that the rate of supply has remained constant but increased with time.
>
> (Eugster, 1970, p. 228)

This calculation, uncertain as it is, gets us much closer to the upper limit of 9000 years given by the C^{14} date of the High Magadi Beds (see Table 15-1). Judging from the large variation in thickness of the Evaporite Series, trona deposition began soon after High Magadi time in a few localized depressions. As the climate became more arid, solute acquisition accelerated, spring waters became more concentrated and trona precipitation spread until it covered the whole lake as it does today. Net yearly accumulation of trona would have been at the rate of no more than 0.4 cm. This is in good agreement with the rates of 0.2–0.3 cm/year calculated by Fahey (1962) for Bed 17 of the Green River Formation. Exceptional events, such as the drying up of Owens Lake, can produce much higher rates for a brief period.

Lake Magadi is a saline lake at its peak of productivity. We can speculate what its ultimate fate might be, provided the present climatic trend continues. Fluoride has already accumulated in the surface brines sufficiently for villiaumite, NaF, to be a common and permanent constituent of the evaporites. Halite forms occasionally on the lake, but its presence is still ephemeral. There is no doubt, however, that the amount of recirculation noticed at Magadi will increase the chloride content of the residual brines, until eventually NaCl will crystallize in large amounts. Perhaps we can expect the formation of trona-halite couplets with trona at the bottom. Eventually, through fractional dissolution and movement to local depressions, pure halite layers might accumulate. The trona-halite distribution might be very similar to that described by Deardorff and Mannion (1971) for Bed 17 of the Green River Formation. The evaporite deposit will be completed when the solute input ceases.

CHERTS AND THE HIGH MAGADI BEDS

The High Magadi Beds were described in detail by Baker (1958) and Eugster (1969). They were deposited during a high stand of the lake some 9,000–10,000 years ago. The shore lines of this lake are still visible on the eastern scarp as a break in vegetation some 13 m above the present lake level. The sediments deposited from this lake consist mainly of bedded clay- and silt-size volcanic debris and cherts. The thickness in the outcrops at the basin margins is rarely more than a few meters, while it may be as much as 80 m in the drill cores (Surdam and Eugster, 1976, Table 7). High Magadi Bed outcrops are best preserved to the NE, E and SE of the present lake. They are also present in the NW, W and SW, but they are more fragmentary there and, because of the complex overlapping with the Oloronga Beds, more difficult to recognize. They not only occur in hummocks (Fig.15-16) but also form the flat floors of the dry valleys surrounding the lake.

Eugster (1969) divided the High Magadi Beds into an upper and lower sequence. The upper sequence consists of an erionite-rich volcanic tuff of uniform thickness (4 m) (see

Fig. 15-16. Tuffaceous silt of the High Magadi Beds. Typical erosional shapes formed by precipitation and dissolution of efflorescent crusts.

Fig. 15-16). In outcrop it appears as a dense, vaguely bedded, brown unit, usually covered by efflorescent crusts, while in the cores it occurs as a very dense, black, plastic mud. It was probably deposited in the lake during a series of rapid volcanic events.

Below this tuff is a horizon rich in organic carbon and fish fossils identified as *Tilapia nilotica* (Baker, 1958), a fish similar to the species living in the more dilute lagoon waters of Magadi at the present time (T. alcalica, see Beadle, 1974). The fossil fish are considerably larger, indicating lower salinities for the High Magadi Lake. Below the Tilapia horizon, which has been dated at 9,100 years B. P. (Butzer et al., 1972), and interbedded with clay-size volcanic deposits, are one or more magadiite ($NaSi_7O_{13}(OH)_3 \cdot 3H_2O$) horizons. In outcrops, the number of horizons are few and the beds are less than 50 cm thick, but judging from the frequency of chert beds in the High Magadi horizon of the drill cores, magadiite was deposited at frequent intervals in the main lake.

The lower sequence of the High Magadi Beds, which consists of the finely bedded volcanic material, the magadiite horizons and the Tilapia bed, rests either on lava rubble or on Oloronga Beds. The top of the latter is formed by a thick caliche horizon and this makes it easy to assign all lake beds above this horizon to the High Magadi sequence.

Surdam and Eugster (1976) have described the mineralogy of the High Magadi Beds in detail. Anorthoclase and amphibole are the main detrital minerals, and both are of volcanic derivation. Some illite is present and it probably represents detrital altered feldspar. Calcite occurs especially in the tuff of the upper sequence as veinlets and aggregates of anhedral crystals replacing silicate matrix. It is probably a late product formed during capillary evaporation of dilute ground water, the same process which leads to the formation of the efflorescent crusts. The most common authigenic mineral is erionite. It occurs in felted masses replacing glass shards. Analcime and chabazite are less abundant and formed later.

Three sodium silicates have been found at Magadi: magadiite, $NaSi_7O_{13}(OH)_3.3H_2O$ (Eugster, 1967), kenyaite, $NaSi_{11}O_{20.5}(OH)_3.H_2O$ (Eugster, 1967), and makatite, $NaSi_2O_3(OH)_3.H_2O$ (Sheppard et al., 1970). A fourth sodium silicate, kanemite, described by Maglione from the Chad basin (Johan and Maglione, 1972) might well be present, but has not yet been encountered. Of these, only magadiite occurs in large amounts. Makatite is confined to a few crystals intergrown with bedded trona of the Evaporite Series, found in a drill core at a depth of 30 m. Kenyaite is common and forms concretions several cm in diameter at certain magadiite horizons in High Magadi outcrops. It may be an intermediate product of the magadiite-chert transformation (Eugster, 1969). The mineralogical properties of magadiite have been summarized by Brindley (1969), and Lagaly et al. (1975).

The petrologic significance of magadiite lies in the fact that it is a precursor of bedded chert. In the High Magadi Beds, the in situ conversion magadiite → chert can be observed anywhere at the perimeter of the dry valley floors surrounding the lake. The bulk of the floors are formed by the brown volcanic tuff of the upper sequence of the High Magadi Beds, which is very dense and hard. At the edges of the basin, the lower units emerge on the surface with primary depositional dips. Where a magadiite horizon emerges, it is now represented by large, flaggy chert plates, 1–5 cm thick. If one follows this chert horizon basinward, by digging a trench, one finds it replaced by a magadiite horizon and the changeover takes place over a distance of less than 1 m and close to the surface (see

Eugster, 1969, Fig. 5). The conversion observed under these conditions has been related to leaching of sodium by more dilute runoff.

$$NaSi_7O_{13}(OH)_3 \cdot 3H_2O + H^+ \rightarrow 7SiO_2 + Na^+ + 5H_2O$$

Hay (1968, 1970) suggested an alternate mechanism, based on the spontaneous crystallization of quartz from magadiite. In this process, Na^+ is expelled and thus it can proceed even in the presence of concentrated brines. Evidence for this process are the trona casts many of the cherts of the drill cores have. For the magadiite horizons buried in the middle of the lake, leaching by more dilute waters is obviously not feasible and the spontaneous conversion is more likely. Isotopic studies by O'Neil and Hay (1973) have demonstrated that both mechanisms are important.

Since the original discovery at Magadi, cherts formed from magadiite ("Magadi-type" cherts) have been found in many basins, with some as old as Jurassic (Surdam et al., 1972). Such cherts are most easily recognized by their soft-sediment deformation features, crystal casts and shrinkage cracks. Characteristic features are pictured in Hay (1968), Eugster (1969), Surdam et al. (1972), Eugster and Chou (1973). The soft-sediment deformation features have been inherited from magadiite, which is a white, soft, plastic putty, as long as it is soaked in brine. Upon drying, it becomes hard and porcellaneous, without losing its sharp, well-defined x-ray pattern. Crystal casts, mainly of trona sprays ("arborescent" chert of Baker, 1958), gaylussite and pirssonite molds, may well be associated with the magadiite-chert transformation itself. The molds now contain calcite. It is interesting to note that the thicker magadiite beds also contain rosettes of euhedral calcite crystals. The magadiite-chert transition involves a 25% reduction in volume. It is not surprising, therefore, that many cherts show shrinkage cracks or surface reticulations. The most dramatic shrinkage effects are represented by the chert polygons, some of which are 50 m in diameter and have ridges up to 2 m high (Eugster, 1969). Magadi-type cherts have also been invoked for some Precambrian iron formations (Eugster and Chou, 1973) on the grounds of textural similarities and the abundant presence of sodium-rich minerals such as riebeckite.

So far, magadiite has been found only in alkaline environments (Magadi, Eugster, 1967; Alkali Valley, Rooney et al., 1969; Trinity County, McAtee et al., 1968; Lake Chad, Maglione, 1970), presumably because the silica activity is high enough only in alkaline waters. Bricker (1969) has measured magadiite solubility and given Free Energies, so that natural waters can now be tested for saturation with respect to magadiite. As Fig.15-17 shows, magadiite is less soluble than amorphous silica and more soluble than quartz.

As the studies at Lakes Magadi and Chad show, magadiite can form either as a lake-wide precipitate or as an authigenic mineral by interaction of alkaline brines with siliceous material, such as volcanic glasses. While authigenic growth requires only that the appropriate levels of sodium and silica activities are reached, lake-wide precipitation is more difficult to account for. A possible mechanism, suggested by Jones et al. (1967), is based on a temporary stratification of the water body. Through evaporative concentration, silica concentration in the brine and pH rise, with many brines close to saturation with respect to amorphous silica. The most effective way to reach supersaturation with magadiite is a drop in pH. However, simply mixing with dilute inflow would not accom-

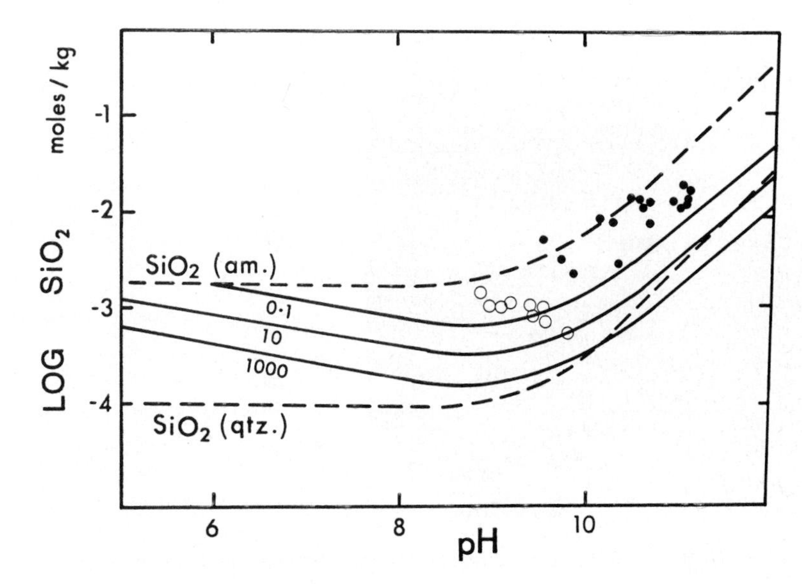

Fig. 15-17. Solubility of magadiite as a function of pH (from Bricker, 1969) shown by solid lines contoured for the sodium activity from 0.1 to 1000. The dashed lines are for quartz and amorphous silica respectively. Dots: Magadi lake brines. Open circles: hot springs.

plish this, because the silica content of the brine would also drop. If the dilute inflow does not mix with the underlying dense brine, magadiite could form at the interface of the two water bodies, where the pH is low and could be kept low through organic activity in the epilimnion. This could account for lake-wide deposition of thick magadiite beds. Eventually, the epilimnion would evaporate or the two water bodies would mix, terminating the magadiite interval. Whatever the detailed mechanism, the appropriate conditions were repeated many times during High Magadi and Oloronga time.

The Green River Formation has a surprising paucity of Magadi-type cherts. Eugster and Hardie (1975) have explained this by pointing out that the Green River lakes had very much shallower gradients than Lake Magadi and wide mud flats surrounding them, and hence it was much less likely that stratification could be achieved before mixing would destroy it.

OLORONGA BEDS

Baker characterized the Oloronga Beds as follows:

> " . . . the Oloronga beds overlie plateau trachyte lava and dip gently to the west. The area of outcrop is not large . . . The large surface of the beds is covered with two kinds of kunkar limestone – a pale grayish white platy variety and a brownish white nodular variety containing large oolitic bodies. Only the upper few feet of the beds are exposed in stream channels . . . " (p. 27).
>
> "The sections described suggest that lacustrine sediments were deposited fairly widely on the plateau trachytes previous to grid-faulting. Such lake beds may be

Fig. 15-18. Typical bedded Oloronga cherts. The resistant bench in the background (out of focus) is the caliche cap.

more widespread than has been indicated and may be present as the lowermost lacustrine series in nearly all the larger basins where they have been protected from erosion." (p. 28).

The Oloronga Beds described by Baker (1958) consist mainly of volcanic ashes, silts and clays. Through careful mapping and measurement of sections it has become possible to separate the cherts assigned by Baker (1958) to the Chert Series into older cherts and younger cherts, with the older cherts being an integral part of the Oloronga Beds (see Fig. 15-18), while the younger cherts are indeed formed from magadiite horizons within the High Magadi Beds, as suggested by Eugster (1969). Unequivocal assignment is possible on almost every outcrop and it is based on the following four general criteria: a) presence of caliche, b) degree of induration, c) fossils, d) color.

Presence of caliche. As mentioned earlier, Baker (1958) noted the presence of several caliche horizons. In the 1958 version, he placed one horizon immediately below the High Magadi Beds and post-dating the Chert Series and minor faulting, while the other occurs directly on top of the Oloronga Beds. In the revision of 1963 (Baker, 1958, 1963) the caliche horizon again appears below the High Magadi Beds and separated from the Oloronga Beds by grid and minor faulting. In the Magadi trough, I have observed two distinct caliche horizons, one occurring within the Oloronga depositional interval and the other capping the youngest Oloronga Beds. The former is thin and very local, as it is restricted to the area immediately SW of Lake Magadi, while the latter is up to 50 cm thick and is present throughout the Magadi trough wherever Oloronga Beds occur, except where it is locally removed by erosion (see Fig. 15-19). The caliche is of the pisolitic type and, though quite porous, it is extremely hard. In fact, it obviously has protected many Oloronga Beds from further erosion, as it caps many steep outcrops. The caliche surface is not necessarily horizontal, but follows the depositional slopes so common for lake beds deposited in narrow, steep-sided troughs. Some of the caliche formation is clearly older than the youngest lava flows, because 10 cm caliche were found to underlie a lava flow N of the NE lagoon. It is not likely that all caliche is that old. On the other hand, no caliche horizon was ever found in the High Magadi internal, the interval so clearly defined by the fish bed, the magadiite horizons and the High Magadi tuff (Eugster, 1969). In other words, any lake beds occurring underneath a caliche cap can definitely be assigned to Oloronga time.

Degree of induration. The caliche argument is fully supported by the observations on the degree of induration of the lake beds. The basic lithologies found in the Oloronga and High Magadi Beds are very similar: principally volcanogenic sediments of various grain sizes and cherts. This is not surprising considering the fact that the depositional environments and sources of debris must have been very similar in the two lakes. In addition, the High Magadi Beds probably contain a substantial amount of reworked Oloronga material. Because of the age difference, however, the Oloronga tuffs and silts are always highly indurated, distinctly more so than the equivalent lithologies in the High Magadi Beds. This difference does not apply to the drill core material, which is always soft (except for the cherts) and soaked in brine.

Fossils. Fossils, where they occur, are of course an unequivocal guide. In the High Magadi Beds the ubiquitous fish bed (*Tilapia nilotica*) mentioned by Baker (1958) can usually be located with great ease and it is the most reliable stratigraphic marker. It occurs underneath the High Magadi tuff and above the highest magadiite horizon. The fish are abundant, not fossilized and consist of very fragile carbonized skeletons. The horizon is

Fig. 15-19. Caliche cap covering Oloronga Beds. Note the nodular top and the more laminated lower part.

rarely more than 10 cm thick. Fossils also occur in the Oloronga Beds. They are the gastropod limestones and cherts (*viviparous*) mentioned by Baker (1958). Recently, fish fossils have also been found in the Oloronga Beds. These are quite different from the High Magadi fish and are being studied now.

Color. Color is not always a reliable guide for the distinction of Oloronga and High Magadi Beds, but there is an overall difference, both with respect to the volcanic sediments and the cherts. Oloronga sediments normally have a strongly olive-green hue, while High Magadi sediments are tan to brown, except for the cherts, which are always white. Oloronga cherts can also be white, but they are more commonly olive-green, except for the silicified *viviparous* horizons, which is tan. It is obviously not safe to use color as the only criterion, but it is useful as an ancilliary guide in mapping.

Occurence. The major areas of Oloronga outcrops are shown in Fig. 15-3. Boundaries between Oloronga and High Magadi outcrops are not sharp, but rather areas of transition. In such areas the detailed relationships may be very complex on a small scale, with patches of High Magadi Beds overlapping Oloronga Beds and, conversely, erosional windows of Oloronga exposures through High Magadi Beds. In other words, areas mapped as Oloronga are principally occupied by Oloronga Beds, but may contain minor amounts of High Magadi Beds and vice versa.

The major Oloronga outcrops include those mapped by Baker (1958, 1963), principally the large area in the Oloronga region, which gave the formation its name. It is located north of Shombole and southwest of Lake Magadi, from which it is separated by Pleistocene trachyte horsts. Baker (1958) also shows small areas of Oloronga outcrops near the wells northeast of Little Magadi and in the trough north of the NE lagoon. The areas newly assigned to the Oloronga formation are as follows: a) the flats surrounding the southwest lagoon to the west, south and east, mapped by Baker (1958) mainly as Chert Series, b) the broad valley east and northeast of Needle Rock is principally Oloronga and High Magadi and both parts were mapped by Baker (1958) as Chert Series, c) a beautifully exposed section on the east side of the trough immediately north of the NE lagoon, d) a small area south of Little Magadi, e) an area in the vicinity of the NW lagoon. In addition, as Baker (1958) pointed out, many smaller valleys in the basin are very flat and obviously contain lake beds, but erosion has not been sufficient to produce outcrops. Most of the valleys lie above the High Magadi shoreline and hence must be floored by Oloronga Beds. Some typical outcrop sections are given in Fig. 15-20.

Lake Oloronga obviously was an alkaline lake which reached magadiite precipitation many times throughout is existence. Its salinity was probably close to that of the High Magadi lake, but much less than that of the lake from which the Evaporite Series precipitated. Judging from the silica storage mechanism necessary for magadiite supersaturation, pH must have been at least 9.5 at its more concentrated intervals and dissolved solids similar to the present hot springs, 30 g/kg. However, there is no indication that the surface waters ever were saturated with respect to trona. The trona casts frequently observed in the cherts are probably related to the magadiite-chert transformation. In fact, most of the time, Lake Oloronga could have been quite fresh. It was certainly deep in places. In drill hole F6, for instance, the base of the Oloronga Beds was not reached and the

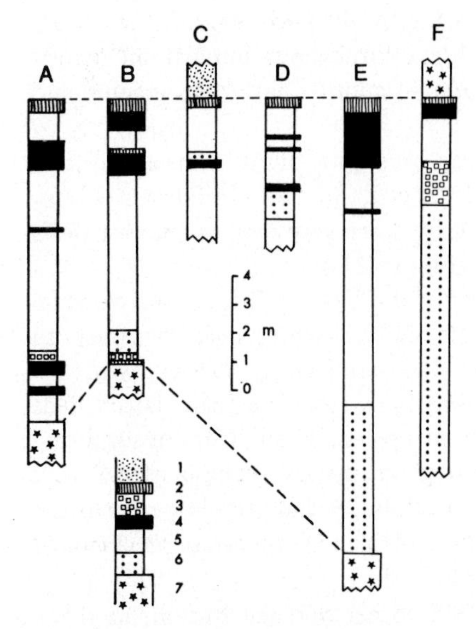

Fig. 15-20. Typical outcrop sections of Oloronga Beds. 1: High Magadi Beds; 2: caliche cap; 3: viviparous limestone or chert; 4: chert; 5: tuffaceous silts and sands, volcanic tuffs; 6: volcanic mudflows; 7: trachyte lava flow. The locations are as follows: a: 5 km SW of the SW Lagoon; B: SW Lagoon; C: Dry Lagoon; D: E of Needle Rock; F: S of Ngejepa; G: E of Enchompoli. The coordinates are: A: AH 751906, B: AH 917806, C: AH 803958, D: BH 003993, E: AJ 994025; F: AJ 999058 on 1:50,000 sheets 160/3, 160/4 and 171/1. For approximate locations see Fig. 15-3.

lowest point is 133 m below the lake surface. Since Oloronga beds are found more than 30 m above the present lake, Lake Oloronga could have been 170–200 m deep in some spots.

Dating and Correlation. Dating of the Oloronga Beds is based on trachyte lava flows extruded onto the lake beds. Two such localities were found. One, a spectacular series of outcrops in part pictured in Fig. 15-21, is located on the E scarp just north of the NE lagoon (see Fig. 15-3). The other is located in a dry stream bed just south of Little Magadi (see Fig. 15-3).

The flow near the North-Eastern Lagoon apparently came from the north or northeast, descended an old lava scarp at an angle of over 45°, filled a trough cut into the Oloronga Beds and then flowed more or less horizontally to the S or SW for at least 300 m. At the contact with the Oloronga Beds the flow was chilled sufficiently to form a 1 m-thick glass selvage. It is this glass which was dated at 780,000 years (Fairhead et al., 1972). Chemical analyses of the glass (225) and the main flow (696 and 698) are given in Surdam and Eugster (1976, Table 3) and the composition is very similar to that of the pre-Oloronga trachytes (see also Baker, 1958 and Baker et al., 1977).

Accordingly, most Oloronga Beds have been deposited at least 800,000 years ago, but we do not know how far back they date. We also do not know how long Lake Oloronga persisted after the lava flow, but we do know that most Oloronga Beds were exposed long enough to accumulate the caliche cap prior to flooding by the High Magadi lake. There could have been in existence a small, saline central lake throughout, but if it did deposit any evaporites, they were probably redissolved, since no evaporites have been encountered at the Oloronga–High Magadi interface.

Some of the lagoons of Lake Oloronga must have been quite dilute. This is indicated by the phillipsite present in the outcrops NE of Little Magadi (Surdam and Eugster, 1976), by unaltered glass found N of the NE lagoon and by some very large viviparous shells present in lake beds below the glassy trachyte flow (Fig. 15-21).

Correlation of Lake Oloronga with other basins of the Eastern Rift is now possible. Oloronga-type rocks were encountered in the Lake Natron basin, near the track to "Blevin's Camp" on the east shore of Lake Natron. They are identical in every way with the Oloronga Beds at Magadi and they can be distinguished from the High Natron Beds (Hay, 1968) by their caliche cap. They have not been mapped in detail. It is not surprising that Oloronga outcrops have been encountered throughout the Magadi basin as well as in the Natron basin. Obviously, Lake Oloronga was much larger and also deeper than the High Magadi lake and it may well have been connected with an Oloronga-age precursor of Lake Natron, which would make it a very large lake indeed.

The Peninj Group, which is preserved on the SW corner of Lake Natron, has been mapped by Isaac (1965) and it spans Oloronga time. It is quite possible that the Moinik Formation, which forms the upper part of the Peninj Group, is contemporaneous with the Oloronga Beds. This is supported by the fact that lobate chert nodules were noted in the Moinik section by Isaac (1965).

Baker and Mitchell (1976) mention a K/Ar age of 400,000 years from a pumice within the diatomaceous lake deposits of the Legemunge basin (the Legemunge Beds of Baker, 1958). It is likely that this lake was contemporaneous with Lake Oloronga.

Fig. 15-21. Lava flow at locality F (Fig.15-20), N of the NE Lagoon. In the foreground are Oloronga Beds (pocket knife for scale) through which hot springs drain (686 of Jones et al., 1977). In the background is the lava flow dated at 780,000 years, descending over Oloronga Beds. Note chilled margin.

AUTHIGENIC MINERAL REACTIONS

All of the lake sediments, from the Recent surface muds to the oldest Oloronga beds, are subject to authigenic reactions when they are in contact with the alkaline brines. Such reactions are promoted by the presence of reactive volcanic material as well as by the high solubility of silica in the brines. The mineralogic nature of these reactions has been discussed in detail by Surdam and Eugster (1976), and their findings will be summarized here briefly.

Zeolites. Erionite, $NaKAl_2Si_7O_{18}.6H_2O$, is the most common zeolite at Magadi. It is present in almost all samples and occurs as felted masses and needles, in some cases clearly replacing glass shards. Surdam and Eugster (1976) concluded that:

> "Assuming conservation of aluminum, erionite can obviously form from trachyte or trachytic glass without addition of any material except water." (p. 1742)

Erionite forms in an environment of high silica and sodium activities and low calcium activities, conditions fulfilled by both the trachytic glass and the brines. It is no surprise, then, that erionite is the dominant zeolite. The waters which participate in the trachyte glass → erionite transformation cannot be rich in calcium, or clinoptilotite or chabazite would form. In other words, they must be brines rather than dilute runoff or ground water.

Under certain conditions, erionite is replaced by analcime through a reaction such as $NaKAl_2Si_7O_{18}.6H_2O + Na^+ \rightarrow 2NaAlSi_2O_6.H_2O + K^+ + 3SiO_2 + 4H_2O$ as reaction which proceeds to the right at high Na and low SiO_2 activities.

> "Lowering the activity of silica and increasing the activity of sodium is also associated with the reaction of magadiite to quartz (Bricker, 1969). In fact, there is evidence that the transition erionite to analcime is coupled to the magadiite to chert conversion: samples containing analcime as a major phase rarely contain even a trace of magadiite." (Surdam and Eugster, 1976, p. 1749)

The transition is accomplished either by the spontaneous crystallization of quartz or by percolating waters leaching sodium. However, these percolating waters cannot be dilute ground waters, because such water would introduce calcium which would favor clinoptilolite over erionite.

Analcime, in addition to forming from a zeolite precursor such as erionite, can also form from the sodium aluminosilicate gels described by Eugster and Jones (1968). Surdam and Eugster (1976) found that

> " . . . there seems to be an antithetical relation between the gel and analcime. In samples of core F6, an evolution of analcime from gel can be demonstrated . . . , clearly indicating that analcime grows at the expense of the gel. This may account for the fact that analcime is more abundant in the deeper parts of the core . . . " (p. 1749)

Clinoptilolite is quite abundant in the Oloronga outcrops and in the High Magadi interval of the drill core. It also represents high silica and sodium activities, but in addition, higher levels of calcium are required. The same holds for mordenite, which was encountered in a single outcrop. Surdam and Eugster (1976) suggested that

"... clinoptilolite and mordenite formed later than erionite, indicating that the alkaline earths may have been supplied by residual glass and other reaction products. The alkaline earths also may have been provided by percolating dilute ground waters." (p. 1748)

Because of the abundance of clinoptilolite in the Oloronga Beds there is some suggestion that Lake Oloronga may have been somewhat more dilute than the High Magadi lake, but since the reactions are post-depositional, this need not have been so.

Chabazite, present in minor amounts, also needs higher calcium levels. It is most abundant in outcrop samples and seems to have formed by a process similar to that responsible for clinoptilolite. Phillipsite, a zeolite indicative of low silica activity, was found only in one outcrop and it was thought to have formed in alkaline soils (Surdam and Eugster, 1976).

The zonation which characterizes so many zeolite deposits (see for instance, Sheppard and Gude, 1968) is present at Magadi only in a very rudimentary form. In the Oloronga Beds, unaltered glass exists in the extreme NE corner and phillipsite in an outcrop to the north, suggesting the sequence unaltered glass → phillipsite → erionite, which could be accounted for by a lateral salinity gradient. However, the zones are very restricted, because Lake Oloronga and High Magadi Lake were both located in narrow fault troughs without the wide mud flats necessary to develop the lateral salinity gradients.

Non-silicates. Gaylussite and fluorite are the most common authigenic non-silicates. Gaylussite forms in many Holocene playas by the interaction of alkaline brines with a calcium source, such as

$$\mathrm{CaCO_3 + 2Na^+ + CO_3^{--} + 5H_2O \rightarrow \underset{\text{gaylussite}}{Na_2CO_3.CaCO_3.5H_2O}}$$

At Magadi it is present in many unconsolidated surface samples in the form of clear crystals up to 5 mm in length.

Fluorite, CaF_2, is present in some outcrops and in many core samples. Surdam and Eugster (1976) noted an antithetical relationship between fluorite and calcite, suggesting a reaction of the type

$$\mathrm{\underset{\text{calcite}}{CaCO_3} + 2F \rightarrow \underset{\text{fluorite}}{CaF_2} + CO_3^{--}}$$

Magadi brines are known to be high in fluoride, with values of up to 2000 mg/kg F (Jones et al., 1977). In the surface deposits, saturation with respect to NaF, villiaumite, occurs and villaumite is found as cubic crystals enclosed in trona (Baker, 1958).

Dilute inflow waters are not unusually high in fluoride (0.1–3 mg/kg), but during brine evolution, calcium is removed as carbonate before saturation with respect to CaF_2 can occur. Hence fluoride accumulates in the brine, since there is no other mineral to remove it. The fluorite which finally does form, is an authigenic product and the result of interaction between fluoride-rich brines and a calcium source. This source could be dilute runoff, detrital minerals, calcite of caliche derivation, gaylussite or volcanic glass.

Gels. Na-Al-silicate gels were first encountered at the N-shore of Little Magadi, where they occur as surface deposits up to 30 cm thick (Eugster and Jones, 1968). They were thought to have formed by interaction of the alkaline hot springs with trachyte debris. Assuming Al is conserved during the reaction, the springs need supply only Na and H_2O, while silica is released. Since the original discovery, such gels have been encountered throughout the Magadi basin, on the surface as well as within the deposits. Surdam and Eugster (1976) noted the presence of gels in many drill core samples, and some of the pits dug into High Magadi beds contained persistent layers, 1–3 mm thick, of translucent rubbery gels. This supports the contention that gels form annually at or near the lake shore and that these accumulations are washed into the lake during the next rainy season. Such gels may eventually crystallize to analcime, forming beds of pure analcime (Hay, 1970).

Calcite pisolites. Interaction of dilute runoff with concentrated lake brines has led to the formation of calcite pisolites near the present lake shore at the south end. The pisolites are button-to-lozenge-shaped, 1–2 cm in diameter, 5–8 mm thick (see Fig. 15-22). They are round with a thicker center and a thin fringe which at times looks corroded. In cross-section they are concentric, but with many cross-cutting layers, indicating repeated precipitation and dissolution events. The pisolites are not cemented and lie on their flat side on unconsolidated sediment. Together with detrital, angular chert chips they form a single-particle pavement with very little overlap. These pavements exist at the interface between the flat, alluvial floor which is flooded during high stage, and the higher chert ridges. The pisolite-chert ratio increases away from the chert ridge, until near the alluvial flat itself the pisolites become much smaller, heavily corroded and eventually disappear altogether.

$CaCO_3$ pisolites have been described from a variety of geologic settings, including saline lagoons (Lucia, 1968; Purser and Loreau, 1973; Scholle, 1974). Perhaps the most com-

Fig. 15-22. Calcite pisolites on the shore of Lake Magadi. Angular fragments are chert chips. Scale in inches.

parable setting is that described by Risacher and Eugster (in press) from the Pastos Grandes Salar in Bolivia, where spherical pisolites, some with diameters of up to 20 cm, form in pools fed by hot springs on a playa surface.

The location and mode of occurence of the Magadi pisolites points to mixing of waters as the chief mechanism for precipitation. During the rainy season, dilute runoff carries Ca-rich waters towards the lake, while the lake expands and its alkaline brines flood some of the normally dry mud flats. At the interface where the two waters meet, supersaturation with respect to calcite occurs through a rise in pH and hence a rise in the CO_3^{--} activity. Calcite precipitates around whatever nuclei are available, frequently angular chert chips or older pisolites. As evaporative concentration sets in, the lake waters recede, exposing the pisolite pavement. The corrosion noted on some of the pisolites may well be due to rain water undersaturated with respect to calcite.

DEPOSITIONAL ENVIRONMENTS

Hardie et al. (in press) have analyzed the sedimentological aspects of hydrologically closed basins. A basin can be divided into a number of subenvironments, each characterized by a set of physical, chemical and biological processes and hence by a diagnostic set of sedimentary features. These subenvironments include the following: (1) alluvial fan, (2) sand flat, (3) dry mud flat, (4) saline mudflat, (5) salt pan of the ephemeral salt lake, (6) perennial saline lake, (7) dune field, (8) perennial stream floodplain, (9) ephemeral stream flood plain, (10) springs, (11) shoreline features. The topography and nature of inflow have much to do with the relative importance of a particular subenvironment. Inflow is by a) perennial streams, b) ephemeral streams, c) unchanneled sheet flow during storms, d) perennial or ephemeral springs, e) ground water.

At Magadi, obviously (6), (7), (8) and a) are missing. Also, because the basin is a fault trough with steep sides, extensive mud flats and flood plains can form only to the N or S of the individual arms. Alluvial fans are present, but in the immediate area of the Magadi basins they are relatively small because of the young topography. They reach much larger dimensions on the west slope of the Rift Valley, where they can be several km long. Sand flats, and dry and saline mud flats are very important at Magadi, and they make up most of the flat, normally dry areas surrounding the lake. However, it is not easy to draw boundaries between them. A dry mud flat may become a saline mud flat during a lake expansion and vice versa. Sand may transgress over a mud flat during a particularly severe storm. In fact, there are no areas except the central salt pan which cannot be reached by currents strong enough to carry sand-size particles. The salt pan itself is well defined as the locus of bedded trona accumulation. Dune fields are absent, probably in part because the fine-grained material is cemented by efflorescent crusts.

The plain just N of Lake Magadi represents a classic ephemeral stream flood plain with braided channels and much sand- to silt-size material. Efflorescent crusts are abundant, because the ground water is close to the surface. Springs are of course the only perennial inflow. Because their waters are devoid of Ca and Mg, no tufa and travertine are formed and the springs simply issue through the lake beds. Shoreline features include small deltas at the mouths of canyons and beaches, such as the pisolite pebble beach described earlier.

Lake Magadi, a typical Rift Valley lake, obviously belongs to the "Alluvial Fan-Ephemeral Saline Lake Complex" of Hardie et al. (1978), though the alluvial fans are subdominant because of the subsequent faulting of the trachyte lava flows which filled the valley floor. At the moment, the individual troughs are small and there simply isn't enough relief and runoff to form the vast fans we are used to from the Great Basin of the U.S. In all other respects, the model fits Lake Magadi perfectly.

It is in the relative importance of the depositional subenvironments that Magadi differs significantly from other well-known alkaline lakes and their deposits. The Eocene Lakes Gosiute and Uinta from which the Green River Formation of Wyoming, Utah and Colorado deposited and for which Lake Magadi has been used as a prototype, fit the Magadi model well in terms of hydrochemistry. However, their gradients were much more gentle and consequently the central lake was surrounded by wide carbonate mud flats. The alluvial fan-sand flat subenvironment was well developed and played a crucial role in the carbonate production (Smoot, 1978). The shoreline moved rapidly and over large distances in response to climatic fluctuations. Sediments are primarily dolomitic mud stones, oil shales and evaporites as well as siliciclastic and carbonate sands. In contrast, at Magadi carbonate production is in the form of cements and cannot lead to carbonate mud flats. What mud flats there are can be drowned by volcanic silts and sands at any time. Hence the sediment packages within the lakes consist mainly of volcanic debris and evaporites. Transgressive-repressive cycles are restricted because of the steep shorelines.

CONCLUSIONS

The faulting of the Rift Valley trachytes formed a trough which has been occupied by several lakes for the last million years or so. Water compositions have always been alkaline and sediments have been principally of volcanic derivations in addition to the cherts formed from magadiite and the trona. The present Lake Magadi, an ephemeral saline lake, is more concentrated than its precursors and it is at its maximum trona productivity. Brine evolution can be documented quantitatively and Lake Magadi is the classic example of an alkaline sodium carbonate lake. Lake Magadi is also the classic example of a Rift Valley lake with steep gradients and high input of clastic material. Authigenic mineral formation is extensive, because of the highly reactive volcanic glass in contact with high-pH brines. The Pleistocene-Holocene history of the Magadi basin correlates well with that of other basins in the Rift Valley. Many problems remain unsolved, but meanwhile Magadi has taught us much about hydrochemistry, brine evolution, mineral precipitation, authigenic reactions and sedimentation in an active closed basin.

SUMMARY

Lake Magadi is one of a number of closed basins located in the Eastern Rift Valley of Africa. It contains a thick deposit of trona, $NaHCO_3.Na_2CO_3.2H_2O$, forming at the present time. It is located within block faulted Pleistocene trachyte flows and it was preceded by two earlier lakes: High Magadi some 10,000–20,000 years ago and Lake Oloronga, at least 800,000 years old. Both earlier lakes were fresher and formed lake beds consisting largely of volcanic sand, silt and clay in addition to numerous Magadi-type chert beds. High Magadi and Oloronga beds overlap in many areas around the lake and they can be most easily distinguished by a thick caliche cap terminating Oloronga deposition.

Brine evolution has been well documented at Magadi. Perennial alkaline springs at the perimeter of the active lake are fed by a saline, hot ground water reservoir which in turn is recharged by dilute runoff and rim streams entering the valley. No perennial streams reach the lake. Precipitation of alkaline earth carbonates and silica and dissolution of efflorescent crusts are some of the important mechanisms responsible for brine evolution. Because of the presence of reactive volcanic glass, authigenic mineral reactions abound. Zeolites as well as silicate gels and non-silicates such as gaylussite and fluorite are common products. Sedimentologically, Lake Magadi is characterized by a large ephemeral salt pan, saline and dry mudflats, some of which are old lake bottoms, ephemeral stream complexes and small alluvial fans. The steep shorelines prevent wide mudflats from forming. Lake Magadi is a type example of (1) a Rift Valley lake, (2) an alkaline sodium carbonate lake and (3) a saline lake in full production of evaporites.

REFERENCES

Baker, B. H., 1958. Geology of the Magadi area. Geol. Surv. Kenya Rept. 42, 82 p.

Baker, B. H., 1965. Geology of the area South of Magadi. Geol. Surv. Kenya Rept. 61, 27 p.

Baker, B. H., Goles, G. G., Leeman, W. P. and Lindstrom, M. M., 1977. Geochemistry and petrogenesis of a basalt-benmoreite-trachyte suite from the southern part of the Gregory Rift, Kenya, Contrib. Mineral. Petrol. 64, 303–332.

Baker, B. H. and Mitchell, J. G., 1976. Volcanic stratigraphy and geochronology of the Kedong – Olorgesailie area and the evolution of the South Kenya rift valley. Journ. Geol. Soc. London 132, 467–484.

Baker, B. H., Mohr, P. A. and Williams, L. A. J., 1972. Geology of the eastern rift system of Africa. Geol. Soc. Am. Spec. Paper 136, 67 p.

Baker, B. H., Williams, L. A. J., Miller, T. A. and Fitch, F. J., 1971. Sequence and geochronology of the Kenya rift volcanics. Tectonophysics 11, 191–215.

Beadle, L. C., 1974. The inland waters of tropical Africa. Longman, London, 365 p.

Bradley, W. H. and Eugster, H. P., 1969. Geochemistry and paleolimnology of the trona deposits and associated authigenic minerals of the Green River Formation of Wyoming. U.S. Geol. Survey Prof. Paper 496–B, 71 p.

Bricker, O. P., 1969. Stability constants and Gibbs Free Energies of Formation of magadiite and kenyaite: Am. Mineral. 54, 1026–1033.

Brindley, G. W., 1969. Unit cell of magadiite in air, vacuo and under other conditions. Am. Mineral. 54, 1583–1591.

Butzer, K. W., Isaac, G. L., Richardson, J. L. and Washbourn-Kamau, C., 1972. Radiocarbon dating of East African lake levels. Science 175, 1069–1076.

Deardorff, D. L. and Mannion, L. E., 1971. Wyoming trona deposits. Wyoming Univ. Contr. Geology 10, 25–37.

Eugster, H. P., 1966. Sodium carbonate-bicarbonate minerals as indicators of PCO_2. J. Geophys. Res. 71, 3369–3377.

Eugster, H. P., 1967. Hydrous sodium silicates from Lake Magadi, Kenya; Precursors of bedded chert. Science 157, 1177–1180.

Eugster, H. P., 1969. Inorganic bedded cherts from the Magadi area, Kenya. Contrib. Mineral. Petrol. 22, 1–31.

Eugster, H. P., 1970. Chemistry and origin of the brines of Lake Magadi, Kenya. Mineral. Soc. Am. Spec. Paper 3, 215–235.

Eugster, H. P., 1971. Origin and deposition of trona. Wyoming Univ. Contr. Geology 10, 49–55.

Eugster, H. P. and Chou, I-Ming, 1973. The depositional environment of Precambrian bedded iron-formations. Econ. Geol. 68, 1144–1168.

Eugster, H. P. and Hardie, L. A., 1975. Sedimentation in an ancient playa-lake complex: the Wilkins Peak Member of the Green River Formation of Wyoming. Bull. Geol. Soc. Am. 86, 319–334.

Eugster, H. P. and Hardie, L. A., (in press). Saline Lakes. In Lerman, A. (ed.) Physics, Chemistry and Geology of Lakes. Springer-Verlag.

Eugster, H. P. and Jones, B. F., 1968. Gels composed of sodium aluminum silicate, Lake Magadi, Kenya. Science 161, 160–164.

Eugster, H. P. and Jones, B. F., 1977. The behavior of potassium and silica during closed-basin evaporation. 2nd IACC Symp. Water-Rock Interaction, Strasbourg, II, 1–12.

Fahey, J. J., 1962. Saline minerals of the Green River Formation. U. S. Geol. Survey Prof. Paper 405, 50 p.

Fairhead, J. D., Mitchell, J. G. and Williams, L. A. J., 1972. New K-Ar determinations on the rift volcanics of Kenya and their bearing on the age of the rift faulting. Nature 238, 66–69.

Garrels, R. M. and Mackenzie, F. T., 1967. Origin of the chemical composition of some springs and lakes. In: Equilibrium concepts in natural water systems. Am. Chem. Soc. Advances in Chemistry, 67, 222–242.

Haines, D. V., 1959. Core logs from Searles Lake, San Bernadino County, California. U.S. Geol. Surv. Bull. 1045–E, 139–317.

Hardie, L. A. and Eugster, H. P., 1970. The evolution of closed-basin brines. Min. Soc. Amer. Spec. Publ. 3, 273–290.

Hardie, L. A., Smoot, J. P. and Eugster, H. P. (in press). Saline Lakes and their deposits: A sedimentological approach. Int. Ass. Sedimentol. Special Publ.

Hay, R. L., 1968. Chert and its sodium-silicate precursors in sodium-carbonate lakes of East Africa. Contr. Mineral. Petrol. 17, 225–274.

Hay, R. L., 1970. Silicate reactions in three lithofacies of a semi-arid basin, Olduvai Gorge, Tanzania. Mineral. Soc. Am. Spec. Paper 3, 237–255.

Hsu, K. T. and Siegenthaler, C., 1969. Preliminary experiments on hydrodynamic movement induced by evaporation and their bearing on the dolomite problem. Sedimentology 12, 11–25.

Hunt, C. B., Robinson, T. W., Bowles, W. A. and Washburn, A. L., 1966. Hydrologic basin, Death Valley, California. U.S. Geol. Surv. Prof. Paper 494 B, 138 p.

Isaac, G. Ll., 1965. The stratigraphy of the Peninj Beds and the provenance of the Natron Australopithecine mandible. Quarternaria 7, 101–130.

Isaac, G. L., Merrick, H. V. and Nelson, C. M., 1972. Stratigraphic and archeological studies in the Lake Nakuru Basin, Kenya. In: Paleoecology of Africa, the surrounding islands and Antarctica (E. M. van Zinderen Bakker, Ed.) Balkema, Cape Town, 225–232.

Johan, Z. and Maglione, G., 1972. La Kanemite, nouveau silicate de sodium hydraté. Bull. Soc. Fr. Mineral. Cristall. 95, 371–382.

Jones, B. F., 1966. Geochemical evolution of closed basin waters in the western Great Basin. Ohio Geol. Soc. Symp. on Salt, 2nd, Ohio Geol. Soc. Cleveland, 1, 181–200.

Jones, B. F., Eugster, H. P. and Rettig, S. L., 1977. Hydrochemistry of the Lake Magadi basin, Kenya. Geochim. Cosmochim. Acta 41, 53–72.

Lagaly, G., Beneke, K. and Weiss, A., 1975. Magadiite and H-magadiite: I. Sodium magadiite and some of its derivatives. Am. Mineral. 60, 642–649.

Lucia, F. J., 1968. Recent sediments and diagenesis of South Bonaire, Netherlands, Antilles. Jour. Sed. Pet. 38, 845–858.

Maglione, G., 1970. La magadiite, silicate sodique de néoformation des facies évaporitiques du Kanem. Bull. Serv. Carte. Géol. Als. Lorr. 23, 177–189.

McAtee, J. L., House, R. and Eugster, H. P., 1968. Magadiite from Trinity County, California. Am. Mineral. 53, 2061–2069.

O'Neil, J. R. and Hay, R. L., 1973. $^{18}O/^{16}O$ ratios in cherts associated with the saline lake deposits of East Africa. Earth Planet. Sc. Letters 19, 257–266.

Parkinson, J., 1914. The East African trough in the neighbourhood of the Soda Lakes. Geogr. Journ. XLIV, 33–49.

Purser, B. H. and Loreau, J. P., 1973. Aragonitic, supra-tidal encrustations on the Trucial Coast, Persian Gulf. In: Purser, B. H. (ed.), The Persian Gulf. Springer-Verlag, N. Y., 343–376.

Richardson, J. L. and Richardson, A. E., 1972. History of an African Rift Lake and its climatic implications. Ecological Monographs 42, 499–534.

Risacher, F. and Eugster, H. P. in press. Holocene pisolites and encrustations associated with spring-fed surface pools. Pastos Grandes, Bolivia. Sedimentology.

Rooney, T. P., Jones, B. F. and Neal, J. J., 1969. Magadiite from Alkali Lake, Oregon. Am. Mineral. 54, 1034–1043.

Scholle, P. A. and Kinsman, D. J. J., 1974. Aragonitic and high-Mg calcite caliche from the Persian Gulf – a modern analog for the Permian of Texas and New Mexico. Jour. Sed. Petr. 44, 904–916.

Sheppard, R. A. and Gude, A. J., 1968. Distribution and genesis of authigenic silicate minerals in tuffs of Pleistocene Lake Tecopa, Inyo County, California. U.S. Geol. Surv. Prof. Paper 597, 38 p.

Sheppard, R. A., Gude, A. J. and Hay, R. L., 1970. Makatite, a new hydrous sodium silicate mineral from Lake Magadi, Kenya. Am. Mineral. 55, 358–366.

Smith, G. I., (in press). Subsurface stratigraphy and geochemistry of late Quarternary evaporites, Searles Lake, California. U.S. Geol. Survey Prof. Paper.

Smith, G. I. and Friedman, I., 1975. Chemical sedimentation and diagenesis of Pleistocene evaporites in Searles Lake, California. 9th Inter. Sediment. Congress, Nice, 137–140.

Smoot, J. P., (in press). Origin of the carbonate sediments in the Wilkins Peak Member of the lacustrine Green River Formation (Eocene), Wyoming, U.S.A. Inter. Sedimentol. Assoc. Special Publ.

Stevens, J. A., 1932. Lake Magadi and its alkaline springs. Unpublished report, summarized in Baker (1958), Geol. Surv. Kenya Report 42.

Stoertz, G. E. and Ericksen, G. E., 1974. Geology of salars in Northern Chile. U.S. Geol. Surv. Prof. Paper 811, 65 p.

Surdam, R. C. and Eugster, H. P., 1976. Mineral reactions in the sedimentary deposits of the Lake Magadi region, Kenya. Bull. Geol. Soc. Am. 87, 1739–1752.

Surdam, R. C., Eugster, H. P. and Mariner, R. H., 1972. Magadi-type chert in Jurassic and Eocene to Pleistocene rocks, Wyoming. Bull. Geol. Soc. Amer. 83, 689–700.

Temperley, B. N., 1951. Some geological and geophysical investigations in the vicinity of Lake Magadi. Unpubl. Report, summarized in Baker (1958), Geol. Surv. Kenya Report 42.

Chapter 16

RECENT SABKHAS MARGINAL TO THE SOUTHERN COASTS OF SINAI, RED SEA

ELIEZER GAVISH

Tel-Aviv University, Ramat-Aviv, Israel

INTRODUCTION

Accumulation of extensive sedimentary evaporites has been attributed to primary and diagenetic processes involving sea-derived brines. So far there is full agreement on the need for the proper environment in which evaporation may produce the saline brines and that the primary precipitation could be the major process of evaporite formation. There is however still uncertainty in the literature about the exact processes and environments in which these evaporites may form and the major factors influencing them. The major environments of evaporite accumulation considered until now have been the shallow marine shelves, enclosed basins and supratidal sabkhas. In recent years, studies of modern equivalents of such environments have shown the evaporites to form mainly in supratidal – sabkhas, thus adding more credibility to this type of environment. These studies were carried out on the coasts of arid to semi arid areas where evaporation is intensive and sea water resupply is possible.

A clear distinction has been made in the literature between the continental and coastal sabkhas (Kinsman, 1969). The continental sabkhas form by concentration of airborn salts in local basins with internal drainage and high water table. In the Sinai area such sabkhas are found in the Arava Valley near the Gulf of Elat (Amiel and Friedman, 1971) and in the northern Sinai near the Mediterranean coastal plain (Levy, 1977). In the southern part of the Sinai peninsula, however no continental sabkhas have formed and the reasons for that could be a very low supply of airborn salts and a lack of rains and thus there is no concentration of the salts by drainage. In the southwestern part of Sinai, between A-Tur and Abu-Rudeiz, in a place called "El-Qa" plain, there is some concentration of evaporites in the upper sediments but not enough to justify it being called a sabkha.

The coastal sabkhas, being the "typical" ones, have been studied more intensively than the continental sabkhas. On the basis of their morphology and hydrodynamics the coastal sabkhas can be divided into two types:

a) Open supratidal environment with unrestricted and continuous hydrodynamic flow regime. In these sabkhas brines form a saline groundwater table which extends continuously from the sea edge landwards. The brines get into the sediment by continuous subsurface seepage and sometimes by occasional floods. This type of sabkha typically occurs in the Persian Gulf and has been studied extensively by a number of investigators (Illing Wells and Taylor, 1965; Kinsman, 1964, 1969; Wood and Wolfe, 1969, Purser, 1973).

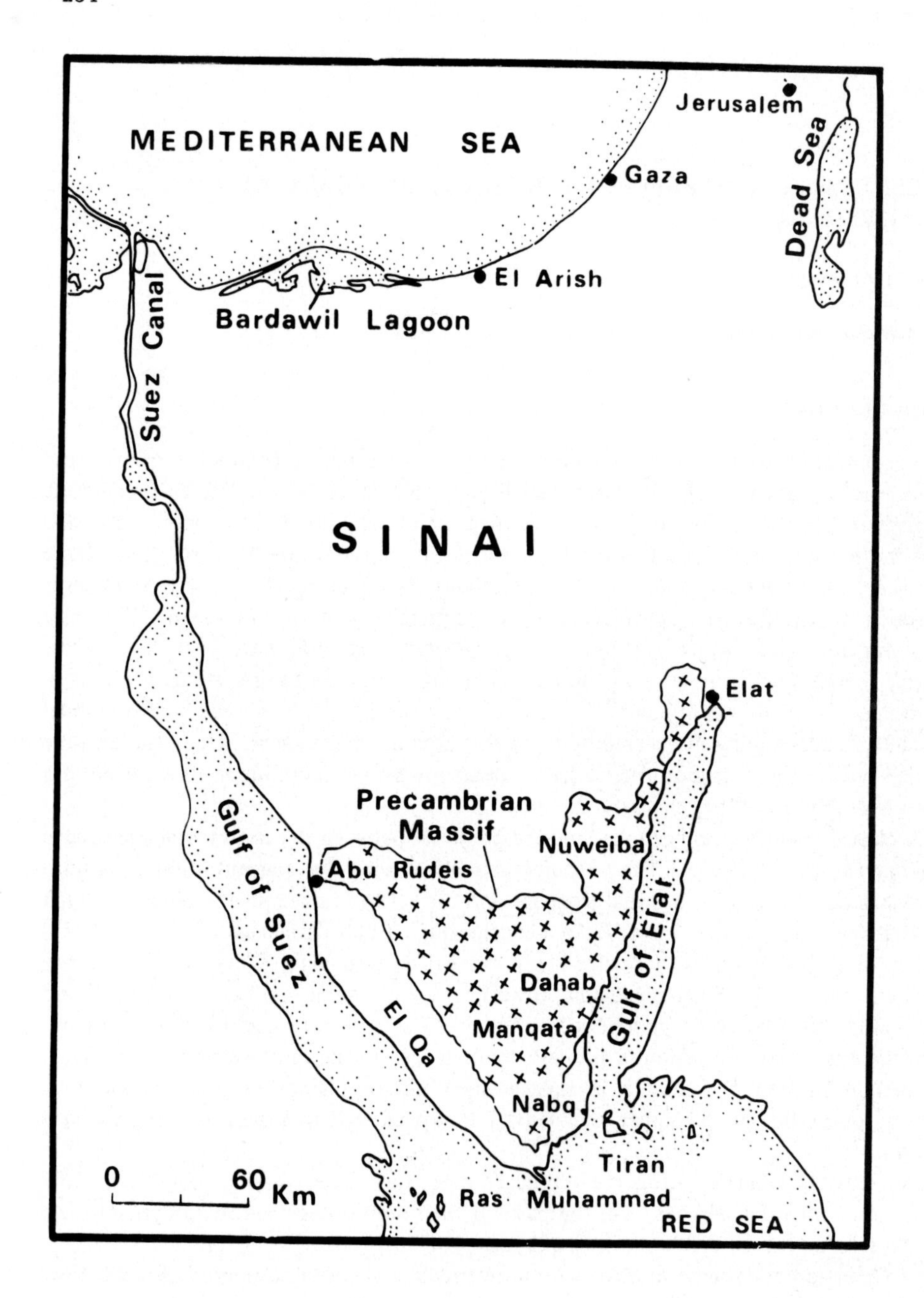

Fig. 16-1a. Index map of the Sinai peninsula showing the major coastal zones along the Gulf of Suez and Elat.

Fig. 16-1b. A satellite photograph of the Sinai Peninsula.

b) Sea marginal brine pans (pools) with restricted and uneven marine water supply. These types of sabkha pools have been studied in the Pekelmeer Lake in Bonaire (Deffeyes, Lucia and Weyl, 1965, Lucia, 1968, Murray, 1969) and in the Ephemeral Lakes of the Coorong district (Skinner, 1963; Alderman, 1965). While in the Pekelmeer the accumulating evaporites are of marine origin, those in the Coorong have contribution by meteoric water. Thus this sabkha could be considered, to a certain extent of continental origin.

In the Sinai peninsula both types of coastal sabkhas are common. The open supratidal environment of evaporite accumulation can be found along the Mediterranean coast, mainly on the Southern margins of the Bardawil lagoon, as well as along the coasts of the Gulf of Suez. Along the margines of the Gulf of Elat, however, open supratidal sabkhas are very few and are typical to that area. Sea marginal brine pans do occur, however, in a few places, thus drawing attention to this unique phenomenon. Fig.16-1a is an index map of the Sinai showing the general locations of the areas referred to in this chapter. Fig.16-1b shows a satellite picture of the Sinai peninsula and the Gulfs of Elat and Suez.

OPEN SUPRATIDAL SABKHA ENVIRONMENT – THE GULF OF SUEZ

The Gulfs of Elat and Suez differ in their geologic history and structure, a fact that causes an extreme difference between their coastal zones. The Gulf of Elat, being a deep and narrow basin, has a coastal zone with a very steep topography and a very narrow coast line. The Gulf of Suez, however, is a relatively shallow basin (average depth about 70 m) and the mountain range flanking it is at a large distance from the coastal zone. As a result the coastal and littoral areas marginal to this Gulf have a very low angle topography and form wide, elongated strips (Fig. 16-1).

Such a wide coastal area where supratidal sabkhas are typically abundant, extends from Abu-Rudeiz to Ras Muhammad along the southern part of the Gulf of Suez (Fig. 16-1). With the exception of a local mountain range near El-Belayim, this area has a flat surface composed of alluvial silicate sediment transported from the Precambrian granitic and metamorphic massif. This area, named El-Qa, is considered to have the most arid and hottest conditions in the Sinai. The annual rainfall is 10–20 mm, mean air temperature in the summer is 29°C, relative humidity is about 50% (Israel Meteorological Service 1970–72) and the evaporation rate is about 3.6 m per year (Griffiths, 1972).

In the subtidal, near-shore zone the water energy is usually low with the waves breaking far from the shoreline, over numerous patch reefs abundant in this area. The general longshore drift is from the norht, producing north-south trending spits which occasionally enclose parts of the shallow littoral zone to form lagoons. One of such lagoons is called El-Belayim (Fig. 16-1) and a part of it is shown in Fig. 16-2. Here an elevated ancient alluvial far is located close to the shoreline making the supratidal zone relatively narrow. Still, even here, sabkha type environments are existent, extending from the water line to the base of the recent alluvial fans (Fig. 16-2). The sediments of the subtidal zone are clastic fine sands composed mainly of silicate minerals transported from the massif, and marine carbonates as biofragments, pellets, ooids and some intraclasts. The proportions of each of the carbonate components varies with the location. Generally in the northern parts of the Gulf of Suez the ooids (Sass, Weiler and Katz, 1972) and intraclasts are

Fig. 16-2. The coast of the Gulf of Suez showing the barrier enclosed El-Belayim lagoon. Note the dark area between the alluvial fan and the shoreline where the supratidal sabkha develops.

relatively abundant, while in the southern parts of this coast, pellets and biofragments are more abundant.

In the intertidal zone the sediments vary from abundant fine marine carbonates on the seaward side of this zone to coarse grained silicates on the landward side. In many places in the upper intertidal zone, there is a somewhat elevated sandy berm (up to 1 m) on a strip of pebbly beachrock (Gavish, 1975a) defining clearly the water edge.

The supratidal zone extends from the water line inland, generally to a distance of a few hundred meters. In places where the berm is nonexistent, water may occasionally flood large parts of the intertidal zone during periods of high tide or storms. In such places a thin upper layer of halite covers the surface during the weeks following the flood, but eventually it disappears. In places where a sandy berm exists, as in El-Belayim (Gavish, 1974), some low areas behind it may be covered by a layer of permanently moist fine sediment. The landward border of the supratidal zone is not as clear and is usually determined at the furthest extent of the domain of authochtonous evaporite minerals.

Fig. 16-3 is a generalized presentation of a typical coastal area in the Gulf of Suez as described above. The values given in this figure are based on many analyses done in a number of places and thus are good representatives of the general conditions along these coasts. The groundwater in the supratidal zone where sabkhas form, originates mainly from the sea and thus the water table remains at about mean sea level and laterally extends up to a few hundred meters landward. The salinity of the groundwater increases landward (Fig. 16-3) unless it is diluted by fresh water as may happen in wadi channels. This gradient of salinity supports the mechanism of marine water seeping through the subsurface toward the land where it gets more concentrated by evaporation through the thin sediment cover (Kinsman, 1969; Gavish, 1974).

The relative distribution of each major evaporite and carbonate component in the upper 0.5 m of these sediments is also shown graphically in Fig. 16-3. These components are, however, unevenly distributed with depth, as can be seen in Figs.16-4 and 5. Fig.16-4 is a photo of a hole dug in a typical sabkha in the supratidal area showing the zonation beneath the surface. Fig. 16-5 is a graphical presentation of the components and their typical distribution in a place similar to the one in which the hole was dug. This typical zonation of the evaporites clearly indicates that they were precipitated from brines migrating upwards from the groundwater. The mechanism of "evaporative pumping" as suggested by Hsu and Siegenthaler (1969) and Hsu and Schneider (1973) could certainly be applied here in order to explain the upward migration of these brines.

The increase in the concentration of the brines as they move upwards causes the precipitation of carbonates first, above them sulphates and finally the halite (Figs.16-3 and 5). The carbonates precipitated are mainly aragonite and Mg-calcite which are difficult to differentiate from the same minerals, but of clastic origin which are abundant in the sediment of the whole coastal zone (Fig. 16-3). Still in a number of places the outhigenic

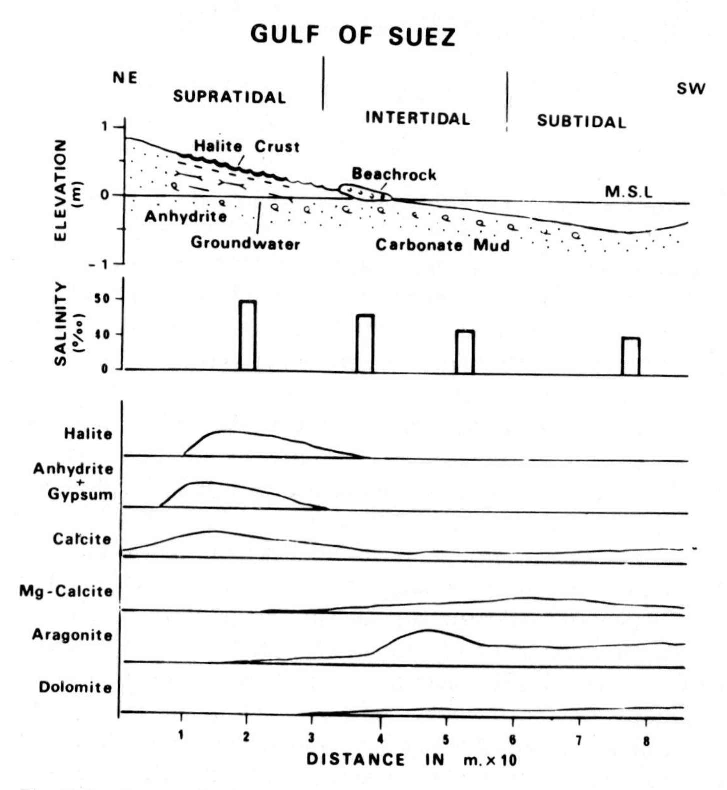

Fig. 16.3 A generalized crossection through the coastal zone along the Gulf of Suez showing typical water geochemistry and sediment composition. The distribution of each major mineral of the sediments is on a relative basis without making a distinction between authigenic and clastic components.

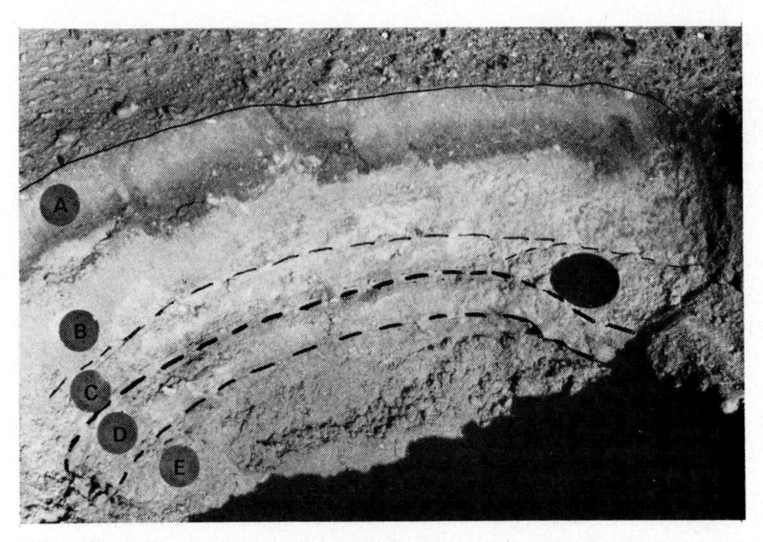

Fig. 16-4. A hole dug in a supratidal sabkha near El-Belayim showing the vertical layering:
a) Lutitic sediment with dispersed halite crystals
b) Sand with halite and anhydrite
c) Lithified layer, mainly by sulphates and carbonates
d) Clean sand.

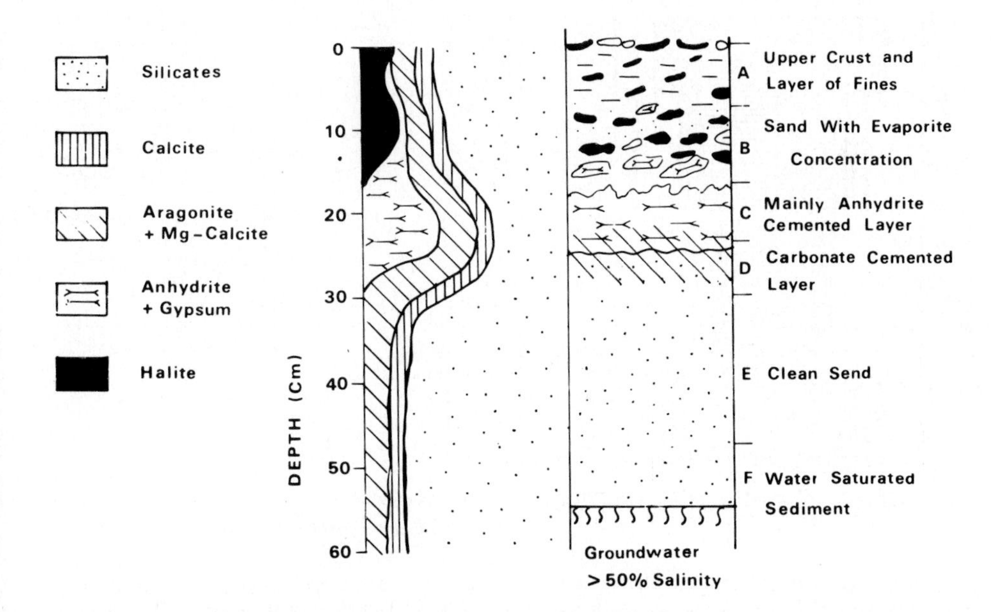

Fig. 16-5. A profile of a typical supratidal sabkha along the Gulf of Suez showing the distribution of the major minerals in its different layers. This profile may well represent the hole shown in Fig. 16-4.

carbonates are abundant enough to form a lithified layer (Fig. 16-5). The sulphates precipitating above the groundwater and above the carbonates, if present, are mainly anhydrite while near the intertidal zone they may contain relatively more gypsum (Fig. 16-3). The anhydrite occurs as dispersed crystalline concentrations or form a lithified layer (Fig. 16-5). The halite under the surface forms scattered crystals with variable concentrations, depending on the texture of their host sediment. On the surface the halite may appear in two forms:

a) As hard polygons on the landward side of the sabkha, where they have accumulated in the past years and are not affected by seasonal changes.

b) As white, thin crusts in the occasionally flooded seaward part of the sabkha where they are usually destroyed after a short time. The fact that not enough halite, relatively to the sulphates, was formed in these sabkhas to account for complete evaporation of the brines, poses still a porblem. But from field observations it seems most reasonable that a large portion of the halite could have been washed down to the groundwater by the infrequent floods or rains.

The process of migration of brines from the groundwater upward gets further support from geochemical analyses of soluble salts in the sabkha sediments as shown in Figs. 16-6 and 7. These figures give the results of analyses done on the salts in 100 gr of sediment

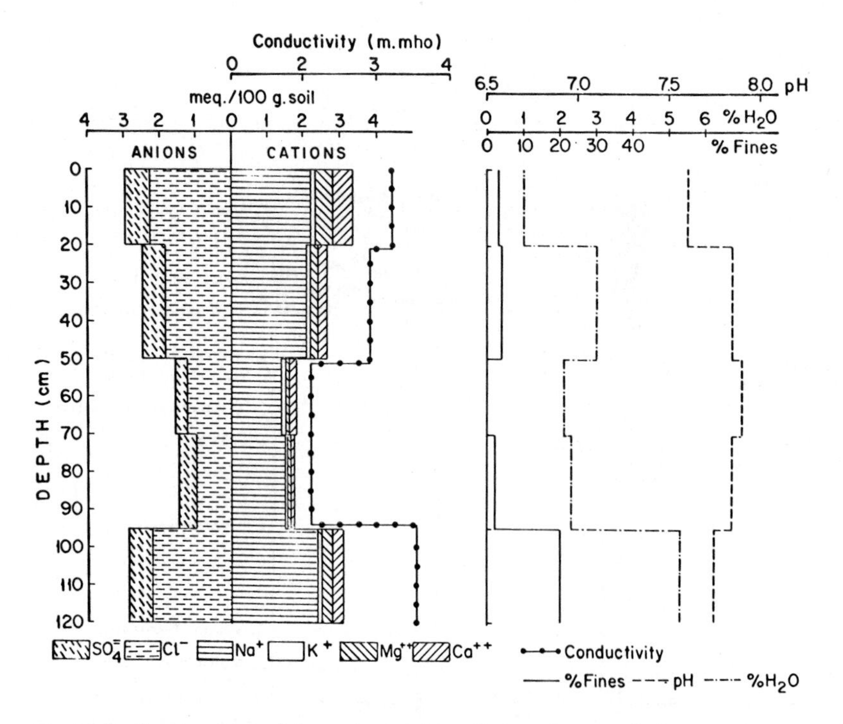

Fig. 16-6. Geochemistry of the major soluble salts in sediments of a berm at the upper intertidal zone along the Gulf of Suez. Sediment water ratio was 100 gr in 100 ml disfilled water. The relatively higher results at the bottom are mainly due to the interstitial brines near the groundwater. (After Gavish, 1974)

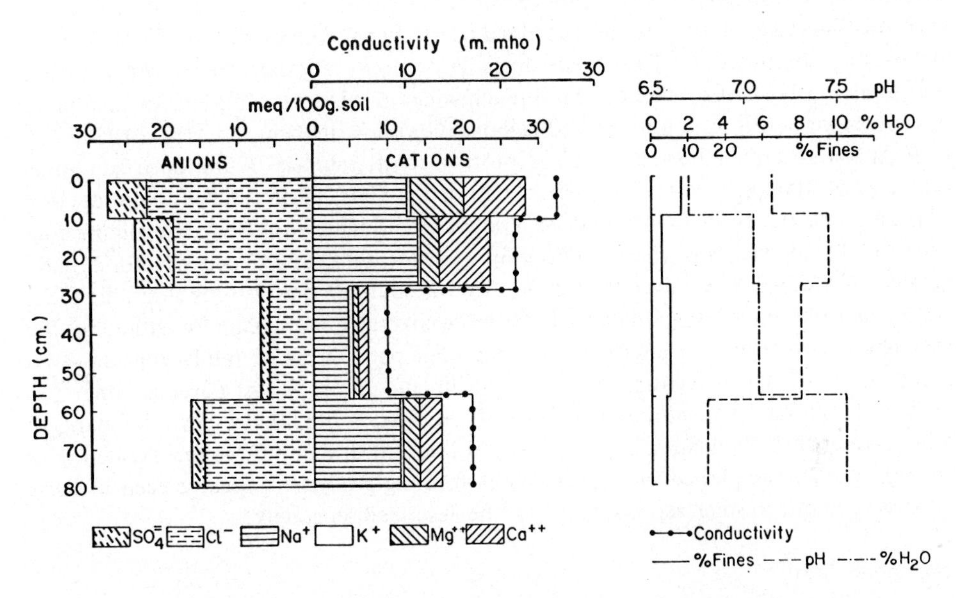

Fig. 16-7. Geochemistry of the major soluble salts in sediments from a supratidal sabkha similar to the one in Figs.16-4 and 5. Sediment-water ratio 1:1. At the bottom contribution of ions from interstitial brines. (After Gavish, 1974)

dissolved by 100 ml distilled water. Fig.16-6 shows the geochemistry of sediments from a berm at the upper part of the intertidal zone, while Fig.16-7 shows results from a hole in a typical sabkha similar to the one in Fig. 16-4. In the berm sediments the ion concentration shows only a minor increase upward and no accumulation of evaporites in the sediments, as can be expected from sediments that are being periodically flushed through. In the sabkha sediments, however, there is a drastic increase in ion concentration upward with the proper proportions to justify the mineral distribution as shown in Fig. 16-5. In areas where the groundwater is deeper than about one meter, brine migration upward apparently does not occur and the sabkha forming processes do not exist.

SEA MARGINAL BRINE PANS – GULF OF ELAT

The coasts of Sinai along the Gulf of Elat have, as stated before, a steep angle topography and are clearly defined by a continuous fringing reef. In most areas, especially between Elat and Manqata (Fig. 16-1), the mountainous granitic and metamorphic massif meets directly the Gulf waters producing a very narrow shore line (Emery, 1963). This shoreline is composed of slump blocks, coarse beachrock plates (Friedman and Gavish, 1971; Gavish, 1975) and narrow reef flat (Friedman, 1968). In a number of places, as

Nuweiba and Dahab, alluvium from wadis has formed large fans which protrude seawards forming a locally wider coastal zone with a less steep topography.

The southern part of this coast, between Manqata and Ras Muhammad (Fig. 16-1), is different in its topography and lithology since the massif range is at a greater distance from the Gulf edge. Here, the backshore is a relatively wide, flat strip with a low angle of inclination, composed of Pleistocene reefs and recent alluvial fans (Gvirtzman and Buchbinder, 1978). Also, the reef flat is much wider in this area and it includes a number of depressions which form open coastal lagoons (Gavish, Krumbein and Tamir, 1978).

In the northern and southern parts of this coast, sabkhas of the open supratidal type do not develop. On the northern coasts of the Gulf, these sabkhas do not develop because under the given topography, backshore supratidal environments are almost non-existent. On alluvial fans where the backshore is wide, sabkhas generally can not develop because of the high rate of sedimentation. On the southern parts of this coast, the topography and rate of sedimentation could be more favorable for sabkha formation but the lithology is not. Here the intertidal zone and large parts of the supratidal zone are composed of lithified subrecent to Pleistocene reefs and the sediment cover is either too shallow or too far from the water to allow the formation of sabkhas. Also, sea-marginal brine pans are not abundant along the coasts of the Gulf of Elat but the few existent ones are very good examples of such local sabkha forming processes. Because each of these sabkhas is unique in some aspects, they shall be described separately.

The "Solar Lake"

The "Solar Lake" is a small pond located about 18 km south of Elat in a typical northern coastal area where the mountains touch the Gulf waters. The pond is about 140 m long, 65 m wide and its maximal water depth reaches about 5 m (Aharon, Kolodny and Sass, 1977). From three sides the pond is surrounded by igneous and metamorphic hills and to the east it is separated from the sea by a 60 m wide and about 3 m high barrier composed of coarse sand and pebbles (Fig. 16-8). This pond has attracted the attention of many investigators in the Earth and Life Sciences and indeed a lot of information has been gathered and many processes studied in it (Friedman, Amiel, Braun and Miller, 1973; Cohen, Krumbein, Goldberg and Shilo, 1977; Aharon et al., 1977).

The peculiarity of this pond is its water temperature which occasionally reaches 57°C at the bottom and about 20–28°C at the surface. The hydrodynamic system, water salinity and the main sediment types of the pond are shown graphically in Fig. 16-9. The water supply into the pond is practically only from the Gulf by seepage through the sediments of the barrier during high tide, a mechanism already established in literature (Adams and Rhodes, 1960; Deffeyes et al., 1965). The high rate of evaporation causes loss of water and makes the pond hypersaline with up to 170 g/l total dissolved solids (Fig. 16-9). According to Krumbein and Cohen (1974) and Aharon et al. (1977) this water body is monomictic and mesothermal with a constant temperature difference between the hypolimnion (57°C) and epilimnion (20–28°C). During the summer when inflow of water is minimal, its salinity is uniformly high, but during the winter the fresh seawater forms an upper layer of relatively low salinity (Fig. 16-9). While this stratification exists, the lower water body absorbs heat and with lack of circulation, its temperature

Fig. 16-8. The "Solar Lake". A brine pond about 18 km south of Elat. Approximate dimensions 140 x 65 m.

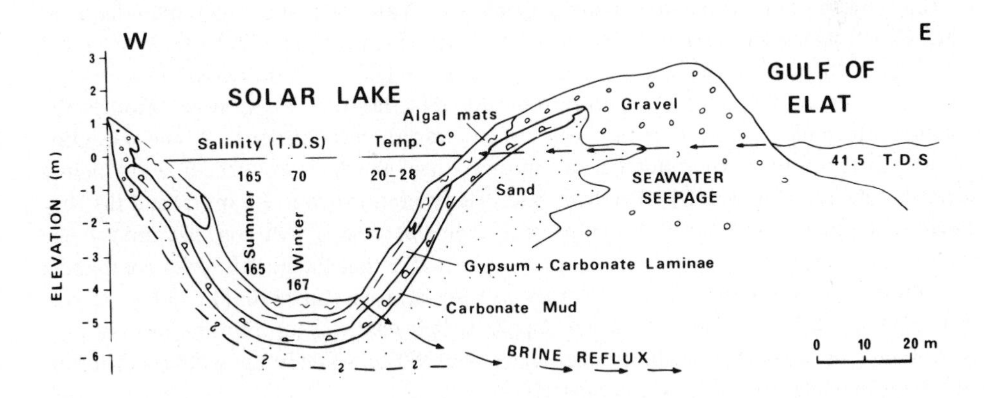

Fig. 16-9. A general crossection through the "Solar Lake" and its barrier showing the major sediment components, water salinity and assumed hydrodynamic system. This figure is a modification of a drawing and data presented by Aharon et al. (1977).

increases due to this green-house effect. The increasing salinity of the upper water layer causes a decrease in the pycnocline instability and a turnover in the water body occurs in the summer.

The pond configuration and its water salinity have existed long enough to form the appropriate evaporite sequence. The margins of the pond are covered by well developed algal mats while its slopes are composed of a layer of white large crystal of gypsum and some aragonite (Fig. 16-9). The carbonate laminae within the algal mats and the gypsum layer are composed of authigenic aragonite Mg-calcite, dolomite and calcite with the dolomite being a replacement product of the aragonite (Friedman et al., 1973; Aharon et al., 1977). The algal and gypsum layers of the pond are underlain by a carbonate mud layer with *cerithium* shells indicating the initial stages of the sabkha type evaporite accumulation.

The geochemical and mineralogical systems in this Lake are similar to those in the Pekelmeer Lakes of the Bonaire (Deffeyes et al., 1965) and therefore also led to the assumption of a possible brine reflux. Indeed the brines of this solar lake have had to precipitate halite and be at a much higher salinity than now, had there not been an escape of heavy brines. Although this negative reasoning is plausible, no positive, direct evidence of this brine reflux has been observed as yet.

The Ras Muhammad Pool

In the most southern tip of the Sinai there is a small peninsula called Ras Muhammad (Fig. 16-1) which is composed mainly of tectonically uplifted Pleistocene reefs (Gavish et al., 1970; Gvirtzman and Buchbinder, 1978). The peninsula is dissected by north-west to south-east trending fault lines that form a number of inlets which are flooded by sea water during high tide. In one place such a tidal inlet intrudes over 1 km inland forming an elongated lagoon. Following this lagoon, there is a separated small depression which probably formed during the late Pleistocene tectonic activity that affected the peninsula. In the center of this depression there is a tiny pool, extending about 200 m in diameter (Fig. 16-10) which shows most of the characteristics of a brine pan sabkha (Friedman and Sanders, 1978, p. 140–141).

The water of the pool originates from the tidal lagoons by seepage through the abundant tectonic crocks, some of which can be noticed on the surface. At the water line around the pool there are a number of "springs" through which seawater flows in mainly during high tide (Fig. 16-10). The high rate of evaporation (over 3.5 m/yr) and the very shallow depth of the pool (0.5 m), cause a rapid increase of salt concentration in the brines. It is only because of the constant supply of water that the pool is never completely dry though its extent and water depth vary with the seasons (Gavish et al., 1978). During the summer and fall, when the water supply is low relatively to evaporation, the pool shrinks in size and its water salinity may reach over 300‰, while in the winter and spring. its size is maximal and salinity is almost 100‰.

Fig. 16-11 is a generalized crossection of this pool showing its major mineral components and their distribution and pool water salinity as measured during spring time. Here we can see that the pool is surrounded by groundwater of about normal Gulf salinity (42‰) and only in the pool itself the water is hypersaline. Maximal salinity is reached

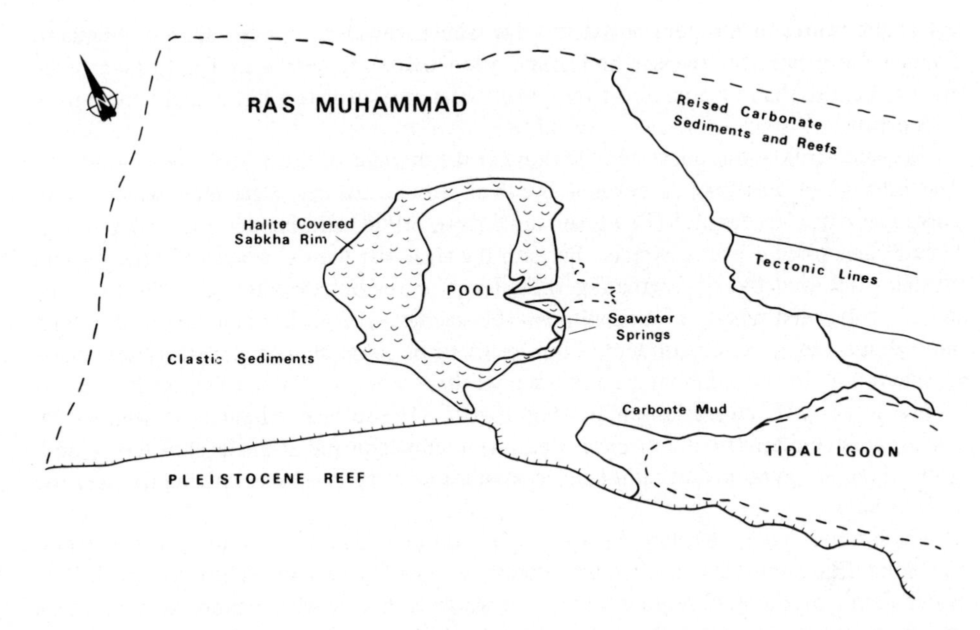

Fig. 16-10. A map showing the hypersaline pool and main features of the Ras-Muhammad peninsula.

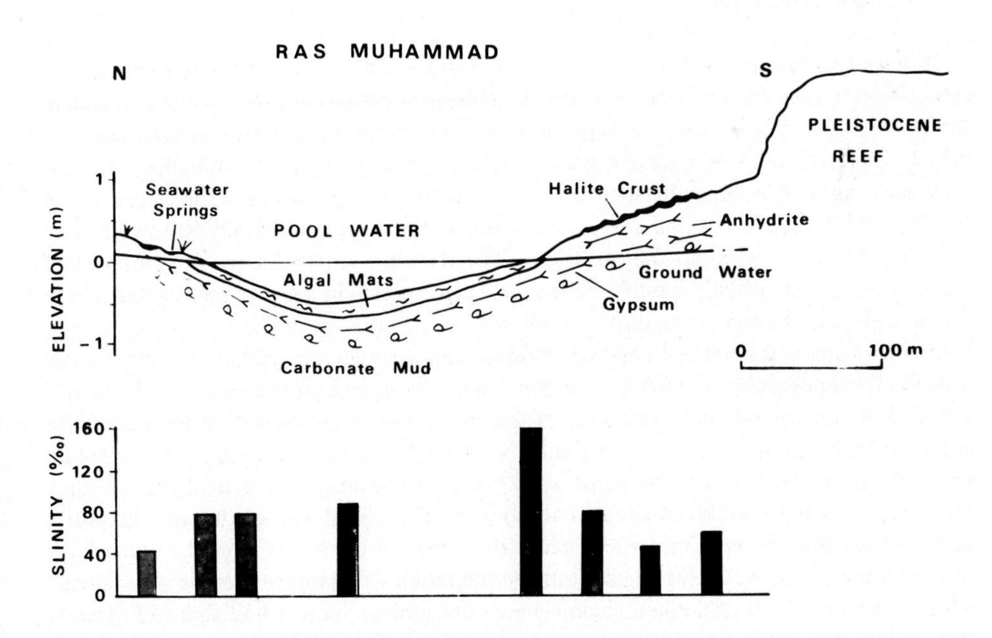

Fig. 16-11. A general crossection through the pool of Ras Muhammad showing the distribution of the major sediment components and water salinity.

not at the center of the pool but at its edge where seawater "springs" are not abundant. Though the process of seepage and evaporation operating here is similar to that in the Solar Lake, this shallow pool shows very small differences in water salinity and temperature with depth.

The sediments at this pool resemble those at the margins of the Solar Lake (Fig. 16-11). The bottom of the pool is covered by thick (up to 30 cm) algal mats which have polygonal crack appearance (Friedman and Brown, 1968) and include thin flakes of carbonates and some gypsum crystals. Beneath the algal mat there is a layer of large gypsum crystals, followed by, or alternating with, layers of carbonate mud (Fig. 16-11). The mud is composed mainly of aragonite and Mg-calcite with smaller amounts of dolomite and calcite. In general, carbonate mud seems to be more abundant in these sediments than gypsum. In the sediment around the pool there is accumulation of evaporites similar to that in an open supratidal sabkha (Fig. 16-10). The surface sediments are lithified by a halite crust, underneath it are concentrations of anhydrite and at the level of the groundwater there are gypsum and carbonate accumulations (Fig. 16-11) resembling those of the Gulf of Suez.

The hydrodynamic mechanism of this pool, though in principal similar to the one in the Solar Lake, must be much more intensive. The high evaporation rate and shallow water depth of the pool require a very fast water turn-over with a short residence time and therefore an intensive reflux. Indeed the pool brines cause extensive accumulation of carbonate mud but not as much as might be expected since at that higher salinity the brines rapidly escape from the system through the abundant tectonic cracks.

The Round Sabkha at Nabq

Perhaps the best example of an enclosed pan-type sabkha with extensive evaporite accumulation is found near Nabq (Fig. 16-1). This area is typical of the southern coasts of the Gulf of Elat; it is a wide, low angle, coastal zone composed of coarse alluvial fan sediments. The sabkha itself is almost symmetrically oval with about 400 m in diameter and is located about the same distance form the waterline (Fig. 16-12). The shape of the sabkha is determined by a well defined inner rim which is topographically its lowest place while the center is by about 0.5 m higher. The elevation of the inner rim is more than 1 m below sea level while the sandy barrier between the sabkha and the sea is about 1.5 m above sea level thus preventing direct flooding.

Fig. 16-13 is a generalized east-west crossection through the sabkha and the barrier showing its topography, the major sediment distribution and groundwater geochemistry. Detailed research which has been done in the last few years by the author on this sabkha caused it to be named locally as "Sabkha Gavish". The data of Fig. 16-13 are based on some of the results of this work (Gavish, 1979) as well as on work still being done today. Water supply into the sabkha is almost entirely from the sea by seepage through the coarse sediments of the barrier (Fig. 16-13). Since the sabkha is below sea level, the upper layer of the seepage front meets the surface outside the sabkha, forming numerous springs from which seawater flows in defined channels into the sabkha (Figs. 16-12 and 13). During the winter and spring the flow can be noticed on the surface of the channels while during the summer and fall it is under the surface. The salinity of the water in the channels is

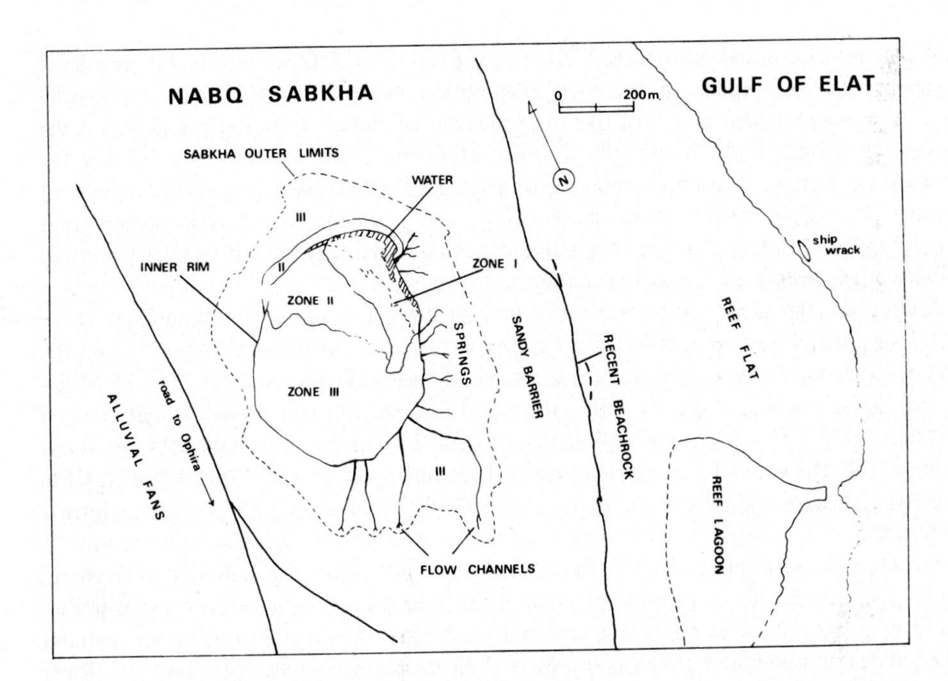

Fig. 16-12. A map of the round sabkha at Nabq, southern coast of Gulf of Elat, showing its major topographic and sedimentary zones.

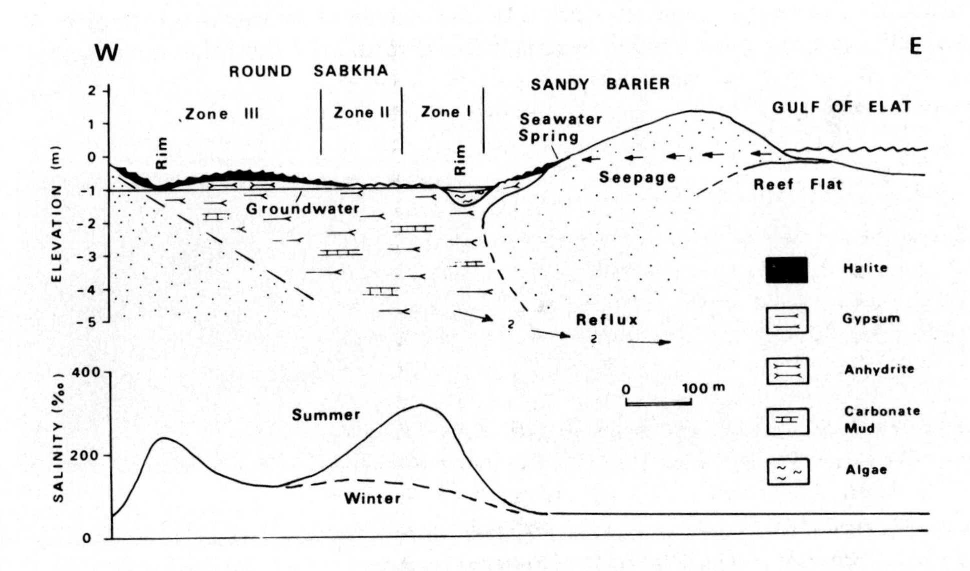

Fig. 16-13. A general crossection through the round sabkha at Nabq showing water salinity and main sediment composition in each topographic zone and the assumed hydrodynamic system.

that of almost normal nearshore Gulf water (41–43‰ T.D.S.) but in the sabkha it increases sharply because of the high evaporation (about 3.5 m/yr) and low rainfall (10–20 mm/yr) of this area. At the inner rim, the sabkha groundwater is exposed at the surface forming a water body only a few centimeters deep. In this zone, during the summer, surface water salinity increases drastically but falls back to a stable salinity of about 120‰ in the center of the sabkha (Fig. 16-13). In the winter, with higher water supply, its level rises slightly and the salinity increases gradually until it reaches the stable salinity characteristic of the sabkha center.

Sedimentologically it is possible to distinguish three zones in this sabkha (Figs. 16-12 and 13). Zone I is topographically the lowest area and it includes the inner rim and the northern part of the sabkha where the surface is at level with groundwater. Fig. 16-14 is a view of this zone showing both parts, the water covered rim and the white salt covered sabkha. At the rim there are abundant green and blue-green algal mats with *cerithium* shells which are studied by microbiologists (Krumbein, Buchholz, Frnake, Giani, Giele and Wonnelberger, submitted for pub.). In some places the algal layer and the sediments underneath it include black carbonate mud but no gypsum, though the water salinity is above the saturation point for this mineral. In the part where the sediment is exposed, there is an upper, thin crust of halite with a thickening layer of gypsum crystals underneath (Fig. 16-13). The halite is a seasonal product formed mainly during the summer when it forms polygonal crusts composed of clean white crystals (Fig. 16-15). These crusts may cover also parts of the muddy rim byt they get largely dissolved during the winter.

Zone II is the central part where the sabkha surface, only a few centimeters above groundwater, is composed of a puffy gypsum crust (Fig. 16-16). Here the sediment under the crust is composed of almost clear gypsum crystals. This type of sediment may reach a depth of a few meters. Zone III which is the highest part of the sabkha, is covered by a well lithified halite crust forming large polygons (Fig. 16-17). This halite is constantly

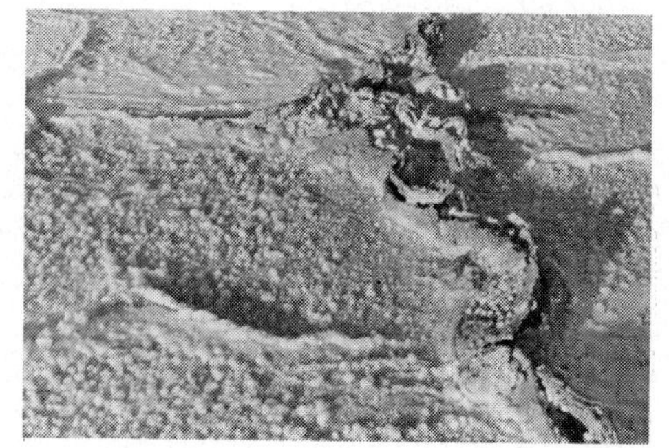

Fig. 16-14. Zone I of the Nabq sabkha, a view from the outer rim looking southward. In the middle is the water covered inner rim and beyond it the halite covered surface.

Fig. 16-15. Zone I of the Nabq sabkha showing the white, seasonally precipitated halite, forming large polygons.

Fig. 16-16. Zone II of the Nabq sabkha showing the puffy surface composed of gypsum and elevated only a few centimeters above groundwater.

Fig. 16-17. Zone III of the Nabq sabkha showing the permanent thick halite crust which forms a rough polygonal surface.

above groundwater and may be up to 30 cm thick. It does not change with the seasons but is a product of constant accumulation by "evaporative pumping" from the groundwater brine. Within the halite and somewhat below it, anhydrite is found in small amounts. The main seidments under the crust are crystalline gupsum and occasional carbonate mud layers composed of aragonite, Mg-calcite, dolomite and calcite. The detailed mineralogical analyses and their distribution with depth are presented by Gavish (1979).

The composition of sediments and the geochemistry of water in this round sabkha indicate very intensive processes of brine concentration. After precipitating gypsum these brines have to escape from the system, and thus their refluxing is inevitable. Water analyses taken from boreholes that are being made now in this sabkha will, hopefully, provide direct evidence for that reflux.

In summary, the three types of sea marginal brine pans shown here, have in common the basic process of seawater seepage, evaporite accumulation from the concentrated brines and their probable reflux. There are however certain differences in the hydrodynamic systems and in the residence time of the brines in these pans which results in different mineralogical and geochemical composition. It is the study of the relationship between these different results as caused by variable conditions, that will make us understand better the sabkha forming process.

ACKNOWLEDGMENT

Parts of the research done by the author on the coasts of Sinai were carried out due to the generous support of the late Mr. Abraham Shapiro of Massachusetts in whose memory this chapter was written. Parts of the research were also supported by the Israel Academy of Science Grant No. 9354.

REFERENCES

Adams, J. E. and Rhodes, M. L., 1960. Dolomitization by seepage refluxion. Am. Ass. Pet. Geol. Bull., 44, 1912–1920.

Aharon, P., Kolodny, Y. and Sass, E., 1977. Recent hot brine dolomitization in the "Solar Lake", Gulf of Elat, Isotopic, chemical and mineralogical study. Jour. Geol., 85, 27–48.

Alderman, A. R., 1965. Dolomitic sediments in their environment in the southeast of South Australia. Geochim. Cosmochim. Acta, 29, 1355–1365.

Amiel, A. J. and Friedman, G. M., 1971. Continental Sabkha in Arava Valley between Dead Sea and Red Sea: Significance for origin of evaporites. A.A.P.G., 55, 581–592.

Cohen, Y., Krumbein, W. E., Goldberg, M. and Shilo, M., 1977. Solar Lake (Sinai): 1: Physical and chemical limnology. Lim. and Ocean. 22, 597–608.

Deffeyes, K. S., Lucia, F. J. and Weyl, P. K., 1965. Dolomitization of recent and Plio-Pleistocene sediments by marine evaporative waters on Bonaire, Netherlands Antilles. Spec. Pub. Soc. Econ. Paleont. Mine., Tulsa, V.B., 71–88.

Emery, K. O., 1963. Sediments of the Gulf of Aqaba (Eilat). In: R. L. Miller (ed.). *Papers in Marine Geology*, MacMillan Co., N.Y., 57–273.

Friedman, G. M., 1968. Geology and Geochemistry of reefs, carbonate sediments, and waters, Gulf of Aqaba (Elat) Red Sea. Jour. Sed. Pet., 38, 895–919.

Friedman, G. M., Amiel, A. J., Braun, M. and Miller, D. S., 1973. Generation of carbonate particles and laminites in algal mats – Example from sea – marginal hypersaline pool, Gulf of Aqaba, Red Sea. Am. Assoc. Pet. Geol. Bull., 57, 541–557.

Friedman, G. M. and Braun, M., 1968. Are "mud-cracked" algal laminate sediments evidence of supratidal origin˄ New evidence from tidal lakes in Sinai peninsula. Geol. Soc. America, Spec. Paper 121, p. 103.

Friedman, G. M., and Gavish, E., 1971. Mediterranean and Red Sea (Gulf of Aqaba) beachrocks. In: Bricker, O. P. (ed.), *Carbonate Cements*, Baltimore and London, The John Hopkins University Press, 376 p.

Friedman, G. M. and Sanders, J. E., 1978. Principles of Sedimentology. New York, John Wiley, 792 p.

Gavish, E., Braun, M., Bein, A., Buchbinder, B. and Friedman, G. M., 1970. Young sediments (recent to Neogene) and their diagenesis in Ras Muhammad (in Hebrew). Proc. Isr. Geol. Soc. Annual Meeting. Central and Southern Sinai. pp. 41–44.

Gavish, E., 1974. Geochemistry and mineralogy of a recent sabkha along the coast of Sinai, Gulf of Suez, Sedimentology, 21, 397–414.

Gavish, E., 1975a. Recent coastal sabkhas marginal to the Gulfs of Suez and Elat, Red Sea, Rapp. Comm. Int. Mer. medit., 23, 129–130.

Gavish, E., 1975b. Recent and Holocene beachrocks along the coasts of Sinai, Gulfs of Elat and Suez; Rapp. Comm. Int. Mer. Medit., 23, 131–132.

Gavish, E., 1979. Mineralogy and groundwater geochemistry as indicators of hydrodynamic systems in a sabkha on the coast of Sinai, Gulf of Elat, Sedimentology (in Press).

Gavish, E., Krumbein, W. and Tamir, N., 1978. Recent clastic (carbonate) sediments and sabkhas marginal to the Gulfs of Elat and Suez. In: Field Excursion Guidebook, I.A.S., Tenth Int. Congress, Part III, 309–332.

Gvirtzman, G. and Buchbinder, B., 1978. Recent and Pleistocene coral reefs and coastal sediments of the Gulf of Elat, In: Field Excursion Guidebook, I.A.S., Tenth Int. Congress, Part III, 163–194.

Hsü, K. J. and Schneider, J., 1973. Progress report on dolomitization – hydrology of Abu-Dhabi sabkha, Arabian Gulf. In: Purser, B. H., (ed.): *The Persian Gulf*, Springer-Verlag, Berlin, 409–422.

Hsü, K. J. and Siegenthaler, C., 1969. Preliminary experiments on hydrodynamic movement induced by evaporation and their bearing on the dolomite problem. Sedimentology, 12, 11–25.

Illing, L. V., Wells, A. J. and Taylor, J. C. M., 1965. Penecontemporary dolomite in the Persian Gulf. In: Pray, L. C. and Murray, R. C. (ed.), Dolomitization and Limestone diagenesis: a symposium – Soc. Econ. Paleontol. Mineral. Spec. Pub. No. 13, 89–111.

Kinsman, D. J. J., 1964. The recent carbonate sediments near Halat el Bahrani, Trucial Coast, Persian Gulf, In: Van Straaten, L. M. J. U., (ed.), *Deltaic and shallow marine deposits.*, Elsevier, Amsterdam, 185–192.

Kinsman, D. J. J., 1969. Modes of formation, sedimentary associations and diagnostic features of shallow water and supratidal evaporites. Bull. Am. Ass. Petrol. Geologists, 53, 830–840.

Krumbein, W. E., Buchholz, H., Franke, P., Giani, D., Giele, C. and Wonneberger, K. O_2 and H_2S coexistance in stromatolites. A model for the origin of mineralogical lamination in stromatolites and banded iron formations. Submitted for publication.

Krumbein, W. E. and Cohen, Y., 1974. Biogene, Klastische und evaporitisch Sedimentation in einem mesothermen monomiktischen ufernahen see (Golf von Aqaba). Geol. Rundschau, Vo. 63, p. 1035–1065.

Levy, Y., 1977. The origin and evolution of brine in coastal sabkhas, Northern Sinai. Jour. Sed. Pet., 38, p. 845–858.

Murray, R. C., 1969. Hydrology of South Bonaire, N. A. – A rock selective dolomitization model. Jour. Sed. Pet., 39, 1007–1013.

Purser, B. H. (ed.), 1973. The Persian Gulf. Springer Verlag Berlin, 471 p.

Sass, E., Weiler, Y. and Katz, A., 1972. Recent sedimentation and oolite formation in the Ras Matarma lagoon, Gulf of Suez. In: Stanley, D. J., (ed.) The Mediterranean Sea, Dowen, Hutchinson and Ross, Inc. Stroudsburg, Pa.

Skinner, h. C. W., 1963. Precipitation of calcium dolomites and magnesian calcites in the south east of South Australia. Am. J. Sci., Vol. 261, p. 449–472.

Wood, G. U. and Wolfe, M. J., 1969. Sabkha cycles in the Arab/Darb formation of the Trucial coast of Arabia. Sedimentology, Vol. 12, p. 165–191.

Chapter 17

TRANSITION FROM OPEN MARINE TO EVAPORITE DEPOSITION IN THE SILURIAN MICHIGAN BASIN[1]

LOUIS I. BRIGGS[2], DAN GILL[3], DARINKA Z. BRIGGS[2], AND R. DOUGLAS ELMORE[2]

SYNOPSIS –

During the later part of the Middle Silurian, the deposition of the Lockport and Guelph formations in the Michigan basin took place in a suite of carbonate depositional environments which delineated continuous circular belts disposed concentrically around the center of the basin. From the outer platforms to the basin interior, the following facies belts are recognized: 1) Platform shelf carbonates containing a diverse suits of lithofacies of predominantly shallow water skeletal wackestone and lagoonal mudstone. 2) Platform margin, shelf edge barrier reef containing an upward shallowing sequence from open marine non-reefal deposition of crinoidal wackestone below wave base to organic reef development with coral-bryozoan packstone, coral-stromatoporoid boundstone and skeletal packstone and wackestone grading upward to lagoonal mudstone. 3) Barrier reef foreslope containing deposits which represent bioclastic debris and lithoclasts transported basinward from the shelf and the barrier reef complex. 4) A belt of isolated pinnacle reefs which developed on the platform to basin transitional slope. The vertical sequence in the pinnacle reefs is very similar to that of the barrier reef, representing an upward shallowing succession from biohermal mounds to organic reefs. 5) A relatively deep water facies of crinoidal and skeletal wackestone in the interior of the basin.

Regional lowering of sea level at the end of the Wenlockian stage (?) was the event that terminated reef growth and initiated the transition from open marine carbonate deposition to restricted, arid, marine evaporites deposition. On the shelf platforms, the shelf-edge barrier reef and the pinnacle reefs, sea level lowering lead to subaerial exposure, hiatuses and disconformities expressed in karstic-vadose diagenetic alterations, green shale partings, dessication cracks, fracturing and solution breccias. In interpinnacle and basin interior areas, the periods of lowered sea level were continuously recorded in the lithofacies of the Cain Formation.

The lowermost member of the Cain Formation, an argillaceous dark micritic limestone, was deposited during the restricted humid phase of the transitional sequence when the exposure of the reef summits produced the extensive karstic vadose alteration. As the basin water changed from penesaline to hypersaline, the deposition of sediments in the basinal environment changed from laminated ostracode limestone, to calcite varvite, calcite-anhydrite varvite, anhydrite and halite. Within the halite bed five thin streaks of calcite-anhydrite varvites are records of minor influexes of sea water when the sea level of

[1] Research for this study was supported by NSF Grant EAR76 – 17410.

[2] *Subsurface Laboratory, University of Michigan, Ann Arbor, Michigan 48109*

[3] *Geological Survey of Israel, Jerusalem, Israel 95501*

the continental platform stood at the silled level of the inlets to the basin, approximately 235–350 ft (67–107 m) below the maximum stand during reef growth in the basin. The water depth in the basin center was about 600 ft (180 m). Evaporative drawdown was rapid so that only a very thin anhydrite bed was deposited at the base of the halite sequence. During deposition of the salts of the A-1 unit of the Salina Group, sea level in the inlets remained at the silled level. The evaporite stage culminated with deposition of potash salts in a greatly restricted sea in the center one-third of the basin. Following deposition of the potash beds, a gradual increased flow of sea water into the basin diluted the basinal brine to the chloride stage. Deposition of the limestones and dolomites of the overlying Ruff Formation took place in shallow penesaline water. The succeeding evaporites of the Salina Group all appear to have been shallow water, penesaline to hypersaline deposits.

STRATIGRAPHIC AND PALEOGEOGRAPHIC SETTING

Near the end of the Ordovician the Taconic Orogeny formed highlands along the eastern seaboard of the continent from which an enormous volume of clastic detritus was eroded and transported into the Appalachian geosyncline and the adjacent continental platform. During the Early Silurian clastic sedimentation gave way to carbonate sedimentation as the volume of eroded clastic detritus diminished.

The beginning of the Middle Silurian Niagaran carbonate deposition coincided with the beginning of a major continental transgression of the interior platform seas, possibly synchronous with a rapid growth of a mid-oceanic ridge, which reached a maximum development during the Wenlockian with the deposition of the Lockport–Guelph carbonates and their stratigraphic equivalents. This was the time of maximum reef development within the Niagaran reef complexes when carbonate platforms and pinnacle reefs formed along the basin margins (Fig. 17-1).

Possibly with the cessation of an oceanic ridge system, the seas regressed from the cratonic platform as the water slowly returned to the ocean basins. This event marked the sedimentological transition from marine carbonates to evaporites, owing to the development of hypersaline waters in the semi-isolated Michigan basin. The first stage of regression temporarily exposed the tops of the pinnacle reefs and the carbonate platforms, and resulted in extensive vadose alteration of the carbonate rocks. The level of the water in the Michigan basin and the adjacent epicontinental seas was lowered such that the ocean water flowed into the basin only through passes or inlets which crossed the carbonate platforms surrounding the basin, possibly at the Georgian Inlet at the northeastern margin of the basin and the Clinton Inlet at the southeastern margin (Alling and Briggs, 1961). The restricted circulation between the waters of the Michigan basin and the open oceans south and northeast increased the salinity of the basinal brine and restricted the biota largely only to algae in the last stage of reef evolution. In the inter-reef and basinal areas this change is recorded in dark gray to black argillaceous limestones representing euxinic penesaline, but non-evaporitic, deposition in a starved basin. Continued restricted sea water flow into the basin combined with a high rate of climatic evaporation led to the deposition of the evaporite rocks of the Salina Group.

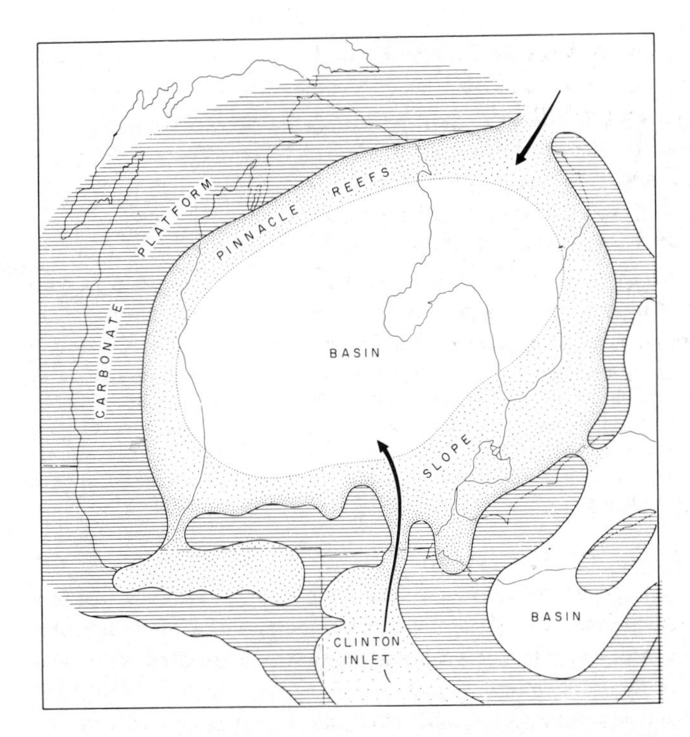

Fig. 17-1. Paleogeographic map of the Michigan basin in the late Silurian at the time of transition from open marine carbonate deposition to restricted marine evaporite deposition. Shown are the carbonate platforms, slope and basinal environments. Two principal inlets transected the platforms, as indicated by the arrows.

A complete and uninterrupted stratigraphic record of the transition from the marine carbonates to evaporites occurs in the inter-reef and basinal sediments that were deposited in the central part of the Michigan basin. In order to appropriately evaluate the stratigraphic and tectonic events that were associated with the transition, it is necessary to examine the contemporaneous facies of the carbonate platforms and the slope environment that contain the pinnacle reefs of the Michigan basin.

PINNACLE REEFS

The reference stratigraphy for the Michigan basin is that established for the pinnacle reefs since they are the most extensively studied and the best known. See Gill (1977) and Huh et al. (1977). The rocks of the pinnacle reefs, excluding their stromatolite caps, represents the Guelph facies of the upper Lockport deposition. Reef accumulation began with a concentration of crinoids and bryozoans on the slope area basinward from an already established carbonate shelf platform. This is the common Lockport biota. The colonies of crinoids and bryozoans apparently trapped the argillaceous carbonate muds washing from the adjacent carbonate shelves so that biohermal mounds were formed. These constitute the lowermost Biohermal stage rocks of the pinnacle reefs (Fig. 17-2).

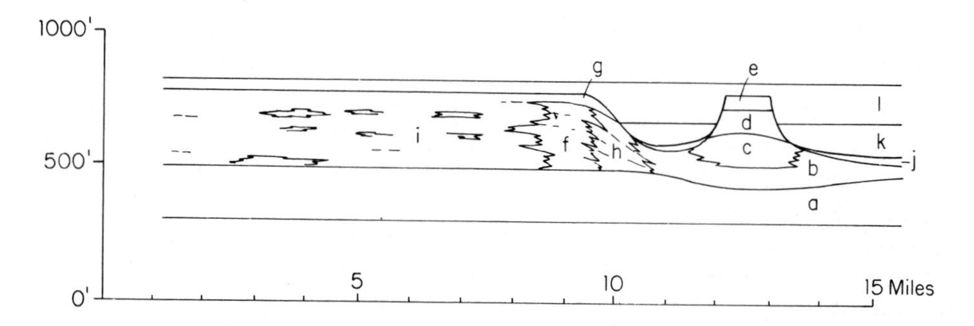

Fig. 17-2. Stratigraphic facies of the Late Wenlockian – Early Ludlovian carbonates and evaporites of the carbonate platform and slope environments, northern Michigan basin. Modified after Meloy (1974). Symbols: a – Lower Lockport Formation, b – Upper Lockport Formation, c – Biohermal stage rocks, d – Organic-reef stage rocks, e – Supratidal-island stage stromatolite, f – Platform margin deposits, g – Lagoonal mudstone, h – Foreslope deposits, i – Shelf carbonates, j – Cain Formation, k – Salina A-1 Salt. 1 – Ruff Formation.

The overlying Organic-reef stage represents the dominant frame-building stromatoporoids and tabulate corals which form the reef-core lithofacies, that is associated with reef-dwellers lithofacies comprised of brachiopods, gastropods, rugose corals, crinoids and bryozoa, and with the reef-detritus lithofacies eroded from the reef cores.

The Biohermal and Organic-reef rocks in the pinnacle reefs reached to a thickness of 300–600 ft (100–200 m), and stood this high above the surrounding sea floor. Debris broken from the reef summits largely during storms accumulated around the base of the reefs as reef-rubble conglomerate. Reef growth ceased with the lowering of sea level and exposure of the upper 30–50 ft (10–15 m) of the reef tops. A karst topography developed and karstic vadose alteration of the upper 30 ft or more (Gill, 1977) of the reefs and the carbonate platforms ensued. This is most readily seen at the quarry in the Maumee reef bank in northeastern Ohio (Kahle, 1974), but it occurs in all of the pinnacle reefs as well. The lowered sea level initiated the restriction of the waters of the Michigan basin, and the transition from the open marine reef development of the Lockport to the restricted marine deposition of the Salina.

When the penesaline waters of the basin again reached the top of the pinnacle reefs, approximately 40 ft (12 m) of algal stromatolite accumulated at the reef crests. These constitute the Supratidal-island stage rocks of the pinnacle reefs (Gill, 1977). Five lithofacies have been recognized (Huh et al., 1977): algal stromatolite, algal-detritus wackestone, lagoonal mudstone, fine-laminated LLH-c algal stromatolite, and flat pebble conglomerate. No fossils other than algae exist in these rocks. There are several disconformities in the stromatolite, indicating a fluctuating sea level during its accumulation. Reef-talus conglomerate at the base of the pinnacle reefs contains fragments of the algal stromatolite lying below the earliest-deposited anhydrite and salt of the Salina Group. According to Huh et al. (1977) the Salina A-1 anhydrite "appears to have been deposited mostly in a sabkha environment. Thus, during deposition of the A-1 anhydrite, the pinnacle reefs probably were completely exposed as islands."

CARBONATE PLATFORMS

The carbonate platform along the northern edge of the Michigan basin in the subsurface rocks has been described by Mesolella et al. (1974) and Meloy (1974). Mesolella et al. (1974) described a section through the shelf margin as a barrier reef, and established the time equivalency of the main reefal phases on the shelf and in the pinnacle reefs. Meloy examined cores from the norhtern carbonate platform and concluded that two major environments were represented, barrier-reef and back-reef platform shelf with a lagoonal environment.

Platform Margin. A vertical succession of the platform margin rocks consists of crinoidal wackestone, containing fragments of stromatoporoids, brachiopods, crinoids, and minor amounts of coral and bryozoa; coral-bryozoan packstone; stromatoporoid wackestone and fossiliferous grainstone. The uppermost unit of the platform of the platform margin succession consists of interbedded mudstone, wackestone and packstone, containing green shale partings, mudcracks, fenestral fabric and algal laminations. The Ruff Formation (Budros and Briggs, 1977) of the Salina Group overlies this unit.

Basinward the platform margin deposits grade into foreslope deposits. The vertical succession in the foreslope beds consists of mudstone, grainstone, and wackestone containing skeletal debris of stromatoporoids, brachiopods, corals, gastropods, and crinoids. The grainstone bedding, some of which is graded, dips to 25° (Meloy, 1974).

The vertical succession through the platform margin deposits probably represents a shallowing from relatively deeper water non-reefal accumulation of crinoidal wackestone below wave base to organic reef development with coral-bryozoan packstone, stromatoporoid wackestone, and fossiliferous grainstone, and finally into restricted lagoonal deposition in the uppermost unit. The mudcracks in this unit indicate that the lagoonal sediments were occasionally exposed and dessicated. The foreslope lithologies represent debris washed from the shelf and marginal environments of the carbonate platform. The grainstone and wackestone probably represent fore-reef or spillover sands from the reef.

Platform Shelf. The basal lithology of the shelf lagoonal carbonate rocks is dark brown crinoidal wackestone containing crinoids, brachiopods, and ostracods. It is overlain by tan wackestone interbedded with mudstone, fossiliferous grainstone, and dark brown wackestone containing mudcracks, decussate anhydrite, burrows (some filled with green shale), pisolites, fine laminations, fenestral fabric, pellets, fossil debris, mottlings, green shale partings and intraclasts. Sparsely distributed fragments of stromatoporoids, brachiopods, gastropods and crinoids are locally abundant. Dolomite and anhydrite infill fenestral vugs, although halite fills some vugs especially in the upper part of the section (Meloy, 1974). Green shale in burrows and in partings also is most abundant in the upper part.

Karstic features also occurs in the rocks throughout the shelf area, though they are most abundant in the upper part of the section. The features include dolomitized fibrous linings in vugs, solution channels filled with anhydrite, vadose silt, green shale, and breccia.

The shelf lithofacies grade into those of the platform margin, but the stratigraphic relationships are not clear. The lowest facies in the shelf area probably represents deeper water deposition that preceeded the buildup of the reef framework of the platform margin, and may correlate with the crinoidal wackestone of the platform edge and with similar rocks of the pinnacle reefs.

The overlying carbonate shelf beds were deposited in shallow lagoons that were periodically dessicated, forming mudcracks and gypsum molds. Many features found in the rocks are indicative of sabkha or supratidal deposition (Kinsman, 1969). Locally, deeper water pools or ponds supported a gastropod-rich fauna which apparently thrived on the organic lagoonal sediments. Fossiliferous grainstones suggest the presence of patch reefs in the lagoon. The cyclic mudstones reported by Meloy (1974) represent minor transgressions and regressions caused by minor sea level fluctuations.

Mudcracks and fenestral fabric indicate short term exposure, but the other vadose features indicate that the shelf area was exposed for a considerable length of time at least once prior to the deposition of the Ruff Formation. The green shale probably represents residue from karstic dissolution of subaerially exposed platform sediments.

A number of correlations between the carbonate platforms and the pinnacle reefs can be suggested on the basis of the faunal evidence and similarity of sequences. The main reef development of the platform margin probably correlated with the Organic-reef stage of the pinnacle reefs (Mesollela et al., 1974). The mudstone overlying the platform margin may be correlative to the Supratidal-island stage of the pinnacle reefs (Huh et al., 1977; Gill, 1977). The platform shelf rocks are even more difficult to correlate because of the sparsity of any distinct lithologic markers. The upper part of the back-reef succession could be equivalent to the mudstone overlying the platform margin on the basis of similar lithologies and stratigraphic position.

The numerous vadose features indicate that the shelf underwent at least one period of substantial subaerial exposure prior to the deposition of the Ruff Formation. The vadose features may be the result of the same periods of subaerial exposure that produced the karst on top of the Organic-reef stage rocks, and to a lesser degree those of the Supratidal-island stage. The fact that the vadose features cannot be related to a specific stratigraphic zone is a problem. Nevertheless, the evidence of vadose alteration is supportive of the drawdown that occurred at the beginning of the evaporite deposition in the Michigan basin.

INTER-REEF AND BASIN CENTER

A very thin sequence of distinctive limestone-evaporite rocks represent the transition from normal marine carbonate deposition of the Lockport Formation to restricted marine evaportie deposition of the Salina Group in the central basinal areas of the Michigan basin. The Lockport Formation, typically bioclastic limestone or dolomite, represents the culminating stage of marine transgression in a clear water open marine environment of the cratonic platform region. Biohermal mounds, patch reefs, and pinnacle reefs developed where waters were shallow, as along basin margins and the fringes of tectonic arches. However, in the basinal regions the carbonate rocks of the Lockport Formation represent detritus eroded and transported from the carbonate platforms along the basin margins into the deeper water of the central basin. In the basinal areas, the Lockport typically consists of pelagic carbonates with argillaceous muds, mixed with bio-

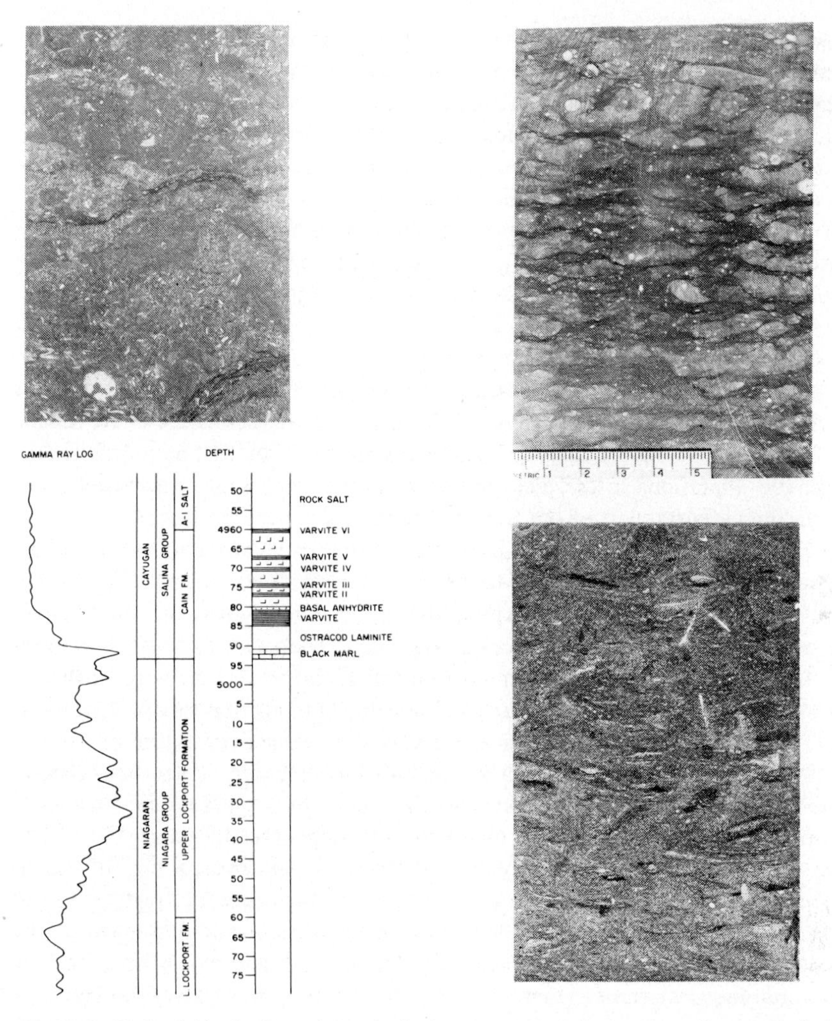

Fig. 17-3. Skeletal, bioclastic, mainly crinoidal wachestone of the Lockport Formation in inter-pinnacle and basinal areas (photomicrograph, width of thin section is 23 mm; sample location: well permit no. 29036 at a depth of 7474 ft.)

Fig. 17-4. Nodular fabric and horse-tail-like dark gray argillaceous seams and wisps, most probably of compactional origin, in the Lockport Formation (sample location: well permit no. 23711 at a depth of 8215 ft.).

Fig. 17-5. The Cain Formation, from type section in core of the Getty Oil Company – Charles A. Cain, et al. no. 1–21, State Permit No. 28866. Located in SW ¼ SW ¼ SE ¼ Section 21, T.31N, R.4E. Montomorency County, Michigan, U.S.A. The core is cataloged and preserved in the Subsurface Laboratory, University of Michigan, Ann Arbor, Michigan, U.S.A. 48109.

Fig. 17-6. Dark gray, dense argillaceous mudstone with rare, scattered crinoid fragments of the euxinic limestone at the base of the Cain Formation. The dark specks and streaks are disseminated pyrite and kerobituminous (?) material (photomicrograph, width of thin section is 23 mm; sample location: well permit no. 23711 at a depth of 8198 ft.)

clastic wackestone containing abundant minute, mainly crinoid fragments but also other skeletal debris (Fig. 17-3). Commonly, the formation displays a distinct nodular fabric and contains abundant dark gray, horse-tail-like argillaceous seams and wisps (Fig. 17-4). These fabrics are most probably of compactional origin. It is noteworthy that similar structures were produced experimentally in the laboratory by compacting a soft, fossiliferous Recent carbonate mud from Florida (Shinn et al., 1977).

Through a transitional vertical interval of about one foot, the Lockport grades upwards into the Cain Formation without any discernable break. The lithologic units of the transitional basal Cain Formation of the Salina Group (Fig. 17-5) are as follows:

Euxinic limestone. – The lowermost member of the Cain Formation is a dark gray, argillaceous dense mudstone with abundant clayey wisps and scattered fossil fragments (Figs. 17-6, 7). This member represents the initial stage of restricted marine water circulation at the beginning of Salina Group time and is correlated to the first period of karstic-vadose diagenetic alterations in the pinnacle reefs and shelf platform. The member occupies only the north central part of the Michigan basin, which apparently was the deepest part of the basin at the end of the Niagaran reef-construction phase of the Lockport (commonly referred to as the Guelph reef facies of the Lockport Formation). The potash facies of the Salina A-1 salt occupies approximately the same position in the basin although it is somewhat more restricted in area (Fig. 17-8). The euxinic limestone averages less than 2 ft (60 cm) in thickness, being up to 6.5 ft (2 m) in the northern slope area near the carbonate shelf platform. Likewise the fabric and composition vary with nearness to the shelf. In the basin center the rock is a very fine-grained admixture of current-bedded detritus and palagic muds, commonly with ostracodes, thin-shelled brachiopods, and trilobite fragments; whereas in the near shelf region the mudstone contains small lithoclasts (Fig. 17-9) and algal (?) coated, diagenetically altered skeletal grains (Fig. 17-10). Adjacent to pinnacle reefs the beds contain admixed reef-derived pebbles of caliche material. The euxinic limestone contains abundant pyrite which occurs as replacement of fossil fragments, as envelops around grians and as fine crystals in the mud matrix. The dark hue is due to the abundance of argillaceous material, pyrite and possibly also carbonaceous (kerobituminous ?) matter (Figs. 17-6, 9). Eleven samples of this lithology were analyzed for organic carbon content. Its abundance was found to range between 0.07%–0.70%, with a mean value of 0.30%.

Ostracode laminite. – The ostracode laminite is a light brown to tan, crudely laminated mudstone with scattered, rather rare, ostracodes (Fig. 17-11). Stylolitic laminae boundaries are common. Near the shelf margin the unit consists of vugular dolomite and contains anhydrite pseudomorphs after gypsum. Some vugs appear to have been formed by the leaching of ostracode tests. Admixed reef derived caliche pebbles are common near the bases of pinnacle reefs. The unit ranges in thickness from 2 ft (60 cm) to 6 ft (180 cm), averaging 3.6 ft (110 cm). It appears to represent the transition from euxinic to penesaline depositional conditions.

Halite–varvite. – The uppermost member of the Cain Formation consists of a lower carbonate-anhydrite varvite and an upper halite salt with 5 streaks of carbonate-anhydrite

Fig. 17-7. Dense, dark gray to black argillaceous mudstone with scattered fossil fragments of the euxinic limestone at the base of the Cain Formation (sample location: Well permit no. 23711 at a depth of 7855 ft., scale bar is 2 cm long).

Fig. 17-8. Map showing areal extent of the potash facies of the Salina A–1 Salt, and the euxinic facies of the basal Cain Formation in the Michigan basin. The potash facies is from Matthews and Eagleson (1974).

Fig. 17-9. Dense argillaceous mudstone of the euxinic limestone of the Cain Formation. Note pyrite – coated dolomitized allochems. The black material in the upper part and the dark specks and streaks are pyrite and organic (kerobituminous ?) matter. (photomicrograph, height of thin section is 23 mm; sample location: well permit no. 29037 at a depth of 6314 ft.).

Fig. 17-10. Algal (?) coated skeletal (some trilobite ? fragments) grains and lithoclasts in the euxinic limestone of the Cain Formation (photomicrograph, height of thin section is 23 mm; sample location: well permit no. 28866 at a depth of 4992 ft.).

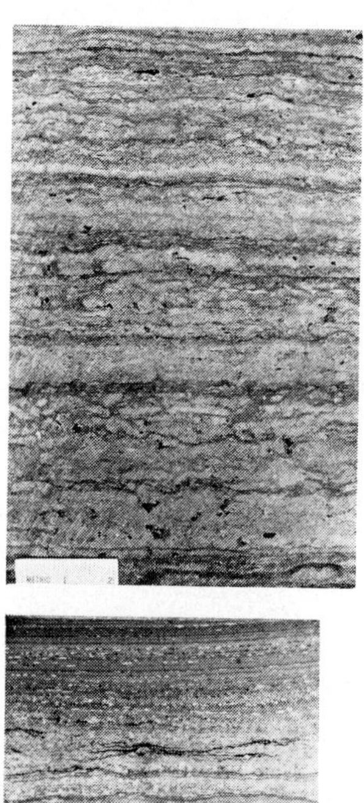

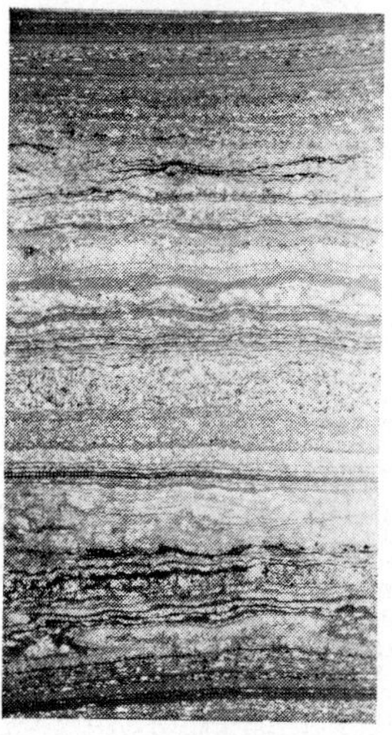

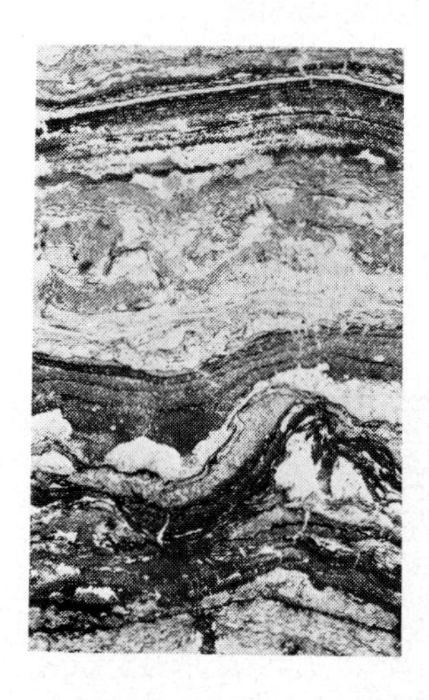

Fig. 17-11. Crudely laminated mudstone with scattered ostracodes, Ostracode laminite lithofacies of the Cain Formation. Note stylolitic laminae boundaries and abundance of pinpoint vugs (sample location: well permit no. 28866 at a depth of 4988 ft.).

Fig. 17-12. Main varvite at base of halite-varvite member of the Cain Formation. The varvite consists of a very orderly succession of planar, submillimeter thick micritic calcite varves alternating with microcrystalline anhydrite varves (sample location: well permit no. 28866 at a depth of 4984 ft.)

Fig. 17-13. The calcite-anhydrite varvite streaks consist of alternations and admixtures of the following components: pure micritic calcite varves; varves consisting of microcrystalline anhydrite; anhydrite laths; aggregates of felted anhydrite forming elliptical micronodules flattened parallel to the bedding; enterolithic anhydrite and seams rich in carbonaceous matter (photomicrograph, width of thin section is 23 mm; sample location: well permit no. 28866 at a depth of 4982 ft.).

Fig. 17-14. Distorted nodules, consisting of felted anhydrite laths (center), displaying hassock flowage fabric due to soft sediment slippage, crudely interlaminated with micritic calcite varves, typical of anhydrite layers bounding the varvite steaks (photomicrograph; width of thin section is 23 mm; sample location: well Dow Chemical no. 8 at a depth of 8593 ft.).

varvite (Fig. 17-3). The thickness of this member ranges from 21.0–25.2 ft (640–768 cm), and averages 23.7 ft (722 cm). The lower, or main varvite ranges in thickness from 1.0 ft (30 cm) to 5.3 ft (160 cm), and averages 2.8 ft (86 cm). The top of the main varvite in the central basin area is a relatively thin (10 cm) crudely laminated and contorted anhydrite bed. The upper part of the member consists mainly of halite salt. Each of the calcite-anhydrite varvite steaks within the salt is approximately 1.0 ft (30 cm) thick and is bounded top and bottom by a thin anhydrite layer, 3–8 cm in thickness. There are 5 such varvite streaks within the salt that can be traced throughout the central basin on gamma ray and electric logs. Near the basin margin, nodular anhydrite takes the place of some or all of the Cain Formation and the Salina A-1 salt, and the carbonates are dolomitized. A brief petrographic description of the various lithologic components of the upper member of the Cain Formation follows.

The varvite streaks contain calcite laminae with thin carbonaceous films, calcite-anhydrite laminae with intercrystalline growth, co-precipitation textures between the two mineral phases and calcite-anhydrite laminae with nodular textures in a progressive but oscillating pattern (Figs. 17-12, 13). Near the basin center the laminae range in thickness from 100 microns to 1000 microns. The boundaries of the laminae are wavy and irregular, a few being microstylolitic; however, most are relatively smooth with dark carbonaceous boundaries and composed of calcite and anhydrite. In the darker monomineralic calcitic laminae the calcite is very fine-grained, approximately 10 microns in diameter, whereas in the anhydrite laminae the calcite measures 20–30 microns. Anhydrite occurs as subhed-

Fig. 17-15. Third varvite streak in the Getty-Cain 1–21 type section of the Cain Formation. A complete depositional cycle displaying a succession of lithologies consisting from bottom upwards of halite, anhydrite, calcite-anhydrite varvite, anhydrite and back to halite (clear crystals at very top 5 mm of slab; sample location: well permit no. 28866 at a depth of 4970 ft.).

Fig. 17-16. The salt between varvite streaks consists of coarse crystalline clear halite, often contaminated by distorted streaks and irregular patches of micritic calcite, anhydrite and carbonaceous matter (sample location: well permit no. 28866 at a depth of 4973 ft.).

ral to anhedral prisms up to 200 microns in maximum diameter, most being about 75 microns, and make up about 5–10 percent of the grains. Where the anhydrite is more abundant a nodular pearly texture prevails within which the anhydrite occurs as fine felted laths. Within thicker anhydritic laminae there are occasional hassock flowage structures apparently owing to soft sediment slippage within the anhydrite-rich layers on a sloping depositional surface. Similar textures appear in the anhydrite layers which bound the varvite steaks (Fig. 17-14). A transitional succession of lithologies that constitutes a complete depositional cycle, the third varvite streak in the Getty-Cain 1–21 type section, is shown in Figure 17-15. Over a vertical distance of 20 cm the lithology, from bottom upwards, passes from halite to anhydrite to calcite-anhydrite varvite to anhydrite and back to halite. The halite consists mainly of clear, very coarse crystals. Between varvite streaks it is often contaminated by distorted streaks and irregular patches of mixtures of micritic calcite, anhydrite and carbonaceous matter (Fig. 17-16).

The halite-varvite member represents deposition of salt by evaporation of concentrated basinal brine while the sea level dropped to the silled level of the principal inlets, the varvites representing temporary small rises in sea levels which flooded the surface of the basin with an influx layer of fresh oceanic seawater. Evaporation of the influx layer produced the varvites in the halite sequence. As the influxing water became more concentrated away from the main inlets, a progression of contemporaneous but petrographically slightly different varvite facies was formed in a pattern shown somewhat diagramatically in Figure 17-17.

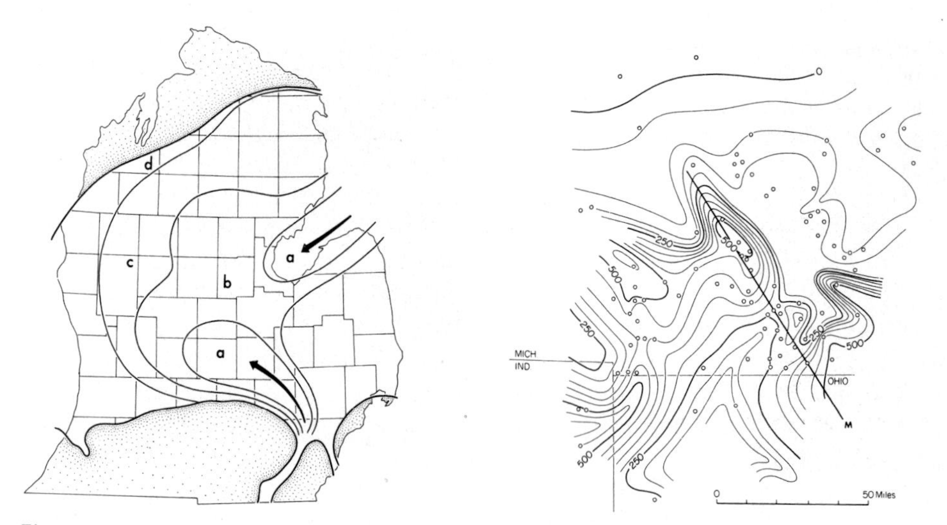

Fig. 17-17. Map showing the distribution of contemporaneous varvite facies during deposition of Cain Formation. The precise boundaries are estimated from a comparison of cores through the same units at different localities. Thus the map is somewhat diagrammatic. a – calcite varvites, b – calcite – anhydrite varvites with co-precipitation textures, c – calcite-anhydrite varvites with nodular anhydrite, d – anhydrite laminites.

Fig. 17-18. The Clinton Inlet is southeastern Michigan and northwestern Ohio, indicated by the thickness contours of the Niagaran reef bank. The line indicates the profile of Fig. 17-19. M is the location of the Maumee quarry, the circles are well locations.

DISCUSSION

The Cain Formation of the Salina Group represents a complete uninterrupted sequence of sedimentation from the open marine limestone deposition of the Lockport Formation to the restricted arid evaporite deposition of the Salina Group. The euxinic limestone member reflects the initial sea level lowering and restricted deposition in the Michigan basin that produced the karstic vadose alteration of the pinnacle reefs and carbonate platforms. It contains a normal marine, but unusual fauna of trilobites and some ostracodes. The sea level had lowered at least 30 ft (10 m), bringing it to below the top of the fringing reef banks and the fringing barriers. Subsequently the basinal sea water became progressively more saline so that laminated carbonates, then anhydrite-carbonate varvites, and lastly halite with interbedded anhydrite-carbonate varvites were deposited within the basin.

The ostracode laminite appears to represent a more saline environment than the euxinic limestone, ostracodes being the dominant form of a sparse fauna. This member was deposited when the Supratidal-island stage algal stromatolite was deposited on the reef tops in penesaline basinal sea water during a somewhat higher sea level stand. The subsequent halite-varvite depositional environment represents hypersaline deposition by surface evaporation of the basinal brine when sea level stood at the silled level of the major inlets to the basin. We estimate this level to be about 230–350 ft below that of the pre-evaporite normal marine conditions from the topographic relief reconstructed for the Clinton inlet in southeastern Michigan and northwestern Ohio (Briggs, 1962; Mantek, 1976). Following deposition of the uppermost varvite in the salt, the sea level dropped to or below the silled level of the inlets, which led to the final evaporative drawdown and deposition of the A-1 salt, including the potash salt in the central part of the Michigan basin.

Although an accurate estimate of the morphology of the inlets through the fringing barrier reefs is difficult to obtain, it can be approximated by determining and mapping the thickness of the Lockport-Guelph rocks along the carbonate platforms. The Clinton inlet (Fig.17-18) in the southern fringing reef bank was discovered by this method and by the distribution of the evaporite facies of the Salina Group in the Michigan basin (Alling and Briggs, 1961). Subsequently, Mantek (1976) showed the same inlet on an isopach map of the Lockport-Guelph rocks.

Preliminary calculations indicate that the Clinton inlet alone could easily provide an adequate flow of oceanic sea water to the evaporite basin. The area of the sea in the Michigan basin within the carbonate platform was approximately 9.4×10^{10} m^2. For a net surface evaporation of one m/yr, the volume lost to evaporation would be 9.4×10^{10} m^3/yr or 10.7×10^6 m^3/hr. Doubling the evaporation rate will double the volume evaporated, but even the higher rates would not greatly alter the conclusions relative to sea level stand during salt deposition.

The topographic profile across the Clinton inlet in southeastern Michigan-northwestern Ohio (Fig. 17-19) shows that it was about 55 mi (90 km) wide at the reef crest and 235 ft (72 m) deep (Briggs, 1962). Mantek (1976) shows the inlet to be approximately 350 ft (107 m) deep. Cross-sectional areas for various steps below the reef crest are shown in Table 17-1.

TABLE 17-1.

Cross-sectional area and rate of flow through the Clinton inlet corresponding to lowering of epicontinental sea level from 60 to 225 ft. The volume of flow to replace water evaporated from the sea in the Michigan basin is estimated to be 10.7 x 10^6 n^3/hr. See Fig. 17-19.

Sea Level Drop Ft	M	Cross-Sectional Area $10^6 m^2$	Flow Through Inlet m/hr
60	18	6.7	1.6
100	30	1.8	6
150	46	0.8	13
200	61	0.19	56
225	69	0.05	214

If the reconstructions are valid, the results indicate that the evaporite deposition began only when the level of the epicontinental sea was no higher than the silled level of the inlets, 235–350 ft (72–107 m) below the crest of the reef banks. Moreover, with this flow configuration there would be little or no reflux of dense concentrated brine from the basin except possibly by means of seepage through the carbonate platforms and banks. Although this may have been an important factor in regional dolomitization by reflux seepage, it is an item of only peripheral concern to this discussion.

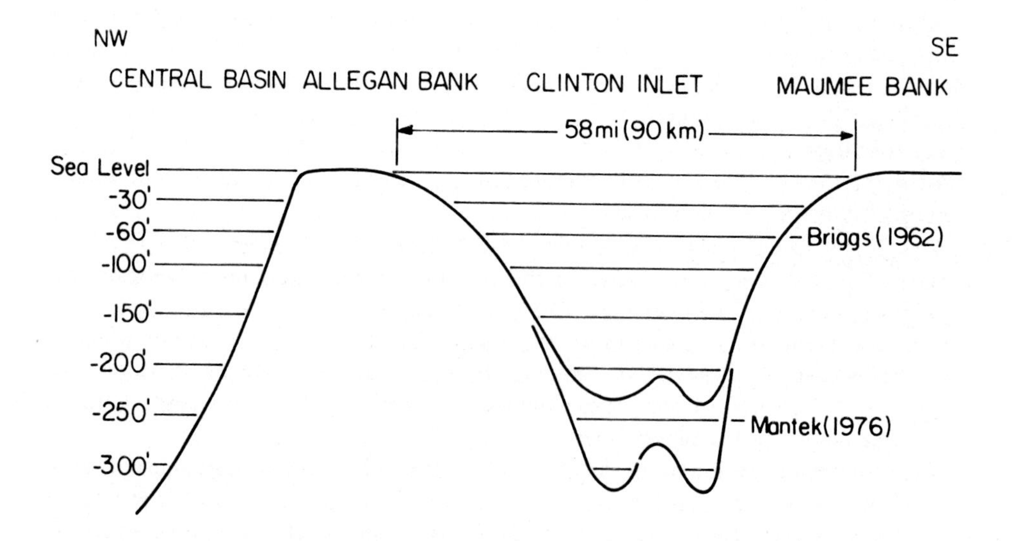

Fig. 17-19. Profile of the Clinton Inlet, shown in Fig. 17-18. The levels corresponding to lowered sea level are shown by the horizontal lines.

The probable depth of water in the basin center can be estimated from the configuration of the carbonate platforms, pinnacle reefs, inlets and their silled depths, and the thickness and distribution of the evaporites. The maximum thicknesses of the margins of the carbonate platforms and of the pinnacle reefs are about 600 ft (180 m), the approximate depth of water along the basin margin. At the basin center the water was probably somewhat deeper. The initial salt and limestone beds of the Salina Group are approximately the same thickness throughout the basinal area, and the two essentially filled the initial basin to the level of the carbonate platform and to the tops of the pinnacle reefs. Thus the water depths in the slope and basinal regions just before the beginning of evaporite deposition probably ranged between 500 and 1000 ft (150–300 m). With the drop in sea level to the silled level of the inlets that initiated evaporite deposition, depths might have ranged from 150 ft (46 m) at the basin margin to a maximum of 800 ft (240 m) at the basin center.

The varvites and the lower part of the salt were deposited when the sea level periodically rose sufficiently to cover the basinal brine with an oceanic influx layer, which evaporated to produce the thin varved sequences (Fig. 17-17). The lower anhydrite of the varvites most likely resulted from mixing of the influx and basinal waters at their interface, and precipitation of gypsum or anhydrite due to oversaturation of the mixed waters with calcium sulfate. The varves and upper anhydrite resulted from precipitation from the influx water owing to surface evaporation; thus the varvites require that the sea level be at or just slightly above the silled level of the inlets.

The volume of salt in the Cain Formation and in the A-1 salt also necessitates essentially continuous slow inflow of oceanic brine into the basin during deposition of the salt. The bromine/chloride ratios in the A-1 salt support the idea of a large volume of brine during salt deposition (Matthews and Eagleston, 1974). The Br/Cl concentration curve shows a very gradual increase to a maximum at the potash facies and a gradual decrease in the upper chloride facies following sylvite deposition. A bromine-concentration curve of this configuration is produced when the potash stage does not occur with dessication of the brine, and there is a relatively large body of brine during the complete depositional sequence. Thus, depths of water even during the potash stage may have been on the order of 100 ft (30 m) in the central basin.

Since the potash salts cover only three-tenths the area of the initial sea in the Michigan basin and since the highly concentrated brines of the potash facies do not evaporate as rapidly as do brines in the chloride and sulfate facies due to a greatly increased surface tension, the amount of evaporation from the surface of the potash-depositing brine might have been only one-tenth of the sea water at the start of evaporite deposition. Thus, the amount of oceanic brine flowing from the open ocean outside the basin through the inlets would have averaged about 10^6 m^3/hr. This is about 2% of the average flow of the Mississippi River, 10% that of the Niger, and 30% that of the Nile. If there were two such inlets they would be equivalent to moderate-sized rivers flowing through the inlets within canyons walled by the limestones of the banks, and the flat salt pans of the outer part of the basin floor to the central potash-depositing sea as large meandering saline streams.

Another consideration relative to the mode and conditions of the initial evaporite deposition involves the very thin anhydrite layer at the base of the salt. It is only 5–10 cm thick. Most large evaporite deposits throughout the world have very thick basal anhydrite beds, which represent a gradual increase in the salinity of the basinal water during sulfate deposition before halite deposition began. The sea water flowing into the Michigan basin during evaporite deposition must have had a normal concentration of calcium sulfate since there is anhydrite in all of the varved sequences and also along the margins of the basins as a facies of the calcite-varvites nearer the inlets. Also the oceanic sea water flowing across the Indiana–Ohio platform through the Clinton inlet and to the Michigan basin supported a normal marine biota that thrived in the northern Indiana platform reefs and inter-reef environments.

An alternative explanation for the thin anhydrite at the base of the salt is that the change from the normal marine Lockport limestone deposition to the restricted arid Cain evaporite deposition was very sudden, and that it involved a very rapid lowering of sea level and evaporative drawdown to the silled level of the inlets. The lower part of the Cain Formation, including the main varvite at the base of the halite-varvite unit would represent gradual concentration of the sea water as it lowered from the reef summits to the base of the inlet, the halite-varvite sequence representing the highly concentrated stage with inflow of oceanic sea water into the basin only during deposition of the varvites in the upper Cain halite sequence. This appears to be the most reasonable explanation for the thin basal anhydrite, since there are no carbonates that might represent bacterial destruction of precipitated calcium sulfate and its replacement by calcium carbonate in the water column above the sea floor, as has been suggested by others (Nurmi and Friedman, 1977).

CONCLUSIONS

The condition that accompanied the transition from normal marine deposition of limestones to restricted arid marine deposition of evaporites during the Late Silurian in the Michigan basin can be reconstructed from the detialed studies of the stratigraphic sequences that were involved in this event. Carbonate platforms built up around the periphery of the Michigan basin where reef organisms flourished in the warm epicontinental seas which accompanied the maximum marine transgressions onto the central continental craton. During this stage in the evolution of the Michigan basin carbonate platforms built out from the shallow water margins to where reefal bioherms, largely of crinoids and bryozoa, had established mounds in the slope environment. Reef framework corals and stromatoporoids established themselves on tops of the mounds and rapidly built up the organic reefs along the platform margins and on the pinnacle reefs in the slope environment to an elevation of 300–600 ft (90–180 m) above the sea floor.

Sea level lowered at least 30 ft (10 m), exposing the tops of the platform reef banks and pinnacle reefs to karstic vadose alteration. This was associated with restriction in the flow of oceanic sea water across the reef platforms to two major inlets, and produced the restricted humid starved basin. Regional lowering of sea level was the event that initiated the transition of open marine carbonate deposition to restricted arid evaporite deposition. The periods of lowered sea level, reflected in the carbonate platforms and pinnacle reefs by hiatuses and disconformities, were continuously recorded in the rocks of the Cain Formation in the basinal areas.

The lowermost member of the Cain Formation, an argillaceous dark micritic limestone, was deposited during the restricted humid phase of the transitional sequence when the exposure of the reef summits produced the extensive karstic vadose alteration. As the basin water changed from penesaline to hypersaline, the deposition of sediments in the basinal environment changed from laminated ostracode limestone, to calcite varvite, calcite-anhydrite varvite, anhydrite and halite. Within the halite bed thin varvites are records of minor influxes of sea water when the sea level of the continental platform stood at the silled level of the inlets to the basin, approximately 235–350 ft (67–107 m) below the maximum stand during reef growth in the basin. The water depth in the basin center was about 600 ft (180 m). Evaporative drawdown was rapid so that only a very thin anhydrite bed was deposited at the base of the halite sequence. During deposition of the salts of the A-1 unit of the Salina Group, sea level in the inlets remained at the silled level. The evaporite stage culminated with deposition of potash salts in a greatly restricted sea in the center one-third of the basin. At no time did the basin water reach dessication. Following deposition of the potash beds, a gradual increased flow of sea water into the basin diluted the basinal brine to the chloride stage. Deposition of the limestones and dolomites of the overlying Ruff Formation took place in shallow penesaline water (Budros and Briggs, 1977). The succeeding evaporites of the Salina Group all appear to have been shallow water, penesaline to hypersaline deposits.

Thus the initial stage of evaporite deposition in the Michigan basin during the Late Silurian began with a starved basin, surrounded by carbonate platforms and banks transected by two major inlets. Evaporite deposition began with a sudden lowering of sea level to the base of the inlets, approximately 235–350 ft (67–107 m) below the reef crests. Water depth in the basin was approximately 600 ft (190 m). The sedimentological history of the complete transition from open marine to restricted arid environments is recorded in the Cain Formation, deposited in the basinal environment of the Michgian basin.

REFERENCES

Alling, H. L., and Briggs, L. I., 1961. Stratigraphy of upper Silurian Cayugan evaporites: A.A.P.G. Bull., v. 45, p. 515–547.

Briggs, L. I., 1962. Niagaran-Cayugan sedimentation in Michigan basin: Michigan Basin Geol. Soc. Ann. Field Conference Guidebook, p. 58–60.

Budros, R. and L. I. Briggs, 1977. Depositional environment of Ruff Formation (Upper Silurian) in southeastern Michigan. p. 57–71 in: Fisher, J. H., (ed.), Reefs and Evaporites – Concepts and Depositional Models: A.A.P.G. Studies in Geology No. 5, 196 p.

Gill, D., 1977. The Belle River Mills gas field: productive Niagaran reef encased by Sabkha deposits, Michigan basin: Michigan Basin Geol. Soc. Spec. Pub. No. 2, 188 p.

Huh, J. M., L. I. Briggs and D. Gill, 1977. Depositional environments of pinnacle reefs, Niagara and Salina Groups, northern shelf, Michigan basin. p. 2–21 in: Fisher J. H. (ed.), Reefs and Evaporites – Concepts and Depositional Models: A.A.P.G. Studies in Geology no. 5, 196 p.

Kahle, C. F., 1974. Nature and significance of Silurian rocks at Maumee Quarry, Ohio. p. 31–54 in: Kesling, R. V., (ed.), Silurian reef-evaporite relationships: Mich. Basin Geol. Soc. Field Conf., 111 p.

Kinsman, D. J., 1969. Modes of formation, sedimentary association and diagenetic features of shallow water and supratidal evaporites: A.A.P.G. Bull., v. 53, p. 830–840.

Mantek, W., 1976. Recent exploration and activity in Michigan: Ontario Petroleum Inst., Proc. 15th Ann. Conf., 29 p.

Matthews, R. D. and G. C. Eagleson, 1974. Origin and implications of a midbasin potash facies in the Salina salt of Michigan. in: Coogan, A. H. (ed.), Fourth Symposium on Salt, vol. 1, Northern Ohio Geol. Soc., Inc., Cleveland, Ohio, p. 16–34.

Meloy, D. U., 1974. Depositional history of the Silurian northern carbonate bank of the Michigan basin: Unpubl. M.Sc. thesis, Univ. of Michigan, 78 p.

Mesolella, K. J., J. D. Robinson, L. M. McCormick and A. R. Ormiston, 1974. Cyclic deposition of Silurian carbonates and evaporites in Michigan basin: A.A.P.G. Bull., v. 56, p. 34–62.

Nurmi, R. D. and G. M. Friedman, 1977. Sedimentology and depositional environments of basin-center evaporites, Lower Salina Group (Upper Silurian), Michigan basin. p. 23–52 in: Fisher, J. H. (ed.), Reef and Evaporites – Concepts and Depositional Models: A.A.P.G. Studies in Geology No. 5, 196 p.

Shinn, E. A., R. B. Halley, J. H. Hudson, and B. H. Lidz, 1977. Limestone compaction: An enigma. Geology, v. 5, p. 21–24.